Introduction to Classical and Modern Analysis and Their Application to Group Representation Theory

Introduction to Classical and Modern Analysis and Their Application to Group Representation Theory

Debabrata Basu

World Scientific

NEW JERSEY · LONDON · SINGAPORE · BEIJING · SHANGHAI · HONG KONG · TAIPEI · CHENNAI

Published by

World Scientific Publishing Co. Pte. Ltd.

5 Toh Tuck Link, Singapore 596224

USA office: 27 Warren Street, Suite 401-402, Hackensack, NJ 07601

UK office: 57 Shelton Street, Covent Garden, London WC2H 9HE

British Library Cataloguing-in-Publication Data
A catalogue record for this book is available from the British Library.

ISBN-13 978-981-4273-29-9
ISBN-10 981-4273-29-5
ISBN-13 978-981-4273-30-5 (pbk)
ISBN-10 981-4273-30-9 (pbk)

Typeset by Stallion Press
Email: enquiries@stallionpress.com

Printed in Singapore by World Scientific Printers.

Dedicated to My Departed Guru

Late Professor Sudhansu Datta Majumdar
Who Taught Me the Fundamentals of
Classical Analysis and Representation Theory

Foreword

This book, by Debabrata Basu, is novel and valuable in several different ways. In this work, the author offers a quality introduction and review of important analytical methods and tools in Part I, and in Part II applies them to an important set of Lie groups presenting them in a modern analytical language that also makes close contact with the language of the well known coherent states. The author effectively uses numerous examples to guide the reader to an ever increasing appreciation of the analytical methods he discusses. As such, this first part already serves as a useful summary of basic fundamental analytical notions. The second part dealing with several examples of locally compact groups offers the reader a smooth introduction into this basic and very important territory. I am pleased to congratulate the author for his significant contribution to this general subject matter.

John Klauder
Gainesville, Florida
January 15, 2009

Preface

During the past few decades analytical methods are being increasingly applied to group representation theory which primarily developed as a branch of algebra in the hands of Frobinius and Schür. Although the application of analytical methods is now the standard approach in Lie groups there is as yet no standard textbook dealing with classical and modern analysis as applied especially to locally compact groups.

It is expected that this gap will be bridged by this book which is essentially an amplification of the lectures of the author to M.Sc. students of the Physics Department of Indian Institute of Technology, Kharagpur. For clarity many standard topics in this book have been treated in a way which substantially differs from traditional treatment and is in a more teachable form.

The author himself does not understand the sophistry of pure mathematics and those who look for elegance and rigour will be sorely disappointed. The book does not provide the most general topological definition of Lie groups, not that the author is unwilling to learn it, but that it is deemed inessential in a preliminary course which this book intends to cover. In a sense, following Ivan Karamazov, discussions are all conducted "as stupidly as possible..." because "the stupider, the more to the point. The stupider, the clearer. Stupidity is brief and artless but intelligence shifts and shuffles and hides itself. Intelligence is a knave, while stupidity is straightforward and honest."

Even a casual reader browsing through this book will not fail to notice the indebtedness of the author to the Russian masters of functional analysis and representation theory. Of course, this does not come anywhere near their magnum opus in depth, breadth of coverage and originality; it is only a modest endeavour to make accessible to the graduate students the fundamentals of the subject created by them.

The first eight chapters of this book may be covered in any traditional graduate course in mathematical physics. In particular later parts of Chapter 8 supplies the mathematical framework of the octet model of Gell-Mann and Neéman which is the foundation of the present day quark model, an inseparable part of standard model. The remaining three chapters deal with infinite dimensional representations of the simplest locally compact groups, namely, $SL(2, R)$, $SL(2, C)$ and the Heisenberg–Weyl group. They are becoming increasingly important in several areas of quantum optics and quantum gravity.

The references at the end of each chapter are those that have been consulted by the author and are a reflection of personal taste rather than anything else.

The author would like to thank Professor J. R. Klauder for not only writing the foreword but also for constructive criticism as well as numerous conceptual and technical corrections, to say nothing of his excellent review submitted to the publishers. The author has immensely benefited from collaboration with Prof. P. Majumdar of Saha Institute of Nuclear Physics, Kolkata in the SERC Winter School held at Benares in several sections of Chapters 8. Finally the author would like to express his heartfelt affection and admiration for the students of Indian Institute of Technology, Kharagpur, who are the main inspiration of this endeavour.

Debabrata Basu

Acknowledgment

Professor Debabrata Basu passed away in November 2009 after a prolonged illness, when the book was still with the publisher. Much of it was written in between his stays at the hospital and the final manuscript was proofread by two of his most favorite students Wrick Sengupta and Dr KV Shajesh. The family of Professor Basu is deeply indebted to Mr Sengupta and Dr Shajesh for having done extremely detailed proofreading of the manuscript. Words are not enough to thank them for their contribution in ensuring that the book sees that the light of the day.

Contents

PART I

Analysis

The first two chapters are of an introductory nature and provide a brief survey of the process of analysis. They touch upon only the most essential points leading to the residue theorem and its applications to the evaluation of definite integrals. The power and scope of this apparently simple theorem has been amply demonstrated in later chapters where it has been employed in the problem of analytic continuation of the hypergeometric series (Chapter 4) on the one hand and in the Clebsch–Gordan problem of $SL(2, R)$ (Chapter 9) on the other.

CHAPTER 1

Basic Analytical Tools

A function of a positive integer designated by $n, n = 1, 2, 3, \ldots$ is called a sequence. This sequence is a set of numbers $u_1, u_2, \ldots$ in a definite order and formed according to a definite rule. The sequence is called finite or infinite according as there are a finite number of terms or not. We shall in general consider infinite sequences only.

A number l is called the limit of an infinite sequence $u_1, u_2, u_3, \ldots$ if for any positive number ϵ, we can find a positive number N depending on ϵ such that $|u_n - l| < \epsilon$ for all $n > $ N. More simply $\lim_{n \to \infty} u_n = l$.

If the limit of a sequence exists, the sequence is called convergent, otherwise it is called divergent.

If a sequence of real numbers does not tend to a limit or to ∞ or to $-\infty$, the sequence is said to oscillate.

1.1 Infinite Series and Its Convergence

Let $u_1, u_2, u_3, \ldots$ be a sequence of numbers, real or complex. We construct a new sequence.

$$S_1 = u_1, \quad S_2 = u_1 + u_2, \ldots, \quad S_n = u_1 + u_2 + \cdots + u_n. \qquad (1.1)$$

S_n called the nth partial sum, is the sum of the first n terms of the sequence. If $\lim_{n \to \infty} S_n = $ S exists, the infinite series $n \to \infty$

$$u_1 + u_2 + u_3 + \cdots \qquad (1.2)$$

is said to be convergent, or to converge to the sum S. In other case the infinite series is said to be divergent.

When the series converges, the expression $S - S_n$, which is the sum of the series

$$u_{n+1} + u_{n+2} + \cdots \qquad (1.3)$$

is called the reminder after n terms, and is frequently denoted by the symbol R_n.

The sum

$$u_{n+1} + u_{n+2} + \cdots + u_{n+p} \qquad (1.4)$$

will be denoted by $S_{n,p}$.

It follows at once that the necessary and sufficient condition for the convergence of an infinite series is that $\lim_{n \to \infty} S_{n,p} = 0$ for every positive value of p.

Since $u_{n+1} = S_{n,1}$ it follows as a particular case that $\lim_{n \to \infty} u_{n+1} = 0$. In other words, the nth term of convergent series must tend to zero as n tends to infinity. But this last condition though necessary, is not sufficient in itself to ensure the convergence of the series, as appears from a study of the series:

$$\frac{1}{1} + \frac{1}{2} + \frac{1}{3} + \frac{1}{4} + \frac{1}{5} + \cdots . \qquad (1.5)$$

In this series

$$S_{n,n} = \frac{1}{n+1} + \frac{1}{n+2} + \frac{1}{n+3} + \cdots + \frac{1}{2n}$$

$$\frac{1}{n+1} > \frac{1}{n+n} = \frac{1}{2n}$$

$$\frac{1}{n+2} > \frac{1}{2n}$$

$$\frac{1}{n+n-1} > \frac{1}{2n} \qquad (1.6)$$

$$S_{n,n} > \frac{1}{2n} + \frac{1}{2n} + \cdots + \frac{1}{2n}$$

$$= \frac{n}{2n} = \frac{1}{2} \qquad (1.7)$$

$$S_{n,n} > \frac{1}{2} \qquad (1.8)$$

$$S_{2^{n+1}} = 1 + S_{1,1} + S_{2,2} + S_{4,4} + S_{8,8} + S_{16,16} \qquad (1.9)$$

$$+ \cdots + S_{2^n,2^n} \qquad (1.10)$$

$$> 1 + \frac{1}{2} + \left(\frac{1}{2} + \frac{1}{2} + \frac{1}{2} + \cdots n \text{ terms} \right) \qquad (1.11)$$

$$\frac{n}{2} + \frac{3}{2} = \frac{1}{2}(n+3) \qquad (1.12)$$

$$\to \infty. \qquad (1.13)$$

So the series is divergent.

We may arrive at a series by some formal process e.g. that of solving a linear differential equation by a series and then to justify the process it will usually have to be proved that the series thus obtained is convergent.

Given an expression S it may be possible to obtain a development

$$S = \sum_{n=1}^{n} u_n + R_n \tag{1.14}$$

valid for all values of n and from the definition of a limit, it follows that if we can prove that $R_n \to 0$ then the series $\sum u_n$ converges and its sum is S.

1.2 Absolute and Conditional Convergence

In order that a series $\sum u_n$ of real or complex terms may converge it is sufficient (but not necessary) that the series of moduli $\sum_{n=1}^{\infty} |u_n|$ should converge. This is a consequence of the inequality

$$|u_1 + u_2 + u_3 + \cdots| \le |u_1| + |u_2| + |u_3| + \cdots \tag{1.15}$$

i.e.

$$\left| \sum_{n=1}^{\infty} u_n \right| \le \sum_{n=1}^{\infty} |u_n|. \tag{1.16}$$

Hence the convergence of the right-hand side ensures that of the left-hand side.

The condition is not necessary for writing $u_n = \frac{(-)^{n+1}}{n}$ we see that

$$1 - \frac{1}{2} + \frac{1}{3} + \cdots \text{ converges,} \tag{1.17}$$

though the harmonic series $1 + \frac{1}{2} + \frac{1}{3} + \cdots$ is known to diverge.

Series which are such that the series formed by the moduli of their terms are convergent, possess special properties of great importance and are called absolutely convergent series. Series which though convergent but not absolutely convergent are said to be conditionally convergent.

1.3 The Geometric Series, and the Series $\sum_{n=1}^{\infty} \frac{1}{n^s}$

The convergence of a particular series is generally investigated not by the direct consideration of the sum $S_{n,p}$ but by a comparison of the given series with some other series which is known to be convergent or divergent. We shall now investigate the convergence of two of the series which are often used as standards of comparison. (a) The geometric series

$$1 + z + z^2 + \cdots .$$

Let us consider the series of moduli

$$1 + |z| + |z|^2 + |z|^3 + \cdots .$$

For this example

$$S_{n,p} = |z|^{n+1} + |z|^{n+2} + \cdots + |z|^{n+p} \tag{1.18}$$

$$= |z|^{n+1} \left[\frac{1 - |z|^p}{1 - |z|} \right] . \tag{1.19}$$

Hence if $|z| < 1$, then $S_{n,p} < \frac{|z|^{n+1}}{1-|z|}$ for all values of p. Thus for all p $\lim_{n\to\infty} S_{n,p} = 0$ and the series is convergent if $|z| < 1$. Therefore the geometric series is absolutely convergent if $|z| < 1$.

(b) The series

$$\frac{1}{1^s} + \frac{1}{2^s} + \cdots = \sum_{n=1}^{\infty} \frac{1}{n^s} . \tag{1.20}$$

Consider

$$n = 2^{p-1} \tag{1.21}$$

$$S_n = \sum_{m=1}^{n} \frac{1}{m^s} \quad \text{for } s > 1. \tag{1.22}$$

$$p = 1 \quad n = 1$$
$$p = 2 \quad n = 2$$
$$p = 3 \quad n = 4$$
$$p = 4 \quad n = 8$$

$$\frac{1}{1^s} = \frac{1}{1^{s-1}} .$$

We have

$$\frac{1}{2^s} + \frac{1}{3^s} < \frac{2}{2^s} = \frac{1}{2^{s-1}} \tag{1.23}$$

$$\frac{1}{4^s} + \frac{1}{5^s} + \frac{1}{6^s} + \frac{1}{7^s} < \frac{4}{4^s} = \frac{1}{4^{s-1}} \tag{1.24}$$

$$\frac{1}{8^s} + \frac{1}{9^s} + \cdots + \frac{1}{15^s} < \frac{8}{8^s} = \frac{1}{8^{s-1}} . \tag{1.25}$$

Hence

$$S_n = \sum_{m=1}^{n=2^{p-1}} \frac{1}{m^s} < \frac{1}{1^{s-1}} + \frac{1}{2^{s-1}} + \frac{1}{4^{s-1}} + \cdots + \frac{1}{2^{(s-1)(p-1)}} \qquad (1.26)$$

$$S_n < \frac{1 - 2^{-(s-1)(p-1)}}{1 - 2^{-(s-1)}} < \frac{1}{1 - 2^{-(s-1)}}. \qquad (1.27)$$

Hence the sum of any number of terms is less than $(1 - 2^{1-s})^{-1}$. Hence the sequence

$$\{S_n\} = \left\{ \sum_{m=1}^{n} m^{-s} \right\} \qquad (1.28)$$

cannot tend to infinity. Therefore the series $\sum_{n-1}^{\infty} \frac{1}{n^s}$ is convergent if $s > 1$, and since the terms are all real and positive the convergence is absolute. The series diverges if $s \leq 1$.

1.4 The Comparison Theorem

A series $u_1 + u_2 + u_3 + \cdots$ is absolutely convergent if $|u_n| < c|v_n|$ where c is a positive number independent of n and v_n is the nth term of another series which is known to be absolutely convergent. For under these conditions, we have

$$|u_{n+1}| + |u_{n+2}| + \cdots + |u_{n+p}| < c(|v_{n+1}| + |v_{n+2}| + \cdots + |v_{n+p}|). \qquad (1.29)$$

Since $\sum v_n$ is absolutely convergent,

$$\lim_{n \to \infty} [|v_{n+1}| + |v_{n+2}| + \cdots + |v_{n+p}|] = 0 \qquad (1.30)$$

for all p,

$$\therefore \quad \lim_{n \to \infty} |u_{n+1}| + |u_{n+2}| + \cdots + |u_{n+p}| \to 0 \qquad (1.31)$$

for all p and $\sum_{n=1}^{\infty} u_n$ is absolutely convergent.

1.5 D' Alembert's Ratio Test for Absolute Convergence

We show that a series $\sum_{n=1}^{\infty} u_n$ is absolutely convergent provided that for all values of n greater than some fixed value r, the ratio $\left| \frac{u_{n+1}}{u_n} \right|$ is less than ρ, where ρ is independent of n and $0 < \rho < 1$. For the terms of the series

$$S_{r,p} = |u_{r+1}| + |u_{r+2}| + \cdots + |u_{r+p}| \tag{1.32}$$

$$= |u_{r+1}|\left(1 + \rho + \rho^2 + \cdots + \rho^p\right) \tag{1.33}$$

$$= \frac{|u_{r+1}|(1 - \rho^p)}{1 - \rho} < \frac{|u_{r+1}|}{1 - \rho}. \tag{1.34}$$

$\lim_{r \to \infty} S_{r;p} \to 0$ for all p and the series $\sum_{n=1}^{\infty} u_n$ is absolutely convergent.

A particular case of the theorem is that $\lim_{n \to \infty} \left|\frac{u_{n+1}}{u_n}\right| < 1$ then the series $\sum_{n=1}^{\infty} u_n$ is absolutely convergent.

By the definition of a limit we can find n such that

$$\left|\left\{\left|\frac{u_{n+1}}{u_n}\right| - l\right\}\right| < \frac{1}{2}(1 - l) \quad \text{when } n > r \tag{1.35}$$

$$\left|\left\{\left|\frac{u_{n+1}}{u_n}\right| - l\right\}\right| > \left|\frac{u_{n+1}}{u_n}\right| - l \tag{1.36}$$

$$\left|\frac{u_{n+1}}{u_n}\right| - l < \frac{1}{2}(1 - l) \tag{1.37}$$

$$\left|\frac{u_{n+1}}{u_n}\right| < \frac{1}{2}(1 + l) < 1 \tag{1.38}$$

when $n > r$.

Example. The series

$$z + \frac{(a - b)}{2!} z^2 + \frac{(a - b)(a - 2b)}{3!} z^3$$

$$+ \frac{(a - b)(a - 2b)(a - 3b)z^4}{4!} + \cdots \tag{1.39}$$

converges absolutely if $|z| < |b|^{-1}$

$$\text{For } \lim_{n \to \infty} \left|\frac{u_{n+1}}{u_n}\right| = \left|\frac{a - n\,b}{n + 1}\right| |z| = |b||z| \tag{1.40}$$

So the condition for absolute convergence is $|b||z| < 1$, i.e. $|z| < |b|^{-1}$.

1.5.1   *A general theorem for which* $\lim_{n\to\infty} \left|\frac{u_{n+1}}{u_n}\right| = 1$

We shall now show that a series $u_1 + u_2 + u_3 + \cdots$ in which $\left|\frac{u_{n+1}}{u_n}\right| = 1$ will be absolutely convergent if a positive number c exists such that,

$$\lim_{n\to\infty} n\left\{\left|\frac{u_{n+1}}{u_n}\right| - 1\right\} = -1 - c. \tag{1.41}$$

Compare the series $\sum |u_n|$ with the absolutely convergent series

$$\sum v_n$$
$$\text{where} \quad v_n = A\, n^{-1-\frac{c}{2}} \tag{1.42}$$

and A is a constant. It has just been shown that the series

$$\sum_{n=1}^{\infty} \frac{1}{n^s}$$

is absolutely convergent whenever $s > 1$. Hence $\sum_{n-1}^{\infty} v_n = A \sum_{n=1}^{\infty} \frac{1}{n^{1+\frac{c}{2}}}$ is absolutely convergent.

We have

$$\frac{v_{n+1}}{v_n} = \left(\frac{n}{n+1}\right)^{1+\frac{c}{2}} = \left(1 + \frac{1}{n}\right)^{-1-\frac{c}{2}}$$
$$= 1 - \frac{1 + \frac{1}{2}c}{n} + \mathcal{O}\left(\frac{1}{n^2}\right).$$

Hence as $n \to \infty$ $n\left\{\frac{v_{n+1}}{v_n} - 1\right\} = -1 - \frac{c}{2}$ and hence we can find m such that, when $n > m$

$$\left|\frac{u_{n+1}}{u_n}\right| \leq \frac{v_{n+1}}{v_n}. \tag{1.43}$$

By a suitable choice of the constant A we can therefore ensure that for all values of n we shall have

$$|u_n| < v_n. \tag{1.44}$$

As $\sum v_n$ is convergent, $\sum |u_n|$ is also convergent and so $\sum u_n$ is absolutely convergent.

1.5.2   *Convergence of the Gauss and generalized hypergeometric series*

The Gauss' hypergeometric series is given by

$$F(a, b; c; z) = \sum_{n=0}^{\infty} \frac{(a)_n\,(b)_n}{(c)_n\,n!} z^n, \quad (a)_n = \frac{\Gamma(a+n)}{\Gamma(a)} \text{ etc.} \qquad (1.45)$$

$$= 1 + \frac{abz}{c} + \frac{a(a+1)b(b+1)}{c(c+1)} \frac{z^2}{2!} + \cdots . \qquad (1.46)$$

In this series

$$\left| \frac{u_{n+1}}{u_n} \right| = \left| \frac{(a+n)(b+n)}{(c+n)(n+1)} \right| |z| \to |z| \qquad (1.47)$$

as $n \to \infty$. Hence the series is absolutely convergent when $|z| < 1$, and divergent when $|z| > 1$. When $|z| = 1$ we have

$$\left| \frac{u_{n+1}}{u_n} \right| = \left| 1 + \frac{a}{n} \right| \left| 1 + \frac{b}{n} \right| \left| 1 - \frac{c}{n} \right| \left| 1 - \frac{1}{n} \right| + \mathcal{O}\left(\frac{1}{n^2} \right) \qquad (1.48)$$

$$= \left| 1 + \frac{a+b-c-1}{n} + \mathcal{O}\left(\frac{1}{n^2} \right) \right|. \qquad (1.49)$$

Let a, b, c be complex numbers so that

$$a = a_1 + ia_2 \quad b = b_1 + ib_2 \quad c = c_1 + ic_2.$$

Hence

$$\left| \frac{u_{n+1}}{u_n} \right| = \left| 1 + \frac{a_1 + b_1 - c_1 - 1}{n} + \frac{i\,(a_2 + b_2 - c_2)}{n} + \mathcal{O}\left(\frac{1}{n^2} \right) \right| \qquad (1.50)$$

$$= \left[\left(1 + \frac{a_1 + b_1 - c_1 - 1}{n} \right)^2 + \left(\frac{a_2 + b_2 - c_2}{n} \right)^2 + \mathcal{O}\left(\frac{1}{n^2} \right) \right]^{\frac{1}{2}} \qquad (1.51)$$

$$= 1 + \frac{-1 - (c_1 - a_1 - b_1)}{n} + \mathcal{O}\left(\frac{1}{n^2} \right). \qquad (1.52)$$

Hence when $|z| = 1$, a sufficient condition for the absolute convergence of the Gauss' hypergeometric series is that the real part of $(c - a - b)$ shall be positive. Bromwich has shown that this condition is also necessary. In a similar manner it can be shown that the generalized hypergeometric series,

$$
{}_3F_2 \begin{bmatrix} a, b, c \\ d, e; z \end{bmatrix} = \sum_{n=0}^{\infty} \frac{(a)_n (b)_n (c)_n}{(d)_n (e)_n \, n!} z^n \tag{1.53}
$$

converges absolutely when $|z| < 1$ and diverges when $|z| > 1$.
It converges absolutely for $|z| = 1$ if

$$
\mathrm{Re}(d + e - a - b - c) > 0. \tag{1.54}
$$

It will be shown later that the generalized hypergeometric series of unit argument appears as the Clebsch Gordan coefficient of $SL(2, R)$ in the $SO(2)$ basis.

1.6 Analytic Functions

Let us consider the function

$$
f(x, y) = u(x, y) + iv(x, y). \tag{1.55}
$$

It is often convenient to introduce the complex numbers

$$
z = x + iy \quad \bar{z} = x - iy.
$$

Then instead of using the differential operators $\frac{\partial}{\partial x}$ and $\frac{\partial}{\partial y}$ one uses derivatives with respect to z and $\bar{z}$, namely $\frac{\partial}{\partial z}$ and $\frac{\partial}{\partial \bar{z}}$. These new operators are defined with the requirement that the ordinary rules of differentiation, in particular the chain rule hold for them, so that

$$
\frac{\partial}{\partial z} = \frac{1}{2} \left(\frac{\partial}{\partial x} - i \frac{\partial}{\partial y} \right) \tag{1.56}
$$

$$
\frac{\partial}{\partial \bar{z}} = \frac{1}{2} \left(\frac{\partial}{\partial x} + i \frac{\partial}{\partial y} \right). \tag{1.57}
$$

Consider

$$
f(x, y) = u(x, y) + iv(x, y).
$$

Put $x = \frac{1}{2}(z + \bar{z}) y = \frac{1}{2i}(z - \bar{z})$. If the $\bar{z}$ dependence cancels out from the two terms then the function $f(x, y) = f_1(z)$ will be called an analytic function of the complex variable $z = x + iy$.

Example. Let us consider

$$f(z) = x^2 - y^2 + 2ixy \tag{1.58}$$

$$= \left(\frac{z+\bar{z}}{2}\right)^2 - \left(\frac{z-\bar{z}}{2i}\right)^2 + 2i\left(\frac{z+\bar{z}}{2}\right)\left(\frac{z-\bar{z}}{2i}\right) \tag{1.59}$$

$$= \frac{1}{4}(z+\bar{z})^2 + \frac{1}{4}(z-\bar{z})^2 + \frac{1}{2}(z^2 - \bar{z}^2) \tag{1.60}$$

$$= \frac{1}{2}(z^2 + \bar{z}^2) + \frac{1}{2}(z^2 - \bar{z}^2) = z^2. \tag{1.61}$$

Hence the function is analytic while

$$f = x^2 - y^2 + 4i(xy) = \frac{3}{2}z^2 - \frac{1}{2}\bar{z}^2 \tag{1.62}$$

is nonanalytic. Hence for analytic functions

$$\frac{\partial f}{\partial \bar{z}} = 0, \quad \text{i.e.} \quad \frac{1}{2}\left(\frac{\partial}{\partial x} + i\frac{\partial}{\partial y}\right)(u + iv) = 0. \tag{1.63}$$

Thus

$$\left(\frac{\partial u}{\partial x} - \frac{\partial v}{\partial y}\right) + i\left(\frac{\partial u}{\partial y} + \frac{\partial v}{\partial x}\right) = 0. \tag{1.64}$$

Hence we obtain the Cauchy–Riemann equations

$$\frac{\partial u}{\partial x} = \frac{\partial v}{\partial y} \qquad \frac{\partial u}{\partial y} = -\frac{\partial v}{\partial x}. \tag{1.65}$$

Problem. Derive the polar form of Cauchy–Riemann equations

$$\frac{\partial u}{\partial r} = \frac{1}{r}\frac{\partial v}{\partial \theta} \tag{1.66}$$

$$\frac{\partial v}{\partial r} = -\frac{1}{r}\frac{\partial u}{\partial \theta}. \tag{1.67}$$

We shall now show that the Cauchy–Riemann equations ensure the following result:

For an analytic function single valued in some region R of the complex plane the derivative of $f(z)$ defined as

$$f'(z) = \lim_{\delta z \to 0} \frac{f(z + \delta z) - f(z)}{\delta z} \tag{1.68}$$

exists and is independent of the manner in which $\delta z \to 0$.

Proof.

$$f'(z) = \lim_{\substack{\delta x \to 0 \\ \delta y \to 0}} \frac{[u\,(x + \delta x, y + \delta y) + iv\,(x, +\delta x, y + \delta y) - \{u(x,y) + iv(x,y)\}]}{(\delta x + i\delta y)}.$$

$$(1.69)$$

Case 1, $\delta y = 0$, $\delta x \to 0$

$$[f'(z)]_1 = \lim_{\delta x \to 0} \frac{u(x + \delta x, y) + iv(x + \delta x, y) - u(x,y) - iv(x,y)}{\delta x}$$

$$= \frac{\partial u}{\partial x} + i\frac{\partial v}{\partial x}.$$

$$(1.70)$$

Case 2, $\delta x = 0$, $\delta y \to 0$

$$[f'(z)]_2 = \frac{u(x, y + \delta y) + iv(x, y + \delta y) - u(x,y) - iv(x,y)}{i\,\delta y}$$

$$= -i\frac{\partial u}{\partial y} + \frac{\partial v}{\partial y}.$$

$$(1.71)$$

Now the Cauchy–Riemann equations demand,

$$\frac{\partial u}{\partial x} = \frac{\partial v}{\partial y}, \quad \frac{\partial v}{\partial x} = \frac{-\partial u}{\partial y}$$

$$(1.72)$$

$$[f'(z)]_1 = [f'(z)]_2 = f'(z).$$

$$(1.73)$$

One of the remarkable theorems valid for analytic functions is the following:

If $f(z)$ is analytic in a region R so $f'(z)$, $f''(z)$ are also analytic in R, i.e. all higher derivatives exist in R. $\qquad\square$

1.7 Singularities of an Analytic Function

A point at which $f(z)$ fails to be analytic is called a singular point or a singularity of $f(z)$. All nontrivial analytic functions have one or more singular points somewhere in the complex z plane. Various types of singularities exists.

1. Poles: If we find a positive integer n such that $\lim_{z \to z_0}(z - z_0)^n f(z) = A \neq 0$ then $z = z_0$ is called a pole of order n. If $n = 1$, z_0 is called a simple pole.

Example 1. $f(z) = \frac{1}{(z-2)^3}$ has a pole of order 3 at $z = 2$.

Example 2. $f(z) = \frac{3z-2}{(z-1)^2(z+1)(z-4)}$ has a pole of order 2 at $z = 1$ and simple poles at $z = -1$ and $z = 4$.

1.8   Branch Points of Multiple Valued Functions

Let $\omega = z^{\frac{1}{2}}$. Let us suppose that corresponding to a particular value $z = z_1$, we have $\omega = \omega_1$. It is easy to verify that if we start at the point z_1 in the z-plane and make one complete circuit counterclockwise around the origin, the value of ω on returning to z_1 is $\omega_1 e^{\pi i} = -\omega_1$.

Proof. We have $z = re^{i\theta}$ so that $\omega = z^{\frac{1}{2}} = r^{\frac{1}{2}} e^{\frac{i\theta}{2}}$.

If $r = r_1$, and $\theta = \theta_1$ then $\omega_1 = r_1^{\frac{1}{2}} e^{\frac{i\theta_1}{2}}$. As θ_1 increases from θ_1 to $\theta_1 + 2\pi$ which happens when one complete circuit counterclockwise around the origin is made, we find

$$\omega = r_1^{\frac{1}{2}} e^{i(\theta_1 + 2\pi)/2} = \omega_1 e^{\pi i} = -\omega_1. \tag{1.74}$$

After two complete circuits the value of ω is $\omega = \omega_1 e^{2\pi i} = \omega_1$. Thus the original value is obtained after 2 revolutions around the origin. If the path does not enclose the origin the increase in $\arg z$ is zero so that increase in $\arg \omega$ is also zero. In this case the value $\omega = \omega_1$ regardless of the number of circuits. It is evident that we can consider ω as a collection of single valued functions called the branches of the multiple valued function by properly restricting θ. Thus for example we can write

$$\omega = r^{\frac{1}{2}} \left(\cos \frac{\theta}{2} + i \sin \frac{\theta}{2} \right) \tag{1.75}$$

where we take two possible intervals of θ given by,

$$0 < \theta < 2\pi, \ \ 2\pi < \theta < 4\pi.$$

The set of values of the function for the first interval $0 < \theta < 2\pi$ is called the principal branch. We start with the principal branch

$$\omega = r^{\frac{1}{2}} \left(\cos \frac{\theta}{2} + i \sin \frac{\theta}{2} \right), \ \ \ 0 < \theta < 2\pi. \tag{1.76}$$

As different values of $f(z)$ are obtained by successively encircling $z = 0$, we call $z = 0$ a branch point. We can restrict ourselves to a particular single valued function, usually the principal branch, by insuring that not more than one complete circuit about the branch point is made, i.e. by suitably restricting θ. In the case of the principal range $0 < \theta < 2\pi$ this is accomplished by constructing an impassable barrier called a branch cut as in the diagram (1.1). □

The circle and the gap is to be understood in the limiting sense, i.e. the radius of the circle is assumed to approach zero and the gap about the real

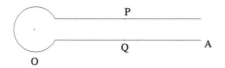

Fig. 1.1

axis is made infinitely narrow. OA is called a branch cut which we agree not to cross. We cite some other functions having a branch point at $z = 0$.

(i) $\omega = \ln z$, $\omega = z^a$, a complex; similarly $w = (1 - z)^a$, $\omega = \ln(z - 1)$ have branch points at $z = 1$. All these functions are infinitely many valued.

Let us take a point P just above the positive real axis in Fig. 1.1. Here

$$\omega_P = z_P^{\frac{1}{2}} = (r^{\frac{1}{2}} e^{\frac{i\theta}{2}})_{\theta=0} = r^{\frac{1}{2}}. \tag{1.77}$$

We now take a point Q just below the real axis

$$\omega_Q = z_Q^{\frac{1}{2}} = (re^{2\pi i})^{\frac{1}{2}} = r^{\frac{1}{2}} e^{\pi i} = -r^{\frac{1}{2}}. \tag{1.78}$$

$$\omega_Q - \omega_P = -2r^{\frac{1}{2}} \tag{1.79}$$

Hence the value of the function changes discontinuously across the cut. In dealing with branch points and branch cuts we shall often need the correspondence between complex numbers and two dimensional vectors. A complex number $z = x + iy$ adds up just like a two dimensional vector, for example in 1.2.

$$\underline{OP} = z_2$$

$$\underline{OQ} = z_1$$

$$\underline{QP} = z_2 - z_1$$

$$\underline{OQ} + \underline{QP} = z_1 + z_2 - z_1 = z_2 = \underline{OP}. \tag{1.80}$$

In almost all of Cauchy's works on analytic functions the single valuedness of the function $f(z)$ is assumed. To incorporate multiple valued functions in this framework all that is needed is to remain only on one branch of such functions. This is achieved just by setting up the branch cut in an appropriate position which we agree not to cross. All standard books use for this purpose the principal branch which is generally defined to be $0 < \arg z < 2\pi$, $-\pi < \arg z < \pi$, etc. For example when origin is the branch point, the branch cut can be chosen as in Fig. 1.1. This choice corresponds to $0 < \arg z < 2\pi$ (principal branch). The branch cut can also be chosen as in Fig. 1.3.

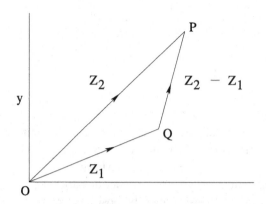

Fig. 1.2

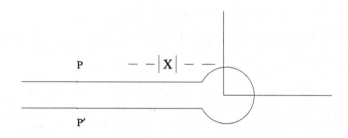

Fig. 1.3

For this choice $-\pi < \arg z < \pi$ (principal branch), another possible choice is Fig. 1.4.

For this choice the principal branch is

$$\frac{-3\pi}{2} < \arg z < \frac{\pi}{2}. \qquad (1.81)$$

The branch cut is chosen according to convenience and there is no rigid rule for its choice. The value of the function across the cut changes discontinuously. For example we take.

$$\omega = f(z) = \ln z$$

$$\text{At P, } \arg z = 0, \quad \omega_P = \ln x.$$

$$\text{At P', } \arg z = 2\pi, \quad \omega_{P'} = \ln(xe^{2\pi i})$$

$$= \ln x + 2\pi i.$$

$$\omega_{P'} - \omega_P = 2\pi i$$

Take the same function and use the cut plane of Fig. 1.3.

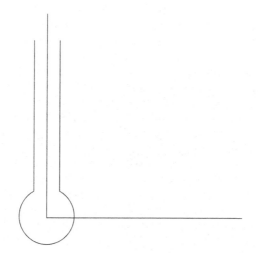

Fig. 1.4

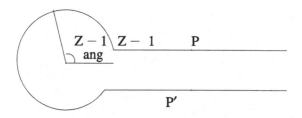

Fig. 1.5

At P,

$$\text{arg}z = \pi, \quad z = |z|e^{i\text{arg}z} = |x|e^{i\pi}$$

$$\omega_p = \ln|x| + i\pi. \tag{1.82}$$

Similarly

$$\omega_{p'} = \ln|x| - i\pi. \tag{1.83}$$

Hence,

$$\omega_{p'} - \omega_p = -2i\pi. \tag{1.84}$$

We now consider a few more complicated examples

$$\omega = f(z) = \ln\left(\frac{z+1}{z-1}\right). \tag{1.85}$$

The branch points of this function are $z = -1, z = 1$.

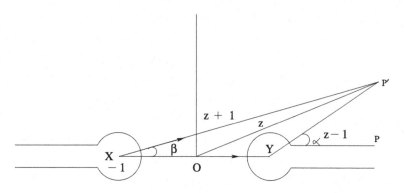

Fig. 1.6

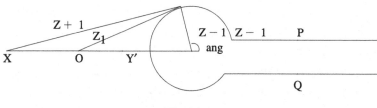

Fig. 1.7

A possible choice of the cut plane is Fig. 1.6.

Take the point P' as shown in Fig. 1.6 At P', $\underline{OP'} = z$.

$$\underline{OY} = 1, \quad \underline{YP'} = z - 1, \quad \arg(z - 1) = \alpha.$$
$$\underline{XP'} = z + 1 \arg(z + 1) = \beta.$$

At P, $\arg(z - 1) = \alpha = 0 = \arg(z + 1) = \beta = 0$.

Hence at P

$$\arg(z + 1) = \arg(z - 1) = 0.$$

As shown in the enlarged diagram Fig. 1.7 when the point P moves along the real axis and comes from P to Q $\arg(z - 1)$ changes from 0 to 2π. Hence at Q, $\arg(z - 1) = 2\pi$. Similarly $\arg(z + 1) = 0$ at Q. Thus moving around the branch point at $z = 1$ does not alter $\arg(z + 1)$ which remains fixed at zero. Since at P $\arg(z + 1) = \arg(z - 1) = 0$ we have

$$\omega_P = \ln\left(\frac{x + 1}{x - 1}\right). \tag{1.86}$$

But at Q no, $\arg(z-1) = 2\pi$, $\arg(z+1) = 0$

$$\omega_Q = \ln\left(\frac{x+1}{x-1}\right) - 2i\pi. \tag{1.87}$$

Hence the discontinuity across the cut

$$\omega_P - \omega_Q = 2i\pi. \tag{1.88}$$

Exercise. Analyze the problem when P, Q are points on the left-hand side cut.

Let us now consider the situation when P is in the region between the branch points. When P_1 is at the location shown in Fig. 1.8, $\arg(z-1) = \pi \arg(z+1) = 0$

$$\omega_p = \ln\left(\left|\frac{1+x}{x-1}\right|\frac{1}{e^{i\pi}}\right) = \ln\left(\frac{1+x}{1-x}\right) - i\pi. \tag{1.89}$$

We now take the same function in a different cut plane Fig. 1.9.

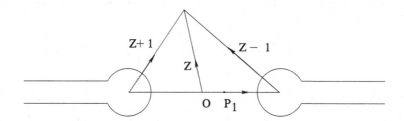

Fig. 1.8

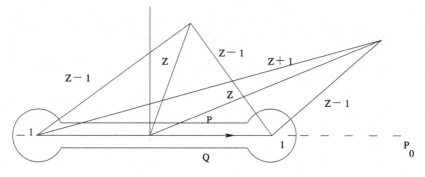

Fig. 1.9

At P_0, $\arg(z+1) = \arg(z-1) = 0$

$$\omega_{P_0} = \ln\left(\frac{x+1}{x-1}\right). \qquad (1.90)$$

But at P, $\arg(z+1) = 0$, $\arg(z-1) = \pi$

$$\omega_P = \ln\left(\frac{1+x}{1-x}\right) - i\pi. \qquad (1.91)$$

As P moves and describes the circuit about $z = -1$ $\arg(z+1)$ changes from 0 to 2π but $\arg(z-1)$ remains fixed at π.

Hence

$$\omega_Q = \ln\frac{|z+1|e^{2\pi i}}{|z-1|e^{\pi i}} = \ln\left(\frac{1+x}{1-x}\right) + i\pi. \qquad (1.92)$$

Thus

$$\omega_Q - \omega_P = 2i\pi. \qquad (1.93)$$

We can come to the point Q by circuiting around the branch point at $z = 1$. When P comes to the point Q by circuiting around $z = 1$ $\arg(z+1)$ remains fixed at zero but $\arg z - 1$ changes from π to $-\pi$. Hence at Q, $\arg(z+1) = 0$, $\arg(z-1) = -\pi$ so that

$$\omega_Q = \ln\left|\frac{z+1}{z-1}\right|\frac{1}{e^{-i\pi}} = \ln\left(\frac{1+x}{1-x}\right) + i\pi. \qquad (1.94)$$

1.9   Essential Singularities

The point $z = z_0$ will be called an essential singularity if we cannot find any positive integer n such that

$$\lim_{z \to z_0} (z - z_0)^n f(z) = A \neq 0.$$

Example. $f(z) = e^{1/(z-a)}$ has an essential singularity at $z = a$.

CHAPTER 2

Complex Integration

2.1 Complex Integration

Let $f(z)$ be continuous at all points of a curve C which we shall assume has a finite length.

Let us subdivide C into n parts as shown in Fig. 2.1 by means of points $z_1, z_2, \ldots, z_{n-1}$ chosen arbitrarily and we call $a = z_0, b = z_n$. On each arc joining z_{k-1} to $z_k (1 \le k \le n)$ let us choose a point ξ_k and we form the sum

$$S_n = \sum f(\xi_k)(z_k - z_{k-1}) = \sum_{k=1}^{n} f(\xi_k)\delta z_k \qquad (2.1)$$

where we have written $\delta z_k = z_k - z_{k-1}$.

Let the number of subdivisions increase in such a way that the largest of the chord length $|\delta z_k|$ approaches zero. Then the sum S_n approaches a limit which does not depend upon the mode of subdivision and we denote

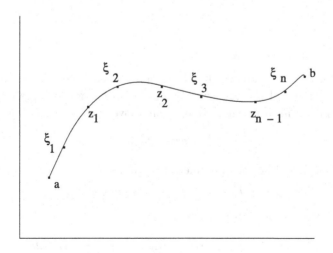

Fig. 2.1

21

this limit by

$$\int_{C_a}^{b} f(z)dz \qquad (2.2)$$

called the line integral of $f(z)$ along the curve C. If this limit exists $f(z)$ is said to be integrable along C. Note that if $f(z)$ is analytic at all points of a region R and if C is a curve lying in R then $f(z)$ is certainly integrable along C.

2.2

If $f(z)$ is integrable along a curve C having a finite length L and if there exists a positive number M such that $|f(z)| \leq M$ on C, then

$$\left| \int_c f(z)dz \right| \leq ML. \qquad (2.3)$$

By definition

$$\int_c f(z)dz = \lim_{n \to \infty} \sum_{k=1}^{n} f(\xi_k)\delta z_k. \qquad (2.4)$$

Now

$$\left| \sum_{k=1}^{n} f(\xi_k)\delta z_k \right| \leq \sum_{k=1}^{n} |f(\xi_k)||\delta z_k| \leq M \sum_{k=1}^{n} |\delta z_k|. \qquad (2.5)$$

Taking the limit of both sides

$$\left| \int_c f(z)dz \right| \leq ML. \qquad (2.6)$$

2.3   Real Line Integrals

If $P(x, y)$ and $Q(x, y)$ are real functions of x and y, continuous at all points of curve C, the real line integral of $Pdx + Qdy$ along the curve C can be defined in a manner similar to that given above and is denoted by

$$\int_C [Pdx + Qdy]. \qquad (2.7)$$

If C is smooth and has the parametric equation

$$x = \phi(t), \ y = \psi(t) \quad \text{where } t_1 \leq t \leq t_2 \qquad (2.8)$$

the value of the above integral is given by

$$\int_{t_1}^{t_2} [P\{\phi(t), \psi(t)\}\phi'(t) + Q\{\phi(t), \ \{\psi(t)\}\psi'(t)\}]dt. \qquad (2.9)$$

If $f(z) = u(x,y) + iv(x,y)$ the complex line integral can be expressed in terms of the real line integral as

$$\int_c f(z)dz = \int_c (udx - vdy) + i\int_c (vdx + udy). \qquad (2.10)$$

2.4 Simply and Multiply Connected Regions

A region R is called simply connected if any simple closed curve which lies in R can be shrunk to a point without leaving R. A region which is not simply connected is called multiply connected. For example:- Suppose R is the region defined by $|z| < 2$ as shown in Fig. 2.2.

If Γ is any simple closed curve lying in R we see that it can be shrunk to a point which lies in R so that R is simply connected. On the other hand if R is the region defined by $1 < |z| < 2$ as shown in Fig. 2.3 then there is a simple closed curve Γ lying in R which cannot be shrunk to a point without leaving R, so that R is multiply connected.

2.5 Convention Regarding Traversal of a Closed Path

The boundary C of a region is said to be traversed in the positive direction (sense) if an observer traveling in this direction (and perpendicular to the plane) has the region to the left. This convention leads to the directions indicated by arrows in Figs. 2.2 and 2.3. We also use the special symbol,

$$\oint f(z)dz$$

to denote the integral around the boundary C in the positive sense. This is called a contour integral.

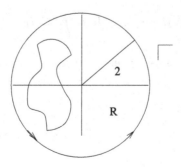

Fig. 2.2

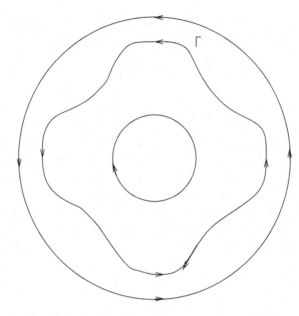

Fig. 2.3

2.6   Green's Theorem in a Plane

Let $P(x,y)$ and $Q(x,y)$ be continuous and have continuous partial derivative in a region R and its closed boundary C. Green's theorem states that

$$\oint_c (Pdx + Qdy) = \int_R \int \left(\frac{\partial Q}{\partial x} - \frac{\partial P}{\partial y} \right) dxdy. \qquad (2.11)$$

We may also state the complex form of Green's theorem

$$\oint_c B(z, \bar{z})dz = 2i \int_R \int \frac{\partial B}{\partial \bar{z}} dxdy. \qquad (2.12)$$

Proof of Green's theorem is entirely elementary and will be omitted.

2.7   Cauchy's Theorem

Let $f(z)$ be analytic in a region R and on its boundary C. Then

$$\oint_C f(z)dz = 0.$$

Proof. We shall prove this theorem, following Cauchy, with the added restriction that $f'(z)$ be continuous in R.

$$\oint_C f(z)dz = \oint_c (u+iv)(dx+idy)$$

$$= \oint_c (udx - vdy) + i\oint (vdx + udy)$$

$$= -\int_R \int \left(\frac{\partial v}{\partial x} + \frac{\partial u}{\partial y}\right) dxdy + i\int_R \int \left(\frac{\partial u}{\partial x} - \frac{\partial v}{\partial y}\right) dxdy$$

$$= 0 \tag{2.13}$$

by Cauchy-Riemann equations. $\qquad\qquad\square$

2.8   Converse of Cauchy's Theorem or Morera's Theorem

Let $f(z)$ be continuous in a simply connected region R and suppose that

$$\oint_C f(z)dz = 0 \tag{2.14}$$

around every simple closed curve C in R. Then $f(z)$ is analytic in R. The proof of this theorem is left as an exercise.

2.9   (a) Indefinite Integrals

If $f(z)$ and $F(z)$ are analytic in a region R and such that $F'(z) = f(z)$ then $F(z)$ is called an indefinite integral or antiderivative of $f(z)$ denoted by

$$\int f(z)dz = F(z) \tag{2.15}$$

the traditional definition of indefinite integral.

2.10   Some Consequences of Cauchy's Theorem

Let $f(z)$ be analytic in a simply connected region R. Then the following theorems hold:

1. If a and z are any two points in R then $\int_a^z f(z)dz$ is independent of the path in R joining a and z.
2. If a and z are any two points in R and $G(z) = \int_a^z f(z)dz$ then $G(z)$ is analytic in R and $G'(z) = f(z)$.

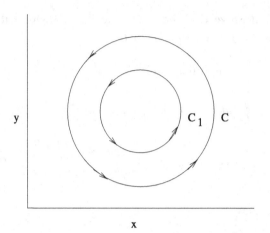

y

x

Fig. 2.4

3. If a and b are any two points in R and $F'(z) = f(z)$ then

$$\int_a^b f(z)dz = F(b) - F(a).$$

4. Let $f(z)$ be analytic in a region bounded by two simple closed curves C and C_1 (where C_1 lies inside C as in Fig. 2.4) and on these curves. Then

$$\oint_C f(z)dz = \oint_{C_1} f(z)dz \qquad (2.16)$$

where C and C_1 are both traversed in the positive sense relative to their interiors.

This shows that the contour of integration can be deformed in any way we like as long as we do not cross any singularity.

Exercise. (a) Let C be any simple closed curve bounding a region having area A. Then

$$A = \frac{1}{2}\oint(+xdy - ydx)). \qquad (2.17)$$

$$\oint(-ydx + xdy) = \int_A \int \left(\frac{\partial x}{\partial x} + \frac{\partial y}{\partial y}\right)dxdy \qquad (2.18)$$

$$= 2A$$

$$A = \frac{1}{2}\oint(xdy - ydx). \qquad (2.19)$$

(b) Use it to compute the area of the ellipse

$$\frac{x^2}{a^2} + \frac{y^2}{b^2} = 1.$$ (2.20)

(c) The parametric equation of the ellipse is given by

$$x = a \cos \psi$$
$$y = b \sin \psi$$ (2.21)

$$A = \frac{1}{2} \left(\oint x\,dy - y\,dx \right)$$ (2.22)

$$= \frac{1}{2} \int_0^{2\pi} d\psi [a \cos \psi (b \cos \psi) + b \sin \psi \, a \sin \psi]$$

$$= \frac{ab}{2} \int_0^{2\pi} d\psi (\sin^2 \psi + \cos^2 \psi) = \pi ab.$$ (2.23)

(d) Evaluate the Fresnel integrals:

$$\int_0^\infty \sin x^2 dx = \int_0^\infty \cos x^2 dx = \sqrt{\frac{\pi}{8}}.$$ (2.24)

Consider

$$\oint_\Sigma e^{iz^2} dz = 0 \quad \text{by Cauchy's theorem}$$ (2.25)

where Σ is the contour shown in Fig. 2.5

$$\int_0^R e^{iz^2} dx + \int_S e^{iz^2} dz + e^{\frac{i\pi}{4}} \int_R^0 e^{i\left(re^{\frac{i\pi}{4}}\right)^2} dr$$ (2.26)

$$= 0.$$

On S $z = Re^{i\theta}$ $0 \le \theta \le \frac{\pi}{4}$ (2.27)

$$\int_S = \int e^{iR^2(\cos 2\theta + i \sin 2\theta)} Re^{i\theta} id\theta$$ (2.28)

$$\left| \int_s e^{iz^2} dz \right| = \left| iR \int^{\frac{\pi}{4}} e^{i\theta} e^{iR^2 \cos 2\theta - R^2 \sin 2\theta} d\theta \right|$$ (2.29)

$$\le R \int_0^{\frac{\pi}{4}} e^{-R^2 \sin 2\theta} d\theta$$ (2.30)

$$= \frac{R}{2} \int_0^{\frac{\pi}{2}} e^{-R^2 \sin \alpha} d\alpha.$$ (2.31)

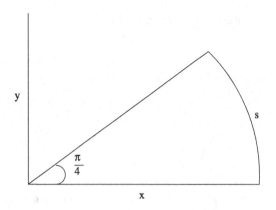

Fig. 2.5

Now $\sin \alpha \geq \frac{2\alpha}{\pi}$ for $0 \leq \alpha \leq \frac{\pi}{2}$ (see Eq. (2.155))

$$\left| \int_s e^{iz^2} dz \right| \leq \frac{R}{2} \times \int_0^{\frac{\pi}{2}} e^{-\frac{R^2\alpha}{\pi}} d\alpha = \frac{\pi}{4R}(1 - e^{-R^2}) \qquad (2.32)$$

$$\to 0 \quad \text{as} \quad R \to \infty$$

$$\int_0^\infty e^{ix^2} dx = e^{\frac{i\pi}{4}} \int_0^\infty e^{-r^2} dr = e^{\frac{i\pi}{4}} \frac{\sqrt{\pi}}{2} \qquad (2.33)$$

$$= \frac{1}{2}(1+i)\frac{\sqrt{\pi}}{\sqrt{2}} \qquad (2.34)$$

$$\int_0^\infty \cos x^2 dx = \sqrt{\frac{\pi}{8}} \qquad (2.35)$$

$$\int_0^\infty \sin x^2 dx = \sqrt{\frac{\pi}{8}}. \qquad (2.36)$$

2.11   Cauchy's Integral Formulas

If $f(z)$ is analytic inside and on a simple closed curve C and a is any point inside C then

$$f(a) = \frac{1}{2\pi i} \oint_C \frac{f(z)dz}{z - a} dz \qquad (2.37)$$

where C is traversed in the positive (counterclockwise) sense.

Also the nth derivative of $f(z)$ at $z = a$ is given by

$$f^{(n)}(a) = \frac{n!}{2\pi i} \oint_C \frac{f(z)dz}{(z - a)^{n+1}}. \qquad (2.38)$$

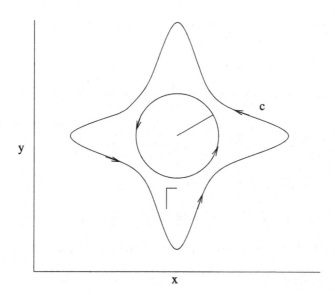

Fig. 2.6

These results are called Cauchy's integral formulas and are quite remarkable because they show that if the function $f(z)$ is known on the simple closed curve C then its value and its derivatives can be found at all points inside C.

Proof. The function $\frac{f(z)}{(z-a)}$ is analytic inside and on C except at the point $z = a$. The contour of integration can, therefore, be deformed to Γ:

$$\frac{1}{2\pi i} \oint_C \frac{f(z)dz}{z-a} = \frac{1}{2\pi i} \oint_\Gamma \frac{f(z)dz}{z-a}. \tag{2.39}$$

On Γ

$$z - a = \epsilon e^{i\theta} \quad dz = i\epsilon e^{i\theta} d\theta. \tag{2.40}$$

Thus

$$\frac{1}{2\pi i} \oint_C \frac{f(z)dz}{z-a} dz = \frac{1}{2\pi\ i}\ i \int_0^{2\pi} f(a + \epsilon e^{i\theta})d\theta. \tag{2.41}$$

Taking the limit $\epsilon \to 0$

$$\frac{1}{2\pi i} \oint_c \frac{f(z)dz}{z-a} = \frac{1}{2\pi} f(a) \int_0^{2\pi} d\theta = f(a). \tag{2.42}$$

We now consider the derivatives of the analytic function

$$f'(a) = \lim_{h \to 0} \frac{f(a+h) - f(a)}{h} \tag{2.43}$$

$$= \lim_{h \to 0} \frac{1}{2\pi i h} \left[\oint \frac{f(z)dz}{(z-a-h)} - \oint \frac{f(z)dz}{(z-a)} \right] \tag{2.44}$$

$$= \lim_{h \to 0} \frac{1}{2\pi i} \frac{1}{h} h \oint \frac{f(z)dz}{(z-a)(z-a-h)} \tag{2.45}$$

$$= \lim_{h \to 0} \frac{1}{2\pi i} \oint \frac{f(z)}{(z-a)^2} \left[1 + \frac{h}{(z-a-h)} \right] \tag{2.46}$$

$$= \frac{1}{2\pi i} \oint_C \frac{f(z)dz}{(z-a)^2} + \lim_{h \to 0} \frac{h}{2\pi i} \oint_C \frac{f(z)dz}{(z-a)^2(z-a-h)}. \tag{2.47}$$

Now on C $f(z)$ is continuous and therefore bounded, and so is $(z-a)^{-2}$. Therefore $\left| \frac{f(z)}{(z-a)^2(z-a-h)} \right|$ is bounded; let its upper bound be K. Then, if l be the length of C,

$$\left| \lim_{h \to 0} \frac{h}{2\pi i} \oint \frac{f(z)dz}{(z-a)^2(z-a-h)} \right| \leq \frac{|h|}{2\pi} \oint \left| \frac{f(z)dz}{(z-a)^2(z-a-h)} \right| \tag{2.48}$$

$$\leq \frac{|h|}{2\pi} Kl \to 0 \tag{2.49}$$

and consequently

$$f'(a) = \frac{1}{2\pi i} \oint_c \frac{f(z)dz}{(z-a)^2}. \tag{2.50}$$

Similarly

$$f''(a) = \frac{2}{2\pi i} \oint \frac{f(z)dz}{(z-a)^2} \tag{2.51}$$

$$f^n(a) = \frac{n!}{2\pi i} \oint \frac{f(z)dz}{(z-a)^{n+1}} \tag{2.52}$$

which can be proved by induction.

Corollary. *If $f(z)$ is an analytic one valued function of z in a ring shaped region bounded by two curves C and C' and a is a point in the region, then*

$$f(a) = \frac{1}{2\pi i} \oint_C \frac{f(z)dz}{z-a} - \frac{1}{2\pi i} \oint_{C'} \frac{f(z)dz}{z-a} \tag{2.53}$$

where C is the outer of the curves and the integrals are taken counterclockwise.

Some further properties of Cauchy's integral formulas

$$\oint_C \frac{dz}{z-a} = 0 \tag{2.54}$$

where C is any simple closed curve and $z = a$ is outside C and

$$\frac{1}{2\pi i} \oint_C \frac{dz}{z-a} = 1 \quad \text{if the point } z = a \text{ is inside } C. \tag{2.55}$$

When $z = a$ is outside C, $\frac{1}{z-a}$ is analytic everywhere within and on C and hence by Cauchy's theorem.

$$\oint_C \frac{dz}{z-a} = 0 \quad \text{for } z = a \text{ outside } C.$$

When the point $z = a$ is inside C.

$$f(a) = \frac{1}{2\pi i} \oint_c \frac{f(z)dz}{z-a}. \tag{2.56}$$

Setting $f(z) = z^0 = 1$ so that $f(a) = a^0 = 1$ we have

$$\frac{1}{2\pi i} \int \frac{dz}{z-a} = 1 \tag{2.57}$$

$$\oint_c \frac{dz}{z-a} = 2\pi i \quad \text{for } z = a \text{ inside } C. \tag{2.58}$$

Corollary.

$$\frac{1}{2\pi i} \oint_c \frac{dz}{(z-a)^n} = 0 \ \textit{because } f'(a) = f''(a) = \cdots = 0 \tag{2.59}$$

for $n = 2, 3, 4, \ldots$.

2.12 Uniform Convergence and Taylor's and Laurent's Series

Let $u_1(z), u_2(z), \ldots, u_n(z)$ denote briefly by $\{u_n(z)\}$ to be a sequence of functions of z defined and single valued in some region of the z plane. We call $u(z)$ the limit of $u_n(z)$ as $n \to \infty$ and write $\lim_{n\to\infty} u_n(z) = u(z)$. Formally

$$|u_n(z) - u(z)| < \epsilon \tag{2.60}$$

for all $n > N$ where N depends in general on both ϵ and z.

In such a case we say that the sequence converges in a region R. We call R the region of convergence of the sequence. A sequence which is not convergent at some value of z is called divergent at z.

2.13 Uniform Convergence of Sequence or Series

In the definition of the limit of a sequence of functions it was pointed out that the number N depends in general on ϵ and the particular value of z. It may happen however, that we can find a number N such that

$$|u_n(z) - u(z)| < \epsilon \tag{2.61}$$

for all $n > N$ where the same number N holds for all z in a region R (i.e. N depends only on ϵ and not on the particular value of z). In such case we say that $u_n(z)$ converges uniformly or is uniformly convergent to $u(z)$.

Example.

$$u_n(z) = z - z^{n+1}. \tag{2.62}$$

If $|z| < 1$, $\lim_{n \to \infty} z^{n+1} = 0$

$$\therefore \quad \lim_{n \to \infty} u_n(z) = z. \tag{2.63}$$

We shall now show that for $|z| \le \frac{1}{2}$ the sequence converges uniformly to z.

$$|u_n(z) - z| = \epsilon \tag{2.64}$$

$$|-z^{n+1}| < \epsilon \tag{2.65}$$

$$n > \frac{|\ln \epsilon|}{(\ln |z|)} - 1. \tag{2.66}$$

When $|z| \le \frac{1}{2}$ the largest value of $\left|\frac{\ln \epsilon}{\ln z}\right|$ occurs when $|z| = \frac{1}{2}$ and given by $\frac{|\ln \epsilon|}{\ln 2} - 1 = N$. It should be noted that for smaller values $|z|$, e.g. $|z| = 3^{-1}$ it is $N' = \frac{|\ln \epsilon|}{\ln 3} - 1 < N$. It therefore follows that $|u_n(z) - z| < \epsilon$ for all $n > N$ where N depends only on ϵ and not on the particular value of z in $|z| \le \frac{1}{2}$. Thus the sequence converges uniformly to z for $|z| \le \frac{1}{2}$. The same argument given above serves to show that sequence converges uniformly to z for $|z| \le 0.9$ or $|z| \le 0.99$ using

$$N = \left|\frac{\ln \epsilon}{\ln .9}\right| - 1 \quad \text{and} \quad N = \left|\frac{\ln \epsilon}{\ln .99}\right| - 1 \tag{2.67}$$

respectively.

It can be similarly proved that the sequence

$$\{u_n(z)\} = \left\{\frac{1}{1 + nz}\right\} \tag{2.68}$$

is uniformly convergent to zero for all z, such that $|z| \ge \delta$, i.e. in any region excluding the point $z = 0$.

These two examples serve to emphasize the essential features of the concept of uniform convergence. If $u_n(z)$ tends to a definite limit for all values of z in a closed interval $|z| \leq R$, we say the sequence converges uniformly to that limit. For example let us take

$$u_n(z) = n \sin \frac{z}{n} \tag{2.69}$$

and we consider the closed interval

$$0 \leq |z| \leq R. \tag{2.70}$$

Then

$$\lim_{n \to \infty} n \sin \frac{z}{n} = z \quad \text{for all } z \text{ in } 0 \leq |z| \leq R \tag{2.71}$$

and the sequence converges uniformly to z when z lies within this interval.

However if $u_n(z) = n \sin \frac{1}{nz}$ then $u_n(z)$ does not tend to a definite limit around $z = 0$. Hence if $0 \leq z \leq R$ the convergence is not uniform.

Similarly if the sequence of partial sums $\{S_n(z)\}$ converges uniformly to $S(z)$ in a given region R, we say that the infinite series.

$$u_1(z) + u_2(z) + \cdots = \sum_{n=1}^{\infty} u_n(z) \tag{2.72}$$

converges uniformly to $S(z)$ in a given region.

If we call

$$R_n(z) = u_{n+1}(z) + u_{n+2}(z) \ldots \tag{2.73}$$

the remainder of the infinite series after n terms, we can equivalently say that the series is uniformly convergent to $S(z)$ in R if given any $\epsilon > 0$, we can find a number N such that for all z in R

$$|R_n(z)| < \epsilon \quad \text{for } n > N \tag{2.74}$$

i.e. for all z in R

$$\lim_{n \to \infty} R_n(z) = 0.$$

We now state some important theorems:

1. If $|u_n(z)| \leq |M_n|$ where M_n is independent of z in a region R and $\sum |M_n|$ converges the series $\sum_{n=1}^{\infty} u_n(z)$ is uniformly convergent in R (Weierstrass M test).

2. If $\{u_n(z)\}$ are analytic in R and $\sum_{n=1}^{\infty} u_n(z)$ is uniformly convergent in R, then

$$S(z) = \sum_{n=1}^{\infty} u_n(z) \tag{2.75}$$

is analytic in R.

3. A power series

$$\sum_{n=1}^{\infty} a_n(z - a)^n \tag{2.76}$$

converges uniformly and absolutely in any region which lies entirely inside its circle of convergence.

2.14 Expansion of Functions in Infinite Series: (a) Taylor's Theorem

Let $f(z)$ be analytic inside and on a simple closed curve C. Let a and $a+h$ be two points inside C, then

$$f(a+h) = f(a) + hf'(a) + \frac{h^2}{z!}f''(a) + \cdots + \frac{h^n}{n!}f^{(n)}(a) + \cdots . \tag{2.77}$$

writing $a + h = z$, $h = z - a$;

$$f(z) = f(a) + (z - a)f'(a) + \frac{(z - a)^2}{z!}f''(a) + \cdots + \frac{(z - a)^n}{n!}f^{(n)}(a) + \cdots . \tag{2.78}$$

The region of convergence of the series is given by $|z - a| < R$ where the radius of convergence R is the distance from a to the nearest singularity of the function $f(z)$. On $|z - a| = R$ the series may or may not converge. For $|z - a| > R$ the series diverges.

Proof. Let $f(z)$ be analytic inside a circle C with center a. Let z be any point inside C. Construct a circle C_1 with center at a and enclosing z.

By Cauchy's integral formula

$$f(z) = \frac{1}{2\pi i} \oint_{c_1} \frac{f(\omega)d\omega}{\omega - z} \tag{2.79}$$

$$\frac{1}{\omega - z} = \frac{1}{(\omega - a) - (z - a)} = \frac{1}{\omega - a}\left[\frac{1}{1 - \frac{z-a}{\omega-a}}\right] \tag{2.80}$$

$$= \frac{1}{\omega - a}\left[\frac{1 - \left(\frac{z-a}{\omega-a}\right)^n}{1 - \frac{z-a}{\omega-a}} + \frac{\left(\frac{z-a}{w-a}\right)^n}{1 - \frac{z-a}{\omega-a}}\right].$$

Setting $\frac{z-a}{\omega-a} = \rho$.

The first term within the bracket is given by

$$\frac{1-\rho^n}{1-\rho} = 1 + \rho + \rho^2 + \cdots + \rho^{n-1}$$

$$= 1 + \left(\frac{z-a}{\omega-a}\right) + \left(\frac{z-a}{\omega-a}\right)^2 + \cdots + \left(\frac{z-a}{\omega-a}\right)^{n-1}. \quad (2.81)$$

Thus

$$\frac{1}{\omega-z} = \frac{1}{\omega-a}\left[1 + \left(\frac{z-a}{\omega-a}\right) + \left(\frac{z-a}{\omega-a}\right)^2 + \cdots + \left(\frac{z-a}{\omega-a}\right)^{n-1}\right] + \frac{\frac{(z-a)^n}{(\omega-a)^n}}{\omega-z}$$

$$= \frac{1}{\omega-a} + \frac{z-a}{(\omega-a)^2} + \frac{(z-a)^2}{(\omega-a)^3}$$

$$+ \cdots + \frac{(z-a)^{n-1}}{(\omega-a)^n} + \frac{\left(\frac{z-a}{\omega-a}\right)^n}{(\omega-z)}. \quad (2.82)$$

Multiplying both sides of this equation by $f(\omega)$ and integrating over C_1,

$$\frac{1}{2\pi i}\oint_{C_1}\frac{f(\omega)d\omega}{\omega-z} = f(z) = \frac{1}{2\pi i}\oint\frac{f(\omega)d\omega}{\omega-a} + \frac{(z-a)}{2\pi i}\int\frac{f(\omega)d\omega}{(\omega-a)}$$

$$+ \frac{(z-a)^2}{2\pi i}\oint\frac{f(\omega)d\omega}{(\omega-a)^3} + \cdots + \frac{(z-a)^{n-1}}{2\pi i}\oint_{c!}\frac{f(\omega)d\omega}{(\omega-a)^n} + R_n \quad (2.83)$$

where

$$R_n = \frac{1}{2\pi i}\int\left(\frac{z-a}{\omega-a}\right)^n\frac{f(\omega)d\omega}{\omega-z}. \quad (2.84)$$

By Cauchy's integral formulas

$$f^{(n)}(a) = \frac{n!}{2\pi i}\int\frac{f(\omega)d\omega}{(\omega-a)^{n+1}}. \quad (2.85)$$

Thus we have

$$f(z) = f(a) + (z-a)f'(a) + \frac{(z-a)^2}{2!}f''(a) + \cdots + \frac{(z-a)^{n-1}}{(n-1)!}f^{(n-1)}(a) + R_n. \quad (2.86)$$

If we can now show that $\lim_{n\to\infty} R_n = 0$, we will have proved the required result. To do that we note that

$$\left|\frac{z-a}{\omega-a}\right| = \frac{|z-a|}{|\omega-a|} < 1 \quad (2.87)$$

$$|\omega-a| > |z-a|$$

$$\frac{r}{R} < 1.$$

Hence

$$|R_n| = \frac{1}{2\pi} \left| \oint \left(\frac{z-a}{\omega-a} \right)^n \frac{f(\omega)}{\omega-z} d\omega \right| \leq \frac{1}{2\pi} \left(\frac{r}{R} \right)^n \oint \frac{|f(\omega)d\omega|}{|\omega-z|}. \qquad (2.88)$$

Now

$$|\omega - z| = |(\omega-a) - (z-a)| \geq |\omega-a| - |z-a| \qquad (2.89)$$

$$\therefore \quad \frac{1}{|\omega-z|} \leq \frac{1}{|\omega-a| - |z-a|} = \frac{1}{R-r} \qquad (2.90)$$

$$|R_n| \leq \frac{1}{2\pi} \left(\frac{r}{R} \right)^n \oint \frac{|f(\omega)|Rd\theta}{R-r}. \qquad (2.91)$$

Since $f(\omega)$ is continuous its modulus cannot exceed some finite number M

$$\therefore \quad |f(\omega)| \leq M \qquad (2.92)$$

$$|R_n| \leq \frac{1}{\cancel{2\pi}} \left(\frac{r}{R} \right)^n \frac{R}{R-r} M \times \cancel{2\pi} \qquad (2.93)$$

$$= \left(\frac{r}{R} \right)^n \frac{RM}{R-r} \to 0$$

as $n \to \infty$.

Thus we obtain Taylor's expansion

$$f(z) = f(a) + (z-a)f'(a) + \frac{(z-a)^2}{z!}f''(a) + \cdots + \frac{(z-a)^n}{n!}f^{(n)}(a) + \cdots . \qquad (2.94)$$

2.14.1   *Laurent's Theorem*

If $f(z)$ is analytic inside and on the boundary of the ring shaped region R bounded by two concentric circles C_1 and C_2 with center at a and respective radii r_1 and $r_2(r_1 > r_2)$ then for all z in R

$$f(z) = \sum_{n=0}^{\infty} a_n(z-a)^n + \sum_{n=1}^{\infty} \frac{a_{-n}}{(z-a)^n} \qquad (2.95)$$

where

$$a_n = \frac{1}{2\pi i} \oint_{c_1} \frac{f(\omega)d\omega}{(\omega-a)^{n+1}} \quad n = 0, 1, 2 \ldots \qquad (2.96)$$

$$a_{-n} = \frac{1}{2\pi i} \oint_{c_2} \frac{f(\omega)d\omega}{(\omega-a)^{-n+1}}. \qquad (2.97)$$

By Cauchy's integral formula

$$f(z) = \frac{1}{2\pi i} \oint_{C_1} \frac{f(\omega)d\omega}{\omega - z} - \frac{1}{2\pi i} \oint_{C_2} \frac{f(\omega)d\omega}{\omega - z}. \tag{2.98}$$

As before for ω on C_1,

$$\frac{1}{\omega - z} = \left[\frac{1}{\omega - a} + \frac{z - a}{(\omega - a)^2} + \frac{(z - a)^2}{(\omega - a)^3} \right.$$
$$\left. + \frac{(z - a)^{n-1}}{(\omega - a)^n} + \left(\frac{z - a}{\omega - a} \right)^n \frac{1}{\omega - z} \right]. \tag{2.99}$$

So that

$$\frac{1}{2\pi i} \oint_c \frac{f(\omega)d\omega}{\omega - z} = \frac{1}{2\pi i} \oint_{C_1} \frac{f(\omega)d\omega}{\omega - a} + \frac{(z - a)}{2\pi i} \oint_{C_1} \frac{f(\omega)d\omega}{(\omega - a)^2}$$
$$+ \frac{(z - a)^2}{2\pi i} \oint \frac{f(\omega)d\omega}{(\omega - a)^3} + \cdots + \frac{(z - a)^{n-1}}{2\pi i} \oint \frac{f(\omega)d\omega}{(\omega - a)^n}$$
$$+ \frac{1}{2\pi i} \oint_{C_2} \left(\frac{z - a}{\omega - a} \right)^n \times \frac{f(\omega)}{\omega - z}$$
$$= a_0 + a_1(z - a) + a_2(z - a)^2$$
$$+ \cdots + (z - a)^{n-1} a_{n-1} + R_n^{(1)} \tag{2.100}$$

where

$$a_{n-1} = \frac{1}{2\pi i} \oint \frac{f(\omega)d\omega}{(\omega - a)^n} \quad n = 1, 2, 3 \ldots \tag{2.101}$$

$$R_n^{(1)} = \frac{1}{2\pi i} \oint \left(\frac{z - a}{\omega - a} \right)^n \frac{f(\omega)d\omega}{\omega - z}. \tag{2.102}$$

Let us consider the second integral. For ω on C_2 we write

$$-\frac{1}{\omega - z} = \frac{1}{z - \omega} = \frac{1}{(z - a) - (\omega - a)}$$
$$= \frac{1}{z - a} \left[\frac{1 - \left(\frac{\omega - a}{z - a} \right)^n}{1 - \left(\frac{\omega - a}{z - a} \right)} + \frac{\left(\frac{\omega - a}{z - a} \right)^n}{1 - \frac{\omega - a}{z - a}} \right]$$
$$= \left[\frac{1}{z - a} + \frac{\omega - a}{(z - a)^2} + \frac{(\omega - a)^2}{(z - a)^3} + \cdots + \frac{(\omega - a)^{n-1}}{(z - a)^n} \right.$$
$$\left. + \left(\frac{\omega - a}{z - a} \right)^n \frac{1}{(z - \omega)} \right]. \tag{2.103}$$

Hence

$$-\frac{1}{2\pi i} \oint_{C_2} \frac{f(\omega)d\omega}{\omega - z} = \frac{a_{-1}}{z - a} + \frac{a_{-2}}{(z - a)^2} + \cdots + \frac{a_{-n}}{(z - a)^n} + R_n^{(2)} \qquad (2.104)$$

where

$$R_n^{(2)} = \frac{1}{2\pi i} \oint_{C_2} \left(\frac{\omega - a}{z - a}\right)^n \frac{f(\omega)d\omega}{z - \omega} \qquad (2.105)$$

and

$$a_{-n} = \frac{1}{2\pi i} \oint_{C_2} (\omega - a)^{n-1} f(\omega)d\omega. \qquad (2.106)$$

Proceeding as before we can show that

$$\lim_{n\to\infty} R_n^{(1)} = 0, \quad \lim_{n\to\infty} R_n^{(2)} = 0. \qquad (2.107)$$

Hence we obtain the Laurent expansion

$$f(z) = \sum_{n=1}^{\infty} \frac{a_{-n}}{(z - a)^n} + \sum_{n=0}^{\infty} a_n(z - a)^n \qquad (2.108)$$

where

$$a_n = \frac{1}{2\pi i} \oint_{C_1} \frac{f(\omega)d\omega}{(\omega - a)^{n+1}} \qquad (2.109)$$

$$a_{-n} = \frac{1}{2\pi i} \oint_{C_2} (\omega - a)^{n-1} f(\omega)d\omega. \qquad (2.110)$$

Example 1. Find the Laurent expansion of the analytic function

$$f(z) = \frac{1}{(z - 1)(z + 2)} \quad \text{about } z = -2. \qquad (2.111)$$

Setting $z + 2 = u$, $z = u - 2$, $z + 1 = u - 1$. We have

$$f(u) = \psi(i) = \frac{u - 2}{(u - 1)u} = \frac{2 - u}{u}[1 + u + u^2 + \cdots]$$

$$= \left(\frac{2}{u} - 1\right)[1 + u + u^2 + \cdots]$$

$$= \left(\frac{2}{u} - 1\right)[1 + u + u^2 + \cdots] - [1 + u + u^2 \cdots]$$

$$= \frac{2}{u}(1 + u + u^2 + \cdots) - (1 + u + u^2 + \cdots)$$

$$= \frac{2}{z + 2} + 1 + (z + 2) + (z + 2)^2 + \cdots. \qquad (2.112)$$

This series converges for $|z + 2| < 1$.

Example 2.

$$\text{Expand} \quad f(z) = e^{\frac{\alpha}{2}(z-\frac{1}{z})} \tag{2.113}$$

about $z = 0$

$$f(z) = \sum_{-\infty}^{\infty} a_n(\alpha) z^n \tag{2.114}$$

$$a_n = \frac{1}{2\pi i} \oint_{C_1} \frac{f(\omega) d\omega}{\omega^{n+1}} \tag{2.115}$$

$$= \frac{1}{2\pi i} \oint_{C_1} \frac{e^{\frac{\alpha}{2}(\omega - \frac{1}{\omega})}}{\omega^{n+1}} d\omega.$$

Taking C_1 to be the unit circle

$$a_n(\alpha) = \frac{1}{2\pi} \int_0^{2\pi} e^{i\alpha \sin\theta - \sin\theta} d\theta$$

$$= \frac{1}{2\pi} \int_0^{2\pi} d\theta [\cos(\alpha \sin\theta - n\theta) + i \sin(\alpha \sin\theta - n\theta)]$$

$$= \frac{1}{2\pi} \int_0^{2\pi} d\theta \cos(\alpha \sin\theta - n\theta) \tag{2.116}$$

because the second term vanishes.

Thus $a_n(\alpha) = J_n(\alpha)$ where the Bessel function

$$J_n(\alpha) = \frac{1}{2\pi} \int_0^{2\pi} \cos(n\theta - \alpha \sin\theta) d\theta. \tag{2.117}$$

2.15   The Residue Theorem

Let $f(z)$ be single valued and analytic inside and on a circle C except at a point $z = a$ chosen as the center of C. Then $f(z)$ has a Laurent series about $z = a$ given by

$$f(z) = \sum_{n=-\infty}^{\infty} a_n(z-a)^n = a_0 + a_1(z-a) + a_2(z-a)^2 + \cdots + \frac{a_{-1}}{z-a} + \frac{a_{-2}}{(z-a)^2} \cdots \tag{2.118}$$

where

$$a_n = \frac{1}{2\pi i} \oint_C \frac{f(z)}{(z-a)^{n+1}}, \quad n = 0, 1, 2, 3 \ldots \tag{2.119}$$

$$a_{-n} = \frac{1}{2\pi i} \oint_C (z-a)^{n-1} f(z) dz, \quad n = 1, 2, 3. \tag{2.120}$$

Setting $n = 1$ we have

$$\oint_C f(z)dz = 2\pi i a_{-1}. \tag{2.121}$$

The coefficient a_{-1} in the Laurent expansion is called the residue of $f(z)$ at $z = a$.

2.16   Cauchy's Residue Theorem

Let $f(z)$ be single valued and analytic inside and on a simple closed curve C except at the poles $a, b, c \ldots$ which have residues given by $a_{-1}, b_{-1}, c_{-1}, \ldots$. then the residue theorem states that,

$$\oint f(z)dz = 2\pi i(a_{-1} + b_{-1} + \cdots). \tag{2.122}$$

Proof. With centers at $a, b, c \ldots$ respectively, let us construct circles $c_1, c_2, c_3 \ldots$ which lie entirely inside C as shown in Fig. 7. Then by Cauchy's theorem.

$$\oint f(z)dz = \oint_{C_1} f(z)dz + \oint_{C_2} f(z)dz + \oint_{C_3} f(z)dz + \cdots. \tag{2.123}$$

But we have already shown

$$\oint_{C_1} f(z)dz = 2\pi i \, a_{-1}. \tag{2.124}$$

$$\oint_{C_2} f(z)dz = 2\pi i \, b_{-1}, \text{ etc.} \tag{2.125}$$

Hence

$$\oint f(z)dz = 2\pi i(a_{-1} + b_{-1} + \cdots \tag{2.126}$$

$$= 2\pi i \sum R \tag{2.127}$$

where $\sum R$ stands for the the sum of residues.   □

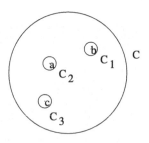

Fig. 2.7

2.17   Calculation of Residues

To obtain the residue of a function $f(z)$ at $z = a$, it may appear that the Laurent expansion of $f(z)$ about $z = a$ must be obtained. however in the case when $z = a$ is a pole of order m there is a simple formula for a_{-1} given by

$$a_{-1} = \lim_{z \to a} \frac{1}{(m-1)!} \frac{d^{m-1}}{dz^{m-1}} [(z-a)^m f(z)]. \tag{2.128}$$

For a simple pole $(m = 1)$ result is especially simple and is given by

$$a_{-1} = \lim_{z \to a} (z-a) f(z). \tag{2.129}$$

Proof. If $f(z)$ has a pole of order m, then the Laurent series of $f(z)$ is given by,

$$f(z) = \frac{a_{-m}}{(z-a)^m} + \frac{a_{-m+1}}{(z-a)^{m-1}} + \cdots + \frac{a_{-1}}{z-a} + a_0 + a_1(z-a) + a_z(z-a)^2 + \cdots . \tag{2.130}$$

Hence

$$(z-a)^m f(z) = a_{-m} + a_{-m+1}(z-a) + a_{-m+2}(z-a)^2$$
$$+ \cdots + a_{-1}(z-a)^{m-1} + a_0(z-a)^m + a_1(z-a)^{m+1} + \cdots .$$

Differentiating both sides with respect to z and taking $\lim_{z \to a}$ we have

$$\lim_{z \to a} \frac{d^{m-1}[(z-a)^m f(z)]}{dz^{m-1}} = a_{-1}(m-1)! \tag{2.131}$$

We therefore obtain

$$a_{-1} = \frac{1}{(m-1)!} \lim_{z \to a} \frac{d^{m-1}[(z-a)^m f(z)]}{dz^{m-1}}. \tag{2.132}$$

2.18   Evaluation of Integrals of the Type

$$I = \int_0^{2\pi} R(\cos\theta, \sin\theta) d\theta \tag{2.133}$$

where the integrand is a rational function of $\sin\theta$ and $\cos\theta$. We write $e^{i\theta} = z$ so that

$$\cos\theta = \frac{1}{2}\left(z + \frac{1}{z}\right) \quad \sin\theta = \frac{1}{2i}\left(z - \frac{1}{z}\right). \tag{2.134}$$

Thus the integral takes the form

$$I = \oint_c S(z) dz \tag{2.135}$$

where $S(z)$ is a rational function of z and c stands for the unit circle

$$\therefore \quad I = 2\pi i \sum R. \tag{2.136}$$

$\sum R$ is the sum of the residues of the poles inside the unit circle.

Example 1.

$$I = \int \frac{d\theta}{1 - 2p\cos\theta + p^2}, \quad 0 < p < 1 \tag{2.137}$$

$$I = \frac{1}{-ip} \oint_c \frac{dz}{z^2 - \frac{(1-p^2)z}{p} + 1} \tag{2.138}$$

$$= -\frac{1}{ip} \oint \frac{1}{(z-p)\left(z - \frac{1}{p}\right)}. \tag{2.139}$$

The integrand has simple poles at $z = p$ and $z = \frac{1}{p}$. Since $0 < p < 1$, the pole at $z = \frac{1}{p}$ lies outside the unit circle C.

$$I = -\frac{2\pi \cancel{i}}{\cancel{i}p\left(p - \frac{1}{p}\right)} = \frac{2\pi}{(1-p^2)}. \tag{2.140}$$

Example 2.

$$I = \int_0^{2\pi} \frac{d\theta}{a + b\sin\theta}, \quad a > |b|. \tag{2.141}$$

Setting $e^{i\theta} = z$ we have

$$I = 2 \oint_c \frac{dz}{bz^2 + 2aiz - b}. \tag{2.142}$$

The roots of the quadratic are

$$\alpha = i\left[\sqrt{\frac{a^2}{b^2} - 1} - \frac{a}{b}\right] \tag{2.143}$$

$$\beta = -i\left[\sqrt{\frac{a^2}{b^2} - 1} + \frac{a}{b}\right] \tag{2.144}$$

$$|\alpha| = \left[\sqrt{\frac{a^2}{b^2} - 1} - \frac{a}{b}\right] \tag{2.145}$$

$$= \left| \frac{\frac{a^2}{b^2} - 1 - \frac{a^2}{b^2}}{\sqrt{\frac{a^2}{b^2} - 1} + \frac{a}{b}} \right| \qquad (2.146)$$

$$= \frac{|b|/|a|}{1 + \sqrt{1 - \frac{b^2}{a^2}}} < 1. \qquad (2.147)$$

Hence the pole $z = \propto$ lies inside the unit circle.

$$I = \frac{2 \times 2\pi i}{(\propto - \beta)} = \frac{2 \times \not{2}\pi \not{i}}{\not{2} \not{i} \sqrt{\frac{a^2}{b^2} - 1}} \qquad (2.148)$$

$$= \frac{2\pi b}{\sqrt{a^2 - b^2}}. \qquad (2.149)$$

2.19 Definite Integrals of the Type

$$I = \int_{-\infty}^{\infty} F(x) dx. \qquad (2.150)$$

If $\lim_{z \to \infty} |z| F(z) \to 0$ then

$$\int_{-\infty}^{\infty} F(x) dx = 2\pi i \sum R \qquad (2.151)$$

where $\sum R$ is the sum of the residues in the upper half plane. The proof of this is left as an exercise.

Example.

$$I = \int_{-\infty}^{\infty} \frac{x^2 dx}{(x^2 + a^2)^2}, \quad a > 0 \qquad (2.152)$$

$$I = \int \frac{(x^2 + a^2 - a^2)}{(x^2 + a^2)^2} dx = \int_{-\infty}^{\infty} \frac{dx}{x^2 + a^2} - a^2 \int_{-\infty}^{\infty} \frac{dx}{(x^2 + a^2)^2} \qquad (2.153)$$

$$= \frac{\pi}{a} + \frac{2\pi \not{i} \times 2a^2}{-8 \not{i} a^3} = \frac{\pi}{2a}. \qquad (2.154)$$

Now we shall prove an inequality:

$$\frac{2\theta}{\pi} \leq \sin \theta \leq \theta, \quad \text{for } 0 \leq \theta \leq \frac{\pi}{2}. \qquad (2.155)$$

We shall give a graphical non rigorous proof. We start from the graph of $\sin \theta$ between $0 \leq \theta \leq \frac{\pi}{2}$.

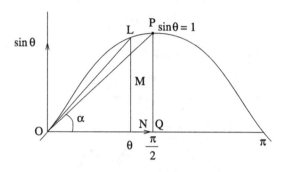

Fig. 2.8

The equation of the straight line OP (Fig. 2.8) is given by

$$y = \tan \alpha \cdot \theta. \qquad (2.156)$$

From the figure

$$\tan \alpha = \frac{PQ}{OQ} = \frac{1}{\frac{\pi}{2}} = \frac{2}{\pi} \qquad (2.157)$$

so that Eq. (2.156) reduces to

$$y = \frac{2\theta}{\pi}.$$

Now taking the point N such that $ON = \theta$, $MN = y NL = \sin\theta$. Thus

$$y = MN \le NL = \sin\theta$$

$$y \le \sin\theta \quad \text{(equality holds only at } P\text{)}.$$

Thus

$$\frac{2\theta}{\pi} \le \sin\theta. \qquad (2.158)$$

Now consider Fig. 2.9,

$$\sin\theta = \frac{d}{r}.$$

Since d is perpendicular, h is hypotenuse

$$d < h$$

$$\frac{d}{r} < \frac{h}{r}\sin\theta \le \frac{h}{r} \quad \left(\text{equality only if } \theta = \frac{\pi}{2}\right)$$

$$\sin\theta < \frac{h}{r}.$$

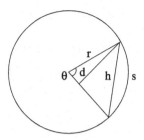

Fig. 2.9

But as the chord h is always less than the arc length s we have

$$h < s$$

$$\frac{h}{r} < \frac{s}{r} = \theta$$

$$\therefore \quad \sin \theta \leq \theta.$$

Hence we have finally

$$\frac{2\theta}{\pi} \leq \sin \theta \leq \theta.$$

We are now in a position to prove Jordan's lemma

$$\lim_{R \to \infty} \left| \int_\Gamma Q(z)e^{imz}dz \right| = 0, \quad m > 0 \tag{2.159}$$

where Γ is a semicircle of radius R and

$$\lim_{R \to \infty} |Q(z)| = \frac{1}{R^k}, \quad k > 0.$$

Now

$$\left| \int Q(z)\frac{e^{imz}}{\pi}dz \right| < R \int_\Gamma |Q(z)||e^{imz}|d\theta$$

$$= \frac{1}{R^{k-1}} \int_0^\pi e^{-2mR\sin\theta}d\theta$$

$$= \frac{2}{R^{k-1}} \int_0^{\pi/2} e^{-mR\sin\theta}d\theta$$

$$< \frac{2}{R^{k-1}} \int_0^{\pi/2} e^{-\frac{2R\theta m}{\pi}}d\theta$$

$$= \frac{\pi}{mR^k}(1 - e^{-mR}) \to 0 \quad \text{as } R \to \infty.$$

Using Jordan's lemma we can evaluate integrals of the form

$$\int_{-\infty}^{\infty} \cos mx \, Q(x) \, dx, \quad Q(x) = Q(-x)$$

$$\int_{-\infty}^{\infty} \sin mx \, Q(x) \, dx, \quad Q(x) = -Q(-x). \tag{2.160}$$

Examples.

(i) $\displaystyle\int_0^{\infty} \frac{\cos x}{x^2 + a^2} dx = \frac{\pi}{2a} e^{-a}, \quad a > 0$ \hfill (2.161)

(ii) $\displaystyle\int \frac{\cos 2ax - \cos 2bx}{x^2} = \pi(b - a) \quad a \geq 0, \; b \geq 0$ \hfill (2.162)

(iii) $\displaystyle\int \frac{x \sin ax}{x^2 + k^2} = \frac{1}{2}\pi \, e^{-ka} \quad k > 0, \; a > 0.$ \hfill (2.163)

2.20 Integrals Involving Branch Points

(a) Integrals of the form

$$\int_0^{\infty} x^{a-1} Q(x) dx. \tag{2.164}$$

Let $Q(x)$ be a rational function of x such that it has no poles on the positive part of the real axis and $x^a Q(x) \to 0$ both when $x \to 0$ and when $x \to \infty$. We take a cut extending from 0 to ∞. Then

$$(1 - e^{2\pi i a}) \int_0^{\infty} x^{a-1} Q(x) dx = 2\pi i \sum R \tag{2.165}$$

where $\sum R$ denote the sum of the residues of $z^{a-1} Q(z)$ at its poles

$$\int_0^{\infty} x^{a-1} Q(x) dx = \frac{2\pi i \sum R}{(1 - e^{2\pi i a})}. \tag{2.166}$$

Example 1.

$$Q(x) = \frac{1}{1 + x} \tag{2.167}$$

$$\sum R = e^{i\pi(a-1)} = -e^{i\pi a} \tag{2.168}$$

$$\therefore \quad \int_0^{\infty} \frac{x^{a-1} dx}{1 + x} = \frac{2\pi i e^{i\pi a}}{(e^{2\pi i a} - 1)} = \frac{\pi}{\sin \pi a}. \tag{2.169}$$

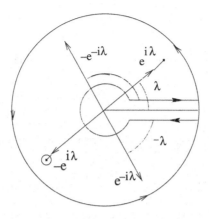

Fig. 2.10

Example 2.

$$\int_0^\infty \frac{x^{-p}dx}{1 + 2x\cos\lambda + x^2} = \frac{\pi}{\sin p\pi}\frac{\sin p\lambda}{\sin\lambda}$$

$$-1 < p < 1, \quad -\pi < \lambda < \pi \qquad (2.170)$$

$$\oint_C \frac{dz\, z^{-p}}{(z + e^{i\lambda})(z + e^{-i\lambda})} = 2\pi i \sum R \qquad (2.171)$$

$$\arg(-e^{i\lambda}) = \pi + \lambda$$
$$\therefore \quad -e^{i\lambda} = e^{i(\pi+\lambda)} \qquad (2.172)$$
$$-e^{-i\lambda} = e^{i(\pi-\lambda)}$$

$$\sum R = \frac{e^{-ip(\pi+\lambda)}}{(e^{i(\pi+\lambda)} + e^{-i\lambda})} + \frac{e^{-ip(\pi-\lambda)}}{(e^{i(\pi-\lambda)} + e^{i\lambda})} \qquad (2.173)$$

$$= -e^{-ip\pi}\frac{\sin p\lambda}{\sin\lambda}.$$

Since the integral over the infinite circle vanishes,

$$(1 - e^{-2\pi ip})\int_0^\infty \frac{x^{-p}dx}{1 + 2x\cos\lambda + x^2} = 2\pi i e^{-\pi ip}\frac{\sin p\lambda}{\sin\lambda} \qquad (2.174)$$

thus

$$\int_0^\infty \frac{x^{-p}dx}{1 + 2x\cos\lambda + x^2} = \frac{\pi}{\sin\pi p}\frac{\sin p\lambda}{\sin\lambda}. \qquad (2.175)$$

Example 3.

$$\int_0^\infty \frac{x^{a-1}}{x^2 + c^2} \sin\left(\frac{\pi a}{2} - bx\right) dx = \frac{\pi}{2} c^{a-2} e^{-bc} \quad \text{(Cauchy)} \quad a < 2, \ b > 0 \tag{2.176}$$

Let us consider

$$\oint_c z^{a-1} \frac{e^{-i(\pi a + bz)}}{z^2 + c^2} dz = -2\pi i \quad \text{(residue at } z = -ic) \tag{2.177}$$

where c is the contour shown in Fig. 2.11.

$$\text{Now, the r.h.s} = -2i\pi (ce^{3i\pi/2})^{a-1} \times \frac{e^{-i(\pi a - ibc)}}{-2ic}$$

$$= i\pi e^{i\pi a/2} c^{a-2} e^{-bc}. \tag{2.178}$$

Since the integral over the infinite semicircle vanishes, the l.h.s. of Eq. (2.177) is given by,

$$-\int_\infty^\epsilon (e^{i\pi} x)^{a-1} e^{-i(\pi a - bx)} \frac{dx}{x^2 + c^2}$$

$$+ \int_\epsilon^\infty (e^{2i\pi} x)^{a-1} e^{-i(\pi a + bx)} \frac{dx}{x^2 + c^2} \tag{2.179}$$

whence

$$\int_0^\infty \frac{x^{a-1}}{x^2 + c^2} \sin\left(\frac{\pi a}{2} - bx\right) dx = \frac{\pi}{2} c^{a-2} e^{-bc}. \tag{2.180}$$

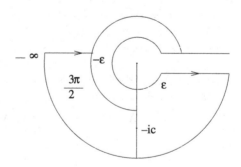

Fig. 2.11

Example 4.

$$\oint_{|z|=10} z^2 \ln \frac{z+1}{z-1} dz = \frac{4i\pi}{3} \tag{2.181}$$

We have

$$\oint_{|z|=10} z^2 \ln \frac{z+1}{z-1} dz = \int_c z^2 \ln \frac{z+1}{z-1} dz$$

$$= \int_{+1}^{-1} x^2 \left[\ln \frac{(1+x)}{(1-x)e^{i\pi}} \right] dx + \int_{-1}^{1} x^2 \left[\ln \frac{(1+x)e^{2i\pi}}{(1-x)e^{i\pi}} \right] dx$$

$$= \int_{+1}^{-1} \left[x^2 \ln \left(\frac{(1+x)}{(1-x)} \right) - i\pi x^2 \right] + \int_{-1}^{1} \left[x^2 \ln \left(\frac{1+x}{1-x} \right) + i\pi x^2 \right] dx$$

$$= 2i\pi \int_{-1}^{1} x^2 dx = \frac{4i\pi}{3}. \tag{2.182}$$

Example 5.

$$\int_0^1 x^{a-1}(1-x)^{-a} dx = \frac{\pi}{\sin \pi a}$$

$$\int_s z^{a-1}(z-1)^{-a} dx = \int_c z^{a-1}(z-1)^{-a} dx. \tag{2.183}$$

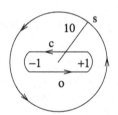

Fig. 2.12

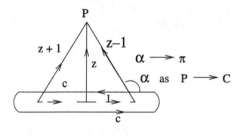

Fig. 2.13

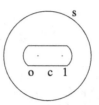

Fig. 2.14

On $S, z = Re^{i\theta}$ with $R \to \infty$

$$\int_S = \int_0^{2\pi} R^{a-\gamma}e^{i\theta(a-\gamma)}R^{-a}e^{-i\theta a}\, R\, e^{i\theta}id\theta$$

$$= i \int_0^{2\pi} d\theta = 2\pi i. \tag{2.184}$$

Hence

$$2\pi i = \int_1^0 x^{a-1}(1-x)^{-a}e^{-i\pi a} + \int_0^1 x^{a-1}e^{2i\pi a}(1-x)^{-a}e^{-i\pi a}. \tag{2.185}$$

Finally

$$2i \sin \pi a \int_0^1 x^{a-1}(1-x)^{-a}dx = 2\pi i. \tag{2.186}$$

Thus

$$\int_0^1 x^{a-1}(1-x)^{-a}dx = \frac{\pi}{\sin \pi a}. \tag{2.187}$$

Example 6.

$$\int_\alpha^\beta \left(\frac{\beta-t}{t-\alpha}\right)^{a-1}\frac{dt}{t} = \frac{\pi}{\sin a\pi}\left[1 - \left(\frac{\beta}{\alpha}\right)^{a-1}\right] \qquad 0 < \alpha < \beta,\ a < 2. \tag{2.188}$$

Consider

$$\int_\Sigma (z-\beta)^{a-1}(z-\alpha)^{1-a}\frac{dz}{z} \tag{2.189}$$

where Σ is the multiply connected region bounding the region A as shown in Fig. 2.15.

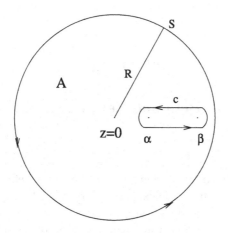

Fig. 2.15

Since the only singularity in A is a simple pole at $z = 0$,

$$\frac{1}{2\pi i} \oint_s (z - \alpha)^{1-a}(z - \beta)^{a-1} \frac{dz}{z} \tag{2.190}$$

$$-\frac{1}{2\pi i} \oint_c (z - \alpha)^{1-a}(z - \beta)^{a-1} \frac{dz}{z} \tag{2.191}$$

$$= \left(\frac{\beta}{\alpha}\right)^{a-1}. \tag{2.192}$$

Setting on S

$$z = Re^{i\theta} \quad \text{with } R \to \infty. \tag{2.193}$$

Now

$$\frac{1}{2\pi i} \int_S \left(\frac{z - \beta}{z - \alpha}\right)^{a-1} \frac{dz}{z} = \frac{1}{2\pi i} 2\pi i = 1. \tag{2.194}$$

Thus finally

$$1 - \left(\frac{e^{i\pi a} - e^{-i\pi a}}{2\pi}\right) \int_\alpha^\beta \left(\frac{\beta - t}{t - \alpha}\right)^{a-1} \frac{dt}{t} = \left(\frac{\beta}{\alpha}\right)^{a-1} \tag{2.195}$$

or

$$\int \left(\frac{\beta - t}{t - \alpha}\right)^{a-1} \frac{dt}{t} = \frac{\pi}{\sin \pi a} \left[1 - \left(\frac{\beta}{\alpha}\right)^{a-1}\right]. \tag{2.196}$$

CHAPTER 3

The Gamma, Beta and Zeta Function of Riemann

The three special functions mentioned above were defined by Euler only on the right half of the complex plane. After Cauchy introduced the concept of analytic function the question arose as to whether Euler's definition could be extended to cover the entire complex plane. This problem was successfully tackled by Cauchy himself for the gamma function and subsequently by Pochhammer and Reimann for the beta and zeta functions respectively. We present here the basic theory in a form which has not so far been presented in any treatise of classical analysis. This formulation will also facilitate the later developments (see Chapters 6 and 9).

3.1   The Process of Analytic Continuation

We start this chapter with a brief introduction to the concept of analytic continuation. Let $F_1(z)$ be a function which is analytic in a region R_1. Let us suppose that we can find a function $F_2(z)$ which is analytic in a region R_2 and which is such that $F_1(z) = F_2(z)$ in the region common to R_1 and R_2. See Fig. 3.1. Then we say that $F_2(z)$ is the analytic continuation of $F_1(z)$. This means that there is a function $F(z)$ analytic in the combined regions R_1 and R_2 such that $F(z) = F_1(z)$ in R_1 and $F(z) = F_2(z)$ in R_2.

The most important description of functions in analysis are either through definite integrals or through power series. We shall first discuss the process of analytic continuation for functions defined by definite integrals of appropriate form. The simplest function defined by a definite integral is the Γ function.

$$\Gamma(z) = \int_0^\infty e^{-t} t^{z-1} dt, \quad \operatorname{Re} z > 0. \tag{3.1}$$

Cauchy recognized that the restriction $\operatorname{Re} z > 0$ is a consequence of the limitation of the above integral but not of the analytic function which it defines. To appreciate this point let us take a simple example. Let us consider the

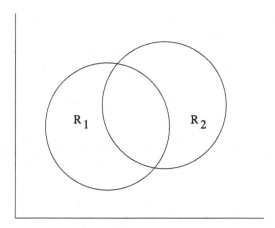

Fig. 3.1

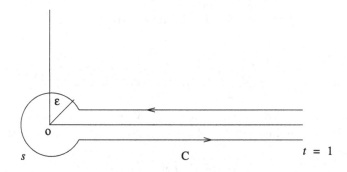

Fig. 3.2

function

$$u(z) \equiv \int_0^1 t^{z-1} dt = \frac{1}{z}, \quad \text{Re}\, z > 0. \tag{3.2}$$

It should be noted that for $\text{Re}\, z \leq 0$ the l.h.s. breaks down but the function $\frac{1}{z}$ is still well defined. Therefore we must modify the integral such that it remains valid for all values of z and reduces to the original line integral extending from 0 to 1 for $\text{Re}\, z > 0$. This is achieved following Cauchy by constructing the contour integral

$$\mu(z) = \int_c t^{z-1} dt \tag{3.3}$$

where C is the contour shown in the Fig. 3.2.

We shall show that the contour integral converges everywhere although the line integral does not.

It should be noted that on the upper linear part of the contour arg $t = 0$ and on the lower linear part arg $t = 2\pi$. Thus

$$\mu(z) = \int_1^{\epsilon} t^{z-1}dt + \int_s t^{z-1}dt + e^{2\pi i z}\int_{\epsilon}^1 t^{z-1}dt$$

$$= (e^{2\pi i z} - 1)\int_{\epsilon}^1 t^{z-1}dt + \int_s t^{z-1}dt. \tag{3.4}$$

Thus

$$\mu(z) = (e^{2\pi i z} - 1)\frac{(1 - \epsilon^z)}{z} + \int_s t^{z-1}dt. \tag{3.5}$$

We now consider the integral over the circle s. Setting $t = \epsilon e^{i\theta}$

$$\int_s t^{z-1}dt = i\epsilon^z \int_0^{2\pi} e^{i\theta z}d\theta = \frac{\epsilon^z}{z}(e^{2\pi i z} - 1). \tag{3.6}$$

Thus

$$\mu(z) = (e^{2\pi i z} - 1)\frac{(1 - \epsilon^z)}{z} + (e^{2\pi i z} - 1)\frac{\epsilon^z}{z} = \frac{(e^{2\pi i z} - 1)}{z}. \tag{3.7}$$

Note that the divergent term cancels exactly and it is not at all necessary to consider only infinitesimal circle. The above analysis shows that:

$$\mu(z) = \int_c t^{z-1}dt \tag{3.8}$$

is finite for all z and is independent of the radius ϵ of the circle s. Next we look for the value of $\mu(z)$ for $\operatorname{Re} z > 0$. For $\operatorname{Re} z > 0$.

$$\int_s t^{z-1}dt = 0. \tag{3.9}$$

Hence

$$\mu(z) = (e^{2\pi i z} - 1)\lim_{\epsilon \to 0}\int_{\epsilon}^1 t^{z-1}dt = (e^{2\pi i z} - 1)u(z). \tag{3.10}$$

Hence generally

$$u(z) = \frac{1}{(e^{2\pi i z} - 1)}\int_c t^{z-1}dt \tag{3.11}$$

defines the function throughout the complex plane.

Exercise.

The integral $\int_{-\infty}^{\infty} e^{-zx^2} dx = f(z)$ defines an analytic function for $\mathrm{Re}\, z > 0$. Analytically continue it to the whole complex plane.

The above example gives us the fundamental method of analytic continuation developed by Cauchy. This method consists of three steps:

1. Replace the line integral by a suitable contour integral which converges everywhere.
2. Examine the contour integral in the restricted region in which the line integral is defined. The contour integral should yield the line integral and some additional factors.
3. Divide the contour integral by these additional factors.
   This procedure will yield the desired analytic continuation of the function beyond the region of convergence of its integral representation.

3.2 Analytic Continuation of the Gamma Function

Let us consider the Euler representation of the gamma function

$$\Gamma(z) = \int_0^\infty e^{-t} t^{z-1} dt, \quad \mathrm{Re}\, z > 0. \tag{3.12}$$

The restriction $\mathrm{Re}\, z > 0$ is a consequence of the requirement of convergence of the Eulerian representation.

Step 1. Following Cauchy's method we consider

$$\mu(z) = \int_c e^{-t} t^{z-1} dt \tag{3.13}$$

where C is the contour shown in Fig. 3.3. We first ensure that the contour integral converges for all z and therefore defines an analytic function

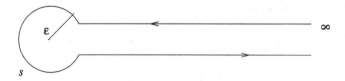

Fig. 3.3

throughout the complex plane.

$$\mu(z) = \int_\infty^\epsilon e^{-t}t^{z-1}dt + \int_s e^{-t}t^{z-1}dt + e^{2\pi i z}\int_\epsilon^\infty e^{-t}t^{z-1}dt$$

$$= (e^{2\pi i z} - 1)\int_\epsilon^\infty e^{-t}t^{z-1}dt + \int_s e^{-t}t^{z-1}dt$$

$$= (e^{2\pi i z} - 1)\int_\epsilon^1 e^{-t}t^{z-1}dt + \int_s e^{-t}t^{z-1}dt$$

$$+ (e^{2\pi i z} - 1)\int_1^\infty e^{-t}t^{z-1}dt.$$

The last integral is finite for all finite z and represents an entire function. Let us call it $E(z)$. Thus

$$\mu(z) = \lambda(z) + (e^{2\pi i z} - 1)E(z) \tag{3.14}$$

where

$$\lambda(z) = \left(e^{2\pi i z} - 1\right)\int_\epsilon^1 e^{-t}t^{z-1}dt + \int_s e^{-t}t^{z-1}dt. \tag{3.15}$$

Now it is easy to see

$$\int_\epsilon^1 e^{-t}t^{z-1}dt = \sum \frac{(-1)^n}{n!}\int_\epsilon^1 t^{n+z-1}dt$$

$$= \sum \frac{(-1)^n}{n!}\frac{(1-\epsilon^{n+z})}{n+z}. \tag{3.16}$$

The integration over the circle can be performed by setting $t = \epsilon e^{i\theta}$ and expanding, as before, the exponential so that

$$\int_s e^{-t}t^{z-1}dt = i\sum \frac{(-1)^n}{n!}\epsilon^{n+z}\int_0^{2\pi} e^{i\theta(n+z)}d\theta$$

$$= (e^{2\pi i z} - 1)\sum \frac{(-1)^n}{n!}\frac{\epsilon^{n+z}}{n+z}. \tag{3.17}$$

Combining these two results

$$\lambda(z) = (e^{2\pi i z} - 1)\sum \frac{(-)^n}{n!}\frac{(1-\epsilon^{n+z})}{(n+z)} + (e^{2\pi i z} - 1)\sum \frac{(-)^n}{n!}\frac{\epsilon^{n+z}}{(n+z)}$$

$$= (e^{2\pi i z} - 1)\sum_{n=0}^\infty \frac{(-)^n/n!}{n+z}. \tag{3.18}$$

Thus

$$\mu(z) = \lambda(z) + (e^{2\pi i z} - 1)\mathrm{E}(z). \tag{3.19}$$

Step 2. The second step is to examine the contour integral for $\mathrm{Re}\, z > 0$:

$$\mu(z) = (e^{2\pi i z} - 1) \lim_{\epsilon \to 0} \int_\epsilon^\infty e^{-t} t^{z-1} dt + \lim_{\epsilon \to 0} \int_s e^{-t} t^{z-1} dt. \tag{3.20}$$

Now since $\mathrm{Re}\, z > 0$

$$\lim_{\epsilon \to 0} \int_s e^{-t} t^{z-1} dt \sim \lim_{\epsilon \to 0} i\epsilon^z \int_0^{2\pi} e^{i\theta z} d\theta = 0. \tag{3.21}$$

Thus for $\mathrm{Re}\, z > 0$

$$\mu(z) = (e^{2\pi i z} - 1)\Gamma(z). \tag{3.22}$$

Thus the formula

$$\Gamma(z) = \frac{1}{(e^{2\pi i z} - 1)} \int_c e^{-t} t^{z-1} dt \tag{3.23}$$

defines the Γ-function everywhere.

The above contour integral representation reveals a very important property of the Γ function. We have already seen

$$\mu(z) = \int_c e^{-t} t^{z-1} dt = \lambda(z) + (e^{2\pi i z} - 1)\mathrm{E}(z)$$

$$= (e^{2\pi i z} - 1) \sum \frac{(-1)^n/n!}{n+z} + (e^{2\pi i z} - 1)\mathrm{E}(z). \tag{3.24}$$

Thus

$$\Gamma(z) = \frac{\mu(z)}{(e^{2\pi i z} - 1)} = \sum_{n=0}^\infty \frac{(-1)^n/n!}{(n+z)} + \mathrm{E}(z). \tag{3.25}$$

Since $\mathrm{E}(z)$ is an entire function the only finite singularities of $\Gamma(z)$ are simple poles at

$$z = -n, \quad n = 0, 1, 2 \ldots$$

and the residue at the simple pole at $z = -n$ is $(-1)^n/n!$. Thus $\Gamma(z)$ is a meromorphic function whose only finite singularities are simple poles at negative integer points.

We shall now prove an important product representation of the gamma function,

$$\Gamma(z) = \lim_{n \to \infty} \frac{n!\, n^z}{z(z+1)\cdots(z+n)}. \tag{3.26}$$

Using the fact that

$$e^{-t} = \lim_{n \to \infty} \left(1 - \frac{t}{n}\right)^n. \tag{3.27}$$

We have

$$\Gamma(z) = \frac{1}{(e^{2\pi i z} - 1)} \int_c e^{-t} t^{z-1} dt \tag{3.28}$$

$$= \frac{1}{(e^{2\pi i z} - 1)} \lim_{n \to \infty} \int_c \left(1 - \frac{t}{n}\right)^n t^{z-1} dt \tag{3.29}$$

where the contour c is given by Fig. 3.4.
We therefore define

$$\pi(z, n) = \frac{1}{(e^{2\pi i z} - 1)} \int_c \left(1 - \frac{t}{n}\right)^n t^{z-1} dt. \tag{3.30}$$

Setting $t = n\tau$ so that $\epsilon \leq \tau \leq 1$ we have

$$\pi(z, n) = n^z (1 - \tau)^n \left.\frac{\tau^z}{z}\right|_\epsilon^1 + \frac{n}{z} n^z \int_\epsilon^1 (1 - \tau)^{n-1} \tau^z d\tau$$

$$+ \frac{1}{(e^{2\pi i z} - 1)} n^z \int_s (1 - \tau)^n \tau^{z-1} d\tau. \tag{3.31}$$

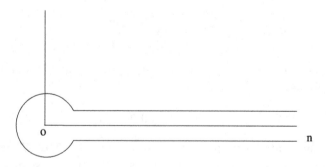

Fig. 3.4

Since the upper limit in the first term vanishes

$$\pi(z,n)n^{-z} = -(1-\epsilon)^n \frac{\epsilon^z}{z} + \frac{n}{z} \int_\epsilon^1 (1-\tau)^{n-1} \tau^z d\tau$$

$$+ \frac{1}{(e^{2\pi iz} - 1)} \int_s (1-\tau)^n \tau^{z-1} d\tau. \qquad (3.32)$$

Carrying out the last integration by parts

$$\pi(z,n)n^{-z} = -(1-\epsilon)^n \frac{\epsilon^z}{z} + \frac{n}{z} \int_\epsilon^i (1-\tau)^{n-1} \tau^z d\tau$$

$$+ \frac{1}{(e^{2\pi iz} - 1)} \frac{n}{z} \int_s (1-\tau)^{n-1} \tau^z d\tau$$

$$+ \frac{1}{(e^{2\pi iz} - 1)} (1 - \epsilon e^{i\theta})^n \frac{(\epsilon e^{i\theta})^z}{z} \Big|_0^{2\pi} \qquad (3.33)$$

$$= \frac{n}{z} \frac{1}{(e^{2\pi iz} - 1)} \int_c (1-\tau)^{n-1} \tau^z d\tau. \qquad (3.34)$$

We therefore obtain

$$\mu_n(z) = n^{-z} \pi(z,n)$$

$$= \frac{n}{z} \mu_{n-1}(z+1)$$

$$= \frac{n(n-1)\mu_{n-2}(z+2)}{z(z+1)}. \qquad (3.35)$$

Proceeding in this way

$$\mu_n(z) = \frac{n!}{z(z+1)\cdots(z+n-1)} \frac{1}{(e^{2\pi iz} - 1)} \int_e \tau^{n+z-1} d\tau$$

$$= \frac{n!}{z(z+1)\cdots(z+n)}. \qquad (3.36)$$

We therefore obtain

$$\pi(z,n) = \frac{n! \, n^z}{z(z+1)(z+2)\cdots(z+n)}. \qquad (3.37)$$

Proceeding to the limit

$$\Gamma(z) = \lim_{n\to\infty} \pi(z,n) = \lim_{n\to\infty} \frac{n! \, n^z}{z(z+1)\cdots(z+n)}. \qquad (3.38)$$

We shall now derive the Wierstrassian product

$$\frac{1}{\Gamma(z)} = z e^{\gamma z} \Pi_{n=1}^\infty \left[\left(1 + \frac{z}{n}\right) e^{\frac{-z}{n}} \right] \qquad (3.39)$$

where

$$\gamma = \lim_{m \to \infty} \left[\frac{1}{1} + \frac{1}{2} + \frac{1}{3} + \cdots + \frac{1}{m} - \ln m \right] = 0.5772157\ldots. \qquad (3.40)$$

The constant γ is known as Euler's or Mascheroni's constant; to prove that it exists we observe that, if

$$u_n = \int_0^1 \frac{dt\, t}{n(n+t)} = \left[\frac{1}{n} - \ln \frac{n+1}{n} \right] \qquad (3.41)$$

$$|u_n| < \int_0^1 \left| \frac{t}{n(n+t)} \right| |dt| < \frac{1}{n(n+1)} \int_0^1 dt < \frac{1}{n^2}. \qquad (3.42)$$

Hence $\sum_1^\infty u_n$ converges and we denote its sum by γ

$$\sum_{n=1}^\infty u_n = \gamma. \qquad (3.43)$$

Now

$$u_n = \frac{1}{n} - \ln \left(\frac{n+1}{n} \right). \qquad (3.44)$$

Hence

$$u_1 + u_2 + \cdots + u_m$$

$$= \sum_{n=1}^m \frac{1}{n} - \ln \left[\frac{\not{2}}{1} \times \frac{\not{3}}{\not{2}} \times \frac{\not{4}}{\not{3}} \times \frac{\not{5}}{\not{4}} \times \cdots \times \frac{\not{m}}{m \not{-1}} \times \frac{m+1}{\not{m}} \right]$$

$$= \sum_{n=1}^m \frac{1}{n} - \ln(m+1). \qquad (3.45)$$

We therefore obtain

$$\sum_{n=1}^m \frac{1}{n} = \sum_{n=1}^m u_n + \ln(m+1). \qquad (3.46)$$

Hence

$$\sum_{n=1}^m \frac{1}{n} - \ln m = \sum_{n=1}^m u_n + \ln \left(\frac{m+1}{m} \right). \qquad (3.47)$$

Thus

$$\lim_{m \to \infty} \left[\sum_{n=1}^m \frac{1}{n} - \ln m \right] = \lim_{m \to \infty} \left[\sum_{n=1}^m u_n + \ln \left(1 + \frac{1}{m} \right) \right]$$

$$= \sum_{n=1}^\infty u_n = \gamma. \qquad (3.48)$$

We now write

$$
\begin{aligned}
\frac{1}{\Gamma(z)} &= \lim_{n\to\infty} \frac{z(z+1)\cdots(z+n)}{n!\, n^z} \\
&= \lim_{n\to\infty} \frac{z(z+1)\cdots(z+n)e^{-z\ln n}}{1.2\cdots n} \\
&= \lim_{n\to\infty} z(1+z)\left(1+\frac{z}{2}\right)\cdots\left(1+\frac{z}{n}\right)e^{-z\ln n} \\
&= \lim_{n\to\infty} z(1+z)e^{-z}\left(1+\frac{z}{2}\right)e^{\frac{-z}{2}}\cdots\left(1+\frac{z}{n}\right)e^{\frac{-z}{n}} \\
&\quad \times e^{z\left[\frac{1}{1}+\frac{1}{2}+\cdots+\frac{1}{n}-\ln n\right]} \\
&= ze^{\gamma z}\Pi_{n=1}^{\infty}\left(1+\frac{z}{n}\right)e^{\frac{-z}{n}}.
\end{aligned}
\tag{3.49}
$$

Exercise. Derive Euler's formula for gamma function.

$$
\Gamma(z) = \frac{1}{z}\Pi_{n=1}^{\infty}\left(1+\frac{1}{n}\right)^z\left(1+\frac{z}{n}\right)^{-1}.
\tag{3.50}
$$

We shall now prove the important property of the gamma function.

$$
\Gamma(z+1) = z\Gamma(z).
\tag{3.51}
$$

We start from the contour integral representation

$$
\Gamma(z+1) = \frac{1}{(e^{2\pi iz}-1)}\int_c e^{-t}\, t^z dt.
\tag{3.52}
$$

Integrating by parts and rearranging the terms we have

$$
\begin{aligned}
\Gamma(z+1) &= -\epsilon^z \cancel{e}^{-\epsilon} + z\int_\epsilon^\infty e^{-t}\, t^{z-1} + \cancel{e}^z e^{-\epsilon} \\
&\quad + \frac{1}{(e^{2\pi iz}-1)}z\int_s e^{-t}\, t^{z-1}dt \\
&= \left(\frac{1}{(e^{2\pi iz}-1)}\right)z\int_c e^{-t}\, t^{z-1}dt.
\end{aligned}
\tag{3.53}
$$

Hence we have

$$
\Gamma(z+1) = z\Gamma(z).
\tag{3.54}
$$

We shall now show that

$$
\Gamma(z)\Gamma(1-z) = \frac{\pi}{\sin\pi z}.
\tag{3.55}
$$

Without any loss of generality we may assume $\mathrm{Re}\, z > 0$. So that in general $\mathrm{Re}(1 - z) < 0$. Thus we have

$$\Gamma(z) = \int e^{-t_1} t_1^{z-1} dt_1 \tag{3.56}$$

$$\Gamma(1 - z) = \frac{1}{(e^{-2\pi i z} - 1)} \int_c e^{-t_2} t_2^{-z} dt_2. \tag{3.57}$$

Hence, setting $t_2/t_1 = x$, $t_1 = y$, we obtain

$$\Gamma(z)\Gamma(1 - z)(e^{-2\pi i z} - 1) = \int_c dx\, x^{-z} \int_0^\infty dy\, e^{-y(1+x)}$$

$$= \int_c \frac{dx\, x^{-z}}{1 + x}. \tag{3.58}$$

It is evident that since the integral is regularized at the origin it is finite for all z. The verification of this is left as an exercise. Since the only singularity on the left of the branch cut is a simple pole at $x = e^{i\pi}$ as shown in Fig. 3.5 by residue theorem

$$\int_c \frac{x^{-z}}{1 + x} dx + \int_s \frac{x^{-z} dx}{1 + x} = -2\pi i e^{-i\pi z}. \tag{3.59}$$

Since the contribution of the infinite circle S is zero we have

$$\int_c \frac{x^{-z} dx}{1 + x} = -2\pi i e^{-i\pi z}. \tag{3.60}$$

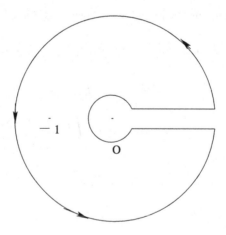

Fig. 3.5

Thus

$$\Gamma(z)\Gamma(1-z) = \frac{-2\pi i e^{-i\pi z}}{(e^{-2\pi i z} - 1)} = \frac{\pi}{\sin \pi z}. \tag{3.61}$$

3.3   The Beta Function

The beta function is defined by the integral

$$\mathrm{B}(x,y) = \int_0^1 t^{x-1}(1-t)^{y-1}dt \quad \mathrm{Re}\,x > 0, \ \ \mathrm{Re}\,y > 0. \tag{3.62}$$

This defines $\mathrm{B}(x,y)$ when the real parts of x and y are both positive. For other values of x and y beta function is defined by analytical continuation.

We first relax the condition $\mathrm{Re}\,x > 0$, but we keep $\mathrm{Re}\,y > 0$. Following Cauchy we define as usual

$$\mu(x,y) = \int_c t^{x-1}(1-t)^{y-1}dt \tag{3.63}$$

where C is the contour shown in Fig. 3.6.

We shall often write

$$\mu(x,y) = \int_1^{0+} t^{x-1}(1-t)^{y-1}dt. \tag{3.64}$$

Repeating the previous arguments one can show that

$$\mu(x,y) = \int_c t^{x-1}(1-t)^{y-1}dt = \int_1^{0+} t^{x-1}(1-t)^{y-1}dt \tag{3.65}$$

$$\mathrm{Re}\,y > 0$$

is finite for all x. Also for $\mathrm{Re}\,x > 0$

$$\mu(x,y) = (e^{2\pi i x} - 1)\mathrm{B}(x,y). \tag{3.66}$$

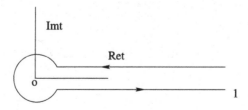

Fig. 3.6

Thus

$$B(x,y) = \frac{1}{(e^{2\pi i x} - 1)} \int_1^{0+} t^{x-1}(1-t)^{y-1}dt$$

$$\operatorname{Re} y > 0.$$

Let us now take $\operatorname{Re} x > 0, \operatorname{Re} y$ arbitrary and define

$$\mu(x,y) = \int_c t^{x-1}(1-t)^{y-1}dt. \tag{3.67}$$

In Fig. 3.7 at P $\arg(1-t) = \propto = 0$ and at Q $\arg(1-t) = 2\pi$. Hence as before

$$\mu(x,y) = -(e^{2\pi i x} - 1)B(x,y). \tag{3.68}$$

Hence we obtain

$$B(x,y) = -\frac{1}{(e^{2\pi i x} - 1)} \int_c t^{x-1}(1-t)^{y-1}dt$$

$$= -\frac{1}{(e^{2\pi i x} - 1)} \int_0^{1+} t^{x-1}(1-t)^{y-1}dt \tag{3.69}$$

y arbitrary, $\operatorname{Re} x > 0$.

We now enquire about the procedure for analytic continuation of the integral for arbitrary values of x and y simultaneously. When x and y are allowed to take arbitrary values we are tempted to modify the contour to the dumb-bell shaped contour shown above.

But we shall see later that such a contour is meaningless.

At every point in any contour the function under consideration must have a unique value. Hence the contour C must be such that if we start

Fig. 3.7

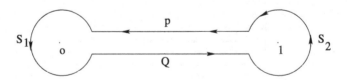

Fig. 3.8

from a particular point P travel along the closed contour C and come back to P the function must return to the original value.

Let us see how we achieve this when the function is multiple valued, i.e. when it has a branch point, say, at the origin. Take for example

$$f(z) = z^\alpha, \quad \alpha = \text{complex}.$$

For this function the contour $\sum$ in Fig. 3.9 is illegal because the function becomes x^α at the start and $x^\alpha e^{2\pi i\alpha}$ at the end. We introduce the branch cut to ensure that the function does come back to its original value. The following contour given by Fig. 3.10 is correct. In fact

$$\begin{aligned}
\text{at} \quad &\text{P} \quad f(z) = x^\alpha \\
\text{at} \quad &\text{Q} \quad f(z) = x^\alpha e^{2\pi i\alpha} \\
\text{at} \quad &\text{R} \quad f(z) = \epsilon^\alpha e^{2\pi i\alpha} \\
\text{at} \quad &\text{S} \quad f(z) = \epsilon^\alpha \\
\text{at} \quad &\text{P} \quad f(z) = x^\alpha.
\end{aligned}$$

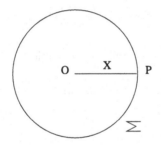

Fig. 3.9

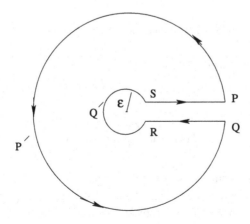

Fig. 3.10

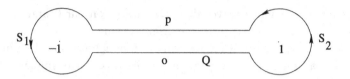

Fig. 3.11

Hence the function returns to the original value and the contour PP′QRQ′SP is correct.

Let us now analyze the dumbbell in Fig. 3.11 for the simple function

$$F(z) = \ln\left(\frac{z+1}{z-1}\right).$$

As explained in Chapter 1 at P, $\arg(z+1) = 0$, $\arg(z-1) = \pi$ so that

$$f(z) = \ln\left(\frac{1+x}{1-x}\right) - i\pi. \tag{3.70}$$

When P comes to Q after describing the circle S_1,

$$\arg(z+1) = 2\pi, \quad \arg(z-1) = \pi. \tag{3.71}$$

Thus at Q,

$$f(z) = \ln\left(\frac{1+x}{1-x}\right) + i\pi. \tag{3.72}$$

As Q moves around S_2 and comes back to P,

$$\arg(z+1) = 2\pi, \quad \arg(z-1) = 3\pi. \tag{3.73}$$

Thus

$$f(z) = \ln\left(\frac{1+x}{1-x}\right) - i\pi. \tag{3.74}$$

so that $f(z)$ does return to the original value.

But this is only a coincidence. We shall see that although the function $f(z) = \ln\frac{z+1}{z-1}$ returns to its original value, the same is not true for all functions having a pair of branch points. Consider for example the beta function integrand over the dumb-bell of Fig. 3.8. At P

$$\arg t = 0, \quad \arg(1 - t) = 0$$
$$f_P(t) = t^{x-1}(1 - t)^{y-1}. \tag{3.75}$$

At Q

$$\arg t = 2\pi, \quad \arg(1 - t) = 0 \tag{3.76}$$
$$f_Q(t) = e^{2\pi i x} t^{x-1}(1 - t)^{y-1}. \tag{3.77}$$

After describing the circle at S_2.

$$\arg t = 2\pi \, \arg(1 - t) = 2\pi. \tag{3.78}$$

So that at P,

$$f'_P(t) = e^{2\pi i x} \, e^{2\pi i y} \, t^{x-1}(1 - t)^{y-1}. \tag{3.79}$$

Thus the function does not return to its original value and the dumb-bell is meaningless.

In fact the above analysis implies that the lines PA and BR do not coincide. If we are to bring back the function to its original value we have to describe two more circles around 0 and 1 in the clockwise direction. This is achieved by the Pochhammer contour in Fig. 3.12: At P,

$$\arg t = 0, \quad \arg(1 - t) = 0 \tag{3.80}$$
$$f_a(t) = t^{x-1}(1 - t)^{y-1}.$$

After describing the 1st circle around zero

$$\arg t = 2\pi, \quad \arg(1 - t) = 0. \tag{3.81}$$

So that

$$f_b(t) = e^{2\pi i x} \, t^{x-1}(1 - t)^{y-1}. \tag{3.82}$$

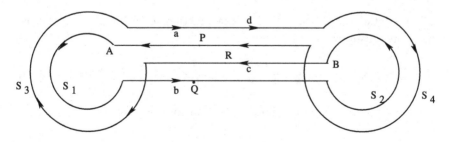

Fig. 3.12

Similarly on c

$$\arg t = \arg(1 - t) = 2\pi \tag{3.83}$$

$$f_c(t) = e^{2\pi i(x+y)}t^{x-1}(1 - t)^{y-1}. \tag{3.84}$$

Finally on d $\arg t = 0$, $\arg(1 - t) = 2\pi$

$$f_d(t) = e^{2\pi iy}t^{x-1}(1 - t)^{y-1}. \tag{3.85}$$

After describing the circle S_4 around 1 we have

$$\arg t = \arg(1 - t) = 0 \tag{3.86}$$

and the function returns to its original value. Let us now consider the following integral

$$\mu(x, y) = \int_P t^{x-1}(1 - t)^{y-1}dt \tag{3.87}$$

where P stands for the Pochhammer contour. We have already seen from the simple examples discussed before that every time we describe the circle around 0 and 1 the corresponding divergence gets cancelled. Thus the function $\mu(x, y)$ is finite for all finite x and y. Now let us find $\mu(x, y)$ for $\operatorname{Re} x > 0$. $\operatorname{Re} y > 0$. If the limit of the vanishing radii is taken with these restrictions the contribution of the circles tend to zero. Thus

$$\mu(x, y) = \int_1^0 f_a(t)dt + \int_0^1 f_b(t)dt + \int_1^0 f_c(t)dt + \int_0^1 f_d(t)dt$$

$$= (e^{2\pi ix} - 1 + e^{2\pi iy} - e^{2\pi i(x+y)} - 1) \int_0^1 t^{x-1}(1 - t)^{y-1}dt$$

$$= -(e^{2\pi ix} - 1)(e^{2\pi iy} - 1)\mathrm{B}(x, y). \tag{3.88}$$

The analytic continuation of the beta function for arbitrary x and y is, therefore, given by

$$\mathrm{B}(x, y) = -\frac{1}{(e^{2\pi ix} - 1)(e^{2\pi iy} - 1)} \int_p t^{x-1}(1 - t)^{y-1}dt. \tag{3.89}$$

The beta function can be expressed in terms of the Γ function

$$\mathrm{B}(x, y) = \frac{\Gamma(x)\,\Gamma(y)}{\Gamma(x + y)}. \tag{3.90}$$

The proof is left as an exercise to the interested reader. We can prove two important formula's for Γ function.

$$\Gamma(z) = \int_1^\infty t^{z-1}e^{-t}dt + \int_\epsilon^1 t^{z-1}\left[e^{-t} - \sum_{r=0}^{n-1}\frac{(-t)^r}{r!}\right]$$

$$= \int_1^\infty e^{-t}t^{z-1}dt + \int_\epsilon^1 t^{z-1}\left[e^{-t} - \left\{1 - t + \frac{t^2}{2!} - \frac{t^3}{3!} + \frac{(-)^n}{(n-1)!}t^n\right\}\right]$$

$$(3.91)$$

valid for $\operatorname{Re} z > -n$. If in addition $-n < \operatorname{Re} z < -n + 1$.

We have

$$\Gamma(z) = \int_0^\infty t^{z-1}\left[e^{-t} - \sum_{r=0}^{n-1}\frac{(-)^r t^r}{r!}\right]dt. \tag{3.92}$$

These two formulas can be proved from the contour integral representation of gamma function and is left as an exercise.

3.4 Asymptotic Expansion of the Gamma Function

3.4.1 *General theory of asymptotic expansion*

Consider the function

$$f(x) = \int_x^\infty e^{x^2 - t^2}dt \tag{3.93}$$

where x is positive and the path of integration is the real axis.

Let us define

$$f_n(x) = \frac{1}{2^n}\int_{x^2}^\infty \frac{e^{x^2 - \tau}}{\tau^{\frac{n}{2}}}d\tau \tag{3.94}$$

so that

$$f_1(x) = f(x). \tag{3.95}$$

Integrating by parts

$$f_n(x) = \frac{1}{2^n x^n} - 2n\, f_{n+2} \tag{3.96}$$

which yields

$$f(x) = f_1(x) = \frac{1}{2x} - 2f_3(x)$$

$$= \frac{1}{2x} - \frac{1}{2^2 x^3} + \frac{1.3}{2^3 x^5} - \frac{1.3.5}{2^4 x^7} + \cdots$$

$$= \sum_n (-)^{n-1} \frac{(2n-2)!}{2^{2n-1}(n-1)! x^{2n-1}}. \tag{3.97}$$

In connection with this function we therefore consider the expression

$$u_n = (-)^{n-1} \frac{(2n-2)!}{2^{2n-1}(n-1)! \, x^{2n-1}} \tag{3.98}$$

and we write

$$\sum_{n=1}^{n} u_n = S_n(x). \tag{3.99}$$

For this sum $\left|\frac{u_{n+1}}{u_n}\right| = \frac{n - \frac{1}{2}}{x^2} \to \infty$.

The series $\sum_{n=1}^{\infty} u_n$ is therefore divergent for all values of x. Inspite of this, however, the series can be used for the calculation of $f(x)$; this can be seen in the following way. Let us take any fixed value for the number n, and calculate S_n. We have

$$|f(x) - S_n(x)| = \frac{(2n)!}{n!} f_{2n+1} \le \frac{2n!}{n! \left(n - \frac{1}{2}\right)} \frac{1}{(2x)^{2n+1}}$$

$$= \frac{\Gamma\left(n - \frac{1}{2}\right)}{2\sqrt{\pi} x^{2n+1}}. \tag{3.100}$$

For values of x which are sufficiently large, the right-hand member of this equation is very small. For example for $n = 3$, and $x = 10$

$$|f(x) - S_3(x)| < 10^{-7}. \tag{3.101}$$

It therefore follows that the value of the function $f(x)$ can be calculated with great accuracy for large values of x, by taking the sum of a suitable number of terms of the series $\sum u_m$.

The series is on this account said to be an asymptotic expansion of the function $f(x)$. The precise definition of the asymptotic expansion which is given below is due to Poincare.

A divergent series

$$A_0 + \frac{A_1}{z} + \frac{A_2}{z^2} + \cdots + \frac{A_n}{z^n} + \cdots \qquad (3.102)$$

in which the first $n+1$ term is $S_n(z)$ is said to be an asymptotic expansion of a function $f(z)$ for a given range of values of $\arg z$, if the expression $R_n(z) = z^n[f(z) - S_n(z)]$ satisfies

$$\lim_{|z| \to \infty} R_n(z) = 0 \quad (n \text{ fixed}) \qquad (3.103)$$

even if $\lim_{n \to \infty}[R_n(z)] = \infty(z \text{ fixed})$. When this is the case we make

$$|z^n[f(z) - S_n(z)]| \to 0 \qquad (3.104)$$

by taking $|z|$ sufficiently large. We denote the series to be the asymptotic expression of $f(z)$ by writing

$$f(z) \sim \sum_{n=0}^{\infty} A_n z^{-n}. \qquad (3.105)$$

This example discussed before satisfies the condition given above

$$|x^n[f(x) - S_n(x)]| < \frac{\Gamma\left(n - \frac{1}{2}\right)}{2\sqrt{\pi}x^{n+1}} \to 0 \quad \text{as } x \to \infty. \qquad (3.106)$$

We shall now state two important theorems.

(a) It is permissible to integrate an asymptotic expansion term by term, the resulting series being the asymptotic expansion of the integral of the function represented by the original series.

$$\int_x^{\infty} f(x)dx \sim \sum_{m=2}^{\infty} \frac{A_m}{(m-1)x^{m-1}} \qquad (3.107)$$

where

$$f(x) \sim \sum_{m=2}^{\infty} \frac{A_m}{x^m}. \qquad (3.108)$$

(b) It is not in general permissible to differentiate an asymptotic expansion. This may be seen by considering $e^{-x}\sin e^x$.

The proof is left as an exercise.

3.5   Watson's Lemma

Suppose $f(t)$ is regular in the neighborhood of $t = 0$ so that it has an expansion for small t of the form.

$$f(t) = \sum_{s=0}^{\infty} a_s t^s. \tag{3.109}$$

Suppose further that corresponding to a given real number $\gamma > 0$ we can find constants K and r_0 such that for all t in the range $0 \le t < \infty$

$$f(t) \le K \exp r_0 t^\gamma. \tag{3.110}$$

Then as $|z| \to \infty$ within the sector $|\arg z| \le \frac{\pi}{2} - \delta$

$$\gamma \int_0^\infty e^{-zt^\gamma} f(t) dt \sim \sum_{s=0}^{\infty} \Gamma\left(\frac{s+1}{\gamma}\right) a_s z^{-(1+S)/\gamma}. \tag{3.111}$$

By our initial assumption

$$F_N(t) = \left| f(t) - \sum_{s=0}^{N-1} a_s t^s \right| \le K_N \, t^N e^{r_0 t^\gamma} \tag{3.112}$$

where K_N is a new constant that in general will depend upon N. Thus

$$\left| \int_0^\infty f(t) e^{-zt^\gamma} dt - \sum_{s=0}^{N-1} a_s \int t^s e^{-zt^\gamma} dt \right| \le K_N \int_0^\infty e^{(r_0-z)r^v} t^N$$

$$= 0 \le K_N |R_N(z)| \tag{3.113}$$

where

$$|R_n(z)| = |K_N| \int e^{-(x-r_0)t^\gamma} t^N dt$$

$$= |K_N| \frac{1}{\gamma} (x - r_0)^{-\frac{(N+1)}{\gamma}} \Gamma\left(\frac{N+1}{\gamma}\right) \to 0 \; x \to \infty \tag{3.114}$$

$$\gamma \int_0^\infty e^{-zt^\gamma} f(t) dt \sim \sum_{s=0}^{\infty} a_s \Gamma\left(\frac{s+1}{\gamma}\right) z^{-\frac{(s+1)}{\gamma}}. \tag{3.115}$$

We now proceed to obtain an asymptotic expansion of $\Gamma(z)$ by first assuming z to be real and positive. The method is based on Watson's discussion of a closely related problem of Ramanujan.

$$\Gamma(z) = \frac{1}{z} \ \Gamma(z+1) = \frac{1}{z} \int t^z e^{-t} dt. \tag{3.116}$$

Setting $t = zu$ so that $dt = z\,du$

$$\Gamma(z) = \frac{1}{\not{z}} z^z \not{z} \int du\, u^z e^{-zu}$$

$$= z^z e^{-z} \int_0^\infty du\, u^z e^{z(1-u)}. \tag{3.117}$$

Thus

$$e^z z^{-z} \Gamma(z) = \int_0^\infty (ue^{1-u})^z du. \tag{3.118}$$

Although we have proved this formula by supposing that z is real and positive it will also hold in any closed region in which $z^{-z} e^z \Gamma(z)$ is regular and the integral is uniformly convergent. The formula is therefore true when $\mathrm{Re}\, z > 0$.

Let us suppose that $\mathrm{Re}\, z > 0$ and that $|z|$ is large. We observe that ue^{1-u} increases as steadily from 0 to 1 as u increases from zero to 1, and then decreases steadily from 1 to 0 as u increases from 1 to infinity. We now write

$$z^{-z} e^z \Gamma(z) = \int_0^1 (ue^{1-u})^z du + \int_1^\infty (Ue^{1-U})^z dU \tag{3.119}$$

it being convenient to use different symbols for the variables in the different parts of the range of integration.

If in the second integral we take the substitution.

$$e^{-t} = Ue^{1-U} \tag{3.120}$$

t increases from zero to infinity as $1 \le U < \infty$. Similarly setting

$$e^{-t} = ue^{1-u} \tag{3.121}$$

t decreases steadily from ∞ to 0 as $0 \le u \le 1$. Thus

$$z^{-z} e^z \Gamma(z) = \int_0^\infty e^{-zt} \left[\frac{dU}{dt} - \frac{du}{dt} \right] dt \tag{3.122}$$

to which we may apply Watson's lemma. Now U and u are two solutions of the transcendental equation.

$$e^{-t} = ue^{1-u}$$

i.e. $t = u - 1 - \ln u = \omega - \ln(1 + \omega)$. \hfill (3.123)

Setting $t = \frac{1}{2}\zeta^2$

$$\frac{1}{\not{2}}\zeta^2 = w - \ln(1+w) = \frac{w^2}{\not{2}}\left[1 - \frac{2}{3}w + \frac{2}{4}w^2\cdots\right]. \tag{3.124}$$

Now ζ regarded as a function of w is two valued in the neighborhood of the origin, its two branches being

$$\zeta = \pm w\left[1 - \frac{2}{3}w + \frac{2}{4}w - \cdots\right]. \tag{3.125}$$

Since each branch is an analytic function of w regular when $|w| < 1$ with a simple zero at $w = 0$ it follows that the equation

$$\zeta = f(w), f(0) = 0 \tag{3.126}$$

has a unique solution

$$w = \sum_{n=1}^{\infty} \frac{\zeta^n}{n!}\left[\frac{d^{n=1}}{dw^{n-1}}\{\psi(w)\}^n\right]_{w=0} \tag{3.127}$$

where

$$f(w) = \frac{w}{\psi(w)}. \tag{3.128}$$

Thus

$$\begin{aligned}
w &= \sum_{n=1}^{\infty} \frac{\zeta^n}{n!}\frac{d^{n-1}}{dw^{n-1}}\left[\frac{w^n}{[f(w)]^n}\right]_{w=0}\\
&= \zeta\left[\frac{w}{f(w)}\right] + \frac{\zeta^2}{2!}\frac{d}{dw}\left[\frac{w^2}{f^2(w)}\right]_{w=0} + \frac{\zeta^2}{3!}\left[\frac{d^2}{dw^2}\frac{w^3}{f^3(w)}\right]_{w=0} + \cdots\\
&= \zeta + \frac{1}{3}\zeta^2 + \frac{1}{36}\zeta^3 - \frac{1}{276}\zeta^4 + \cdots.
\end{aligned} \tag{3.129}$$

This solution is regular in the neighborhood $|\zeta| < \rho$ of the origin. We shall denote this solution by $w_1(\zeta)$. Similarly the solution of

$$\zeta = -w\left(1 - \frac{2}{3}w + \frac{2}{4}w^2 - \cdots\right)^{\frac{1}{2}} \tag{3.130}$$

regular in $|\zeta| < \rho$, is $w = w_2(\zeta) = w_1(-\zeta)$.

We have therefore shown that the function $w(\zeta)$ defined by the equation

$$\frac{1}{2}\zeta^2 = w - \ln(1 + w) \tag{3.131}$$

has two branches $w = w_1(\zeta)$, $w = w_2(\zeta)$ each regular in the neighborhood of the origin:

$$w_{-1}(\zeta) = U - 1 = (2t)^{\frac{1}{2}} + \frac{1}{3}(2t) + \frac{1}{36}(2t)^{3/2} + \cdots \tag{3.132}$$

$$w_{-2}(\zeta) = u - 1 = (2t)^{\frac{1}{2}} + \frac{1}{3}(2t) + \frac{1}{36}(2t)^{3/2} + \cdots . \tag{3.133}$$

Lastly since

$$\frac{dU}{dt} - \frac{du}{dt} = \frac{U}{U-1} + \frac{u}{1-u} = \frac{1}{U-1} + \frac{1}{1-u} \tag{3.134}$$

$\frac{d}{dt}(U - u)$ is bounded when $t \geq \epsilon > 0$. The conditions of Watson's lemma are therefore all satisfied and we have

$$\int_0^\infty e^{-zt} \frac{d}{dt} \left[\sum_n a_n (2t)^{n/2} - \sum (-)^n a_n (2t)^{\frac{n}{2}} \right]$$

$$= \int_0^\infty e^{-zt} \frac{d}{dt} \sum_{n=0}^\infty 2a_{2n+1} (2t)^{n+\frac{1}{2}}. \tag{3.135}$$

After term by term integration

$$\Gamma(z) = e^{-z} z^z \sqrt{\frac{2\pi}{z}} \left[1 + \frac{1}{12z} + \frac{1}{288z^2} + \cdots \right]. \tag{3.136}$$

It can be shown that this asymptotic expansion is valid under the restriction

$$|\arg z| \leq \pi - \delta. \tag{3.137}$$

3.6 Legendre's Duplication Formula

Let us consider the function

$$\psi(z) = \frac{2^{2z} \Gamma(z) \Gamma\left(z + \frac{1}{2}\right)}{\Gamma(2z)}. \tag{3.138}$$

Its only singularities are the poles of the numerator and zeroes of the denominator. Now the numerator has simple poles at the points.

$$z = 0, -\frac{1}{2}, -1, -\frac{3}{2}, \ldots$$

i.e.   at $2z = 0, -1, -2 \ldots$.

But since these points are also simple poles of the denominator they are not singularities of $\psi(z)$. The denominator $\Gamma(2z)$ never vanishes. Hence

$\psi(z)$ has no singularity in the finite part of the z plane and so is an entire function. Again

$$\psi(z+1) = \psi(z) \tag{3.139}$$

by the recurrence formula for the Γ function. Hence $\psi(z)$ is a periodic function of period 1. Thus if we can show that $\psi(z)$ is bounded when $\operatorname{Re} z \geq 1$, say, it will follow by periodicity that it is bounded over the z plane. Now by the asymptotic expansion of $\Gamma(z)$.

$$\psi(z) = 2\left(\frac{\pi}{e}\right)^{\frac{1}{2}}\left(1+\frac{1}{2z}\right)^{z}\left[1+0\left(\frac{1}{|z|}\right)\right] = 2\sqrt{\pi}\left[1+0\left(\frac{1}{|z|}\right)\right]. \tag{3.140}$$

Hence $\psi(z)$ is bounded when $\operatorname{Re} z > 1$. Thus by Liouville's theorem $\psi(z) = c$. Since

$$\lim |z| \to \infty \, \psi(z) = 2\sqrt{\pi} \quad C = 2\sqrt{\pi}. \tag{3.141}$$

Hence we have

$$\frac{2^{2z}\,\Gamma(z)\Gamma\left(z+\frac{1}{2}\right)}{\Gamma(2z)} = 2\sqrt{\pi}. \tag{3.142}$$

Thus

$$\Gamma(2z) = \frac{1}{\sqrt{\pi}\,2^{2z-1}}\Gamma(z)\Gamma\left(z+\frac{1}{2}\right). $$

3.7   The Zeta Function of Riemann

When $\operatorname{Re} s > 1$ the ζ function of Riemann is defined by the definite integral

$$\zeta(s) = \frac{1}{\Gamma(s)}\int_{0}^{\infty}\frac{t^{s-1}dt}{e^{t}-1}. \tag{3.143}$$

It can be easily ascertained

$$\zeta(s) = \sum_{n=1}^{\infty}\frac{1}{n^{s}}. \tag{3.144}$$

This is a uniformly convergent series of analytic functions in any domain in which $\operatorname{Re} s \geq 1+\delta$ with $\delta > 0$. For other values of s the function is defined by analytic continuation.

$$\zeta(s) = \frac{1}{\Gamma(s)(e^{2\pi is}-1)}\int_{\infty}^{0+}\frac{t^{s-1}dt}{e^{t}-1}. \tag{3.145}$$

We shall now state a summation formula originally due to Plana

$$\sum_{n=0}^{\infty} f(n) = \frac{1}{2}f(0) + \int_0^{\infty} f(\tau)d\tau + i\int_0^{\infty} [f(it) - f(-it)]\frac{dt}{e^{2\pi t} - 1}.$$

(3.146)

The condition of validity of this formula are

(1) $f(\zeta)$ is regular for $\operatorname{Re}\zeta \geq 0$.
(2) $\lim_{t\to\infty} e^{-2v|t|}f(\tau + it) = 0$ uniformly for $0 \leq \tau < \infty$.
(3) $\lim_{t\to\infty} \int_{-\infty}^{\infty} e^{-2v|t|}|f(\tau + it)|dt = 0$.

Let us now take

$$f(\zeta) = (\zeta + 1)^{-s}, \quad \operatorname{Re} s > 1$$

(3.147)

Thus

$$\zeta(s) = \sum_{n=0}^{\infty} \frac{1}{(n+1)^s} = \frac{1}{2} + \frac{1}{s-1} + 2\int_0^{\infty} \frac{\sin[s\tan^{-!}t]}{(v^2 + t^2)^{\frac{s}{2}}}$$

(3.148)

which is Hermite's representation of $\zeta(s)$. From this we see that the only singularity $\zeta(s)$ in the finite part of the s plane is a simple pole of unit residue at $s = 1$.

When $s = -m$, the cut in the t plane of the integrand is no longer present. Thus

$$\zeta(-m) = \frac{\Gamma(1+m)(-)^m}{2\pi i} \oint_c \frac{t^{-m-1}dt}{e^t - 1}$$

(3.149)

where C is a closed contour surrounding the point $t = 0$.
Now

$$\frac{t^{-m-1}}{e^t - 1} = \sum(-)^n B_n(1)\frac{t^{n-m-2}}{n!}.$$

(3.150)

Thus

$$\operatorname{Re} s\frac{t^{-m-1}}{e^t - 1} = (-)^{m+1}\frac{B_{m+1}(1)}{(m+1)!}.$$

(3.151)

Thus

$$\zeta(-m) = \frac{m!(-)^m}{2\not\pi i}2\not\pi i(-)^{m+1}\frac{B_{m+1}}{(m+1)!} = -\frac{B_{m+1}(1)}{m+1}.$$

(3.152)

3.8   The Zeta Function for Re $s < 0$

Consider

$$\zeta(s) \quad \text{for} \quad \text{Re } s < 0.$$

Let us now consider the integral

$$\oint \frac{t^{s-1}dt}{e^t - 1} \tag{3.153}$$

over a large circle of radius R. The poles of the integrand are at

$$\pm 2\pi i \cdots \pm 4\pi i + \cdots = 2n\pi e^{\frac{\pi i}{2}} = 2n\pi e^{3\pi i/2}, \quad n = 1, 2, 3 \ldots .$$

Sum of the residues at these poles

$$\sum [2n\pi e^{i\pi/2}]^{s-1} + \sum (2n\pi e^{\frac{3i\pi}{2}})^{s-1} = 2(2\pi)^{s-1} \sin \frac{\pi s}{2} \sum_{n=1}^{\infty} \frac{1}{n^{1-s}}.$$

If S is the large circle of radius R

$$\int_S \frac{t^{s-1}}{e^t - 1}dt + \int_\infty^{0+} \frac{t^{s-1}}{e^t - 1} = 2\pi i \sum \quad \text{Residues}$$

$$= 4\pi i(2\pi)^{s-1} \sin \frac{\pi s}{2} \sum_{n=1}^{\infty} \frac{1}{n^{1-s}}. \tag{3.154}$$

For Re $s < 0$. The integral over S vanishes, thus

$$\int_\infty^{0+} \frac{t^{s-1}}{e^t - 1}dt = 4\pi i(2\pi)^{s-1} \sin \frac{\pi s}{2} \sum \frac{1}{n^{1-s}} \tag{3.155}$$

i.e. $(e^{2\pi i s} - 1)\Gamma(s)\zeta(s) = 4\pi i(2\pi)^{s-1} \sin \dfrac{\pi s}{2} \sum \dfrac{1}{n^{1-s}}.$ (3.156)

We therefore finally obtain

$$\zeta(s) = \frac{2\Gamma(1-s)}{(2\pi)^{1-s}} \sin \frac{\pi s}{2} \sum_{n=1}^{\infty} \frac{1}{n^{1-s}}. \tag{3.157}$$

Since

$$\sum_{n=1}^{\infty} \frac{1}{n^{1-s}} = \zeta(1-s) \tag{3.158}$$

we obtain Riemann's result

$$2^{1-s}\Gamma(s)\zeta(s) \cos \frac{s\pi}{2} = \pi^s \zeta(1-s). \tag{3.159}$$

Since both sides of this equation are analytic functions of s save for isolated values of s at which they have poles the equation proved when Re $s < 0$ is valid for all values of s save at those isolated values.

Exercise.

If m is a positive integer show that

$$\zeta(2m) = 2^{2m-1}\frac{B_m \pi^{2m}}{(2m)!}$$

$$\zeta(-2m) = 0, \quad m = 1, 2.$$

3.9   Euler's Product for $\zeta(s)$

Let $\mathrm{Re}\, s = \sigma \geq 1 + \delta$ and let $2, 3, 5, 7, \ldots p \ldots$ be the prime numbers in order

$$\zeta(s) = \frac{1}{1^s} + \frac{1}{2^s} + \frac{1}{3^s} + \frac{1}{4^s} + \cdots$$

$$= \frac{1}{1^s} + \frac{1}{3^s} + \frac{1}{5^s} + \cdots + \frac{1}{2^s} \times \left[\frac{1}{1^s} + \frac{1}{2^s} + \frac{1}{3^s} + \cdots\right]$$

$$= \frac{1}{1^s} + \frac{1}{3^s} + \frac{1}{5^s} + \cdots + 2^{-s}\zeta(s) \tag{3.160}$$

$$\zeta(s)(1 - 2^{-s}) = \frac{1}{1^s} + \frac{1}{3^s} + \frac{1}{5^s} + \cdots$$

$$= \frac{1}{1^s} + \frac{1}{5^s} + \frac{1}{7^s} + \cdots + 3^{-s}(1 - 2^{-s})\zeta(s) \tag{3.161}$$

$$\zeta(s)(1 - 2^{-s})(1 - 3^{-s}) = \frac{1}{1^s} + \frac{1}{5^s} + \frac{1}{7^s}. \tag{3.162}$$

Proceeding in this way

$$\zeta(z)(1 - 2^{-s})(1 - 3^{-s}) \cdots (1 - p^{-s}) = 1 + \sum' n^{-s} \tag{3.163}$$

where the prime denotes that only those values of n (greater than p) which are prime to $2, 3, \ldots p$ occur in the summation.

Thus

$$\left|\sum' n^{-s}\right| = \sum' n^{-1-\delta} \to 0 \quad \text{as} \quad p \to \infty. \tag{3.164}$$

Thus for $\mathrm{Re}\, s \geq 1 + \delta$ the product

$$\zeta(s)\Pi_p(1 - p^{-s}) \tag{3.165}$$

converges to 1 where the number p assumes prime values $2, 3, 5, \ldots$.

But the product $\Pi_p(1 - p^{-s})$ converges when $\mathrm{Re}\, s \geq 1 + \delta$ for it consists of some of the factors of the absolutely convergent product

$$\Pi_{n=2}(1 - n^{-s}).$$

Consequently we infer that $\zeta(s)$ has no zeros at which $\operatorname{Re} s \geq 1 + \delta$, for if it had any such zeros $\Pi_{p=2}^{\infty}(1 - p^{-s})$ would not converge at them.

Thus for $\operatorname{Re} s \geq 1 + \delta$

$$\frac{1}{\zeta(s)} = \Pi_p(1 - p^{-s}). \tag{3.166}$$

3.10 Riemann's Conjecture Concerning the Zeros of $\zeta(s)$

We have already seen

1. $\zeta(s)$ has no zeros for $\operatorname{Re} s > 1$.
2. $\zeta(-2m) = 0$, $m = 1, 2, 3 \ldots$.

It can be shown that these are the only zeros on the left half of the complex plane. Hence all the zeros of $\zeta(s)$ except at $s = -2, -4, \ldots$ lie in the strip

$$0 < \operatorname{Re} s < 1.$$

It was conjectured by Riemann but it has not yet been proved that all zeros of $\zeta(s)$ in this strip lie on the line.

$$\operatorname{Re} s = \frac{1}{2}.$$

It was proved by Hardy that an infinity of zeros of $\zeta(s)$ actually lie on the line $\operatorname{Re} s = \frac{1}{2}$.

It is highly probable that Riemann's conjecture is correct and its proof will have far-reaching consequences in the theory of prime numbers.

It will be shown in Chapter 6 that the concept of regularization due to Hadamard, Reisz, Schwartz and Gel'fand and coworkers is the natural offspring of Cauchy's theory of analytic continuation as explained above. In particular the Gel'fand–Shilov formula for regularization has been derived from Cauchy's contour integral representation.

CHAPTER 4

The Special Functions Defined by Power Series

4.1 The Hypergeometric Function

The special functions defined by power series can be divided into two categories (a) power series with a finite radius of convergence (b) power series with infinite radius of convergence. The latter will be called entire functions and will have little importance from the standpoint of analytic continuation. However, power series with finite radius of convergence may sometime be continued beyond the circle of convergence. Only then can it be regarded as a bonafide analytic function. The theory of continuation of such power series was developed by Barnes, which can be applied to the entire set of hypergeometric series

$$
{}_{p+1}F_p \left(\begin{matrix} a_1, a_2 \cdots a_{p+1}, \\ b_1, b_2 \cdots b_p \end{matrix} \quad z \right)
$$

the simplest special case being the ordinary hypergeometric series

$$
1 + \frac{ab}{c} \frac{z}{11} + \frac{a(a+1)b(b+1)}{c(c+1)} \frac{z^2}{2!} + \cdots = \sum \frac{(a)_n (b)_n z^n}{(c)_n n!}
$$

where $(a)_n = \frac{\Gamma(a+n)}{\Gamma(a)}$, etc.

It is called Gauss series or the ordinary hypergeometric series. It is represented by the symbol

$$
F(a, b; c; z).
$$

The variable is z and a, b, c are called the parameters of the function. If either of the quantities a or b is a negative integer the series terminates and becomes a polynomial. The series converges absolutely for $|z| < 1$. For $z = 1$ the series converges absolutely if $\text{Re}(c - a - b) > 0$.

It will be clear later that the series can be continued analytically throughout the complex plane so that $F(a, b; c; z)$ can be regarded as an analytic function having a branch point at $z = 1$. Thus if a cut is made from 1 to ∞ along the real axis, the function is analytic and single valued throughout the cut plane.

4.2 The Differential Equation

It is known that the Papperitz equation

$$
w'' + \left(\frac{1-\alpha-\alpha'}{z-\xi} + \frac{1-\beta-\beta'}{z-\eta} + \frac{1-\gamma-\gamma'}{z-\zeta} \right) w'
$$
$$
+ \left[\frac{\alpha\alpha'(\xi-\eta)(\xi-\zeta)}{z-\xi} + \frac{\beta\beta'(\eta-\zeta)(\eta-\xi)}{z-\eta} + \frac{\gamma\gamma'(\zeta-\xi)(\zeta-\eta)}{z-\zeta} \right]
$$
$$
\times \frac{w}{(z-\xi)(z-\eta)(z-\zeta)} = 0 \tag{4.1}
$$

is the most general linear differential equation of the second order whose only singular points are ξ, η, ζ and are all regular. The exponents at these points are $\alpha, \alpha', \beta, \beta'$ and γ, γ' respectively. Since the point at infinity is an ordinary point, it follows that

$$
\alpha + \alpha' + \beta + \beta' + \gamma + \gamma' = 1.
$$

The most important properties of the Eq. (4.1) are

(1) A bilinear transformation preserves the form of the Papperitz equation, transforming the position of the singular points, but preserving the exponents

(2) A transformation of the form $\omega(z) = (z-\xi)^{-\lambda}(z-\eta)^{-\mu}(z-\zeta)^{-\nu}\nu(z)$

where $\lambda + \mu + \nu = 0$, also preserves the form of the Papperitz equation, leaving the singular points unchanged, but increasing the associated exponents by λ, μ and ν respectively.

A combination of these two transformations can therefore transform the singular points to 0, 1 and ∞, and reduce one of the exponents at both 0 and 1 to zero.

If we write as is conventional

$a = \alpha + \beta + \gamma$, $b = \alpha + \beta' + \gamma$, $c = 1 + \alpha - \alpha'$, the Eq. (4.1) becomes

$$
z(1-z)w'' + [c - (a+b+1)z]w' - abw = 0 \tag{4.2}
$$

which is known as hypergeometric equation. A solution is given by $F(a, b; c; z)$ provided $c \neq -n$, $n = 0, 1, 2 \dots$.

Simple algebraic manipulation of the series for $F(a, b; c; z)$ may be used to derive recurrence relations between contiguous functions. Typical relations are

$$cF(a, b+1, c, z) = cF(a, b, c, z) + azF(a+1, b+1; c+1, z) \qquad (4.3)$$

$$c\frac{d}{dz}F(a; b; c; z) = abF(a+1, b+1, c+1, z). \qquad (4.4)$$

If the constant c is not an integer we may construct a second solution of the differential equation using the exponent $\alpha' = 1 - c$. This solution may be written

$$z^{1-c}F(1+a-c, 1+b-c, 2-c, z). \qquad (4.5)$$

4.3 A Representation by an Euler-Type Integral

Gauss's work showed clearly that he was already regarding $F(a, b; c; z)$ as an analytic function of four variables rather than a series in z. To analyze the analyticity in a, b, c we start from the integral representation

$$F(a, b; c; z) = \frac{\Gamma(c)}{\Gamma(b)\Gamma(c-b)} \int_0^1 t^{b-1}(1-t)^{c-b-1}(1-tz)^{-a} dt$$

$$\operatorname{Re} b > 0, \quad \operatorname{Re} c > \operatorname{Re} b. \qquad (4.6)$$

It can be easily verified that the above integral representation yields the power series for $|z| < 1$. For $|z| < 1$ we may expand

$$(1-tz)^{-a} = \sum \frac{(a)_n}{n!} t^n z^n.$$

Thus

$$\frac{\Gamma(c)}{\Gamma(b)\Gamma(c-b)} \int_0^1 t^{b-1}(1-t)^{c-b-1}(1-tz)^{-a}$$

$$= \frac{\Gamma(c)}{\Gamma(b)\Gamma(c-b)} \sum \frac{(a)_n}{n!} z^n \int_0^1 t^{n+b-1}(1-t)^{c-b-1} dt$$

$$= \frac{\Gamma(c)}{\Gamma(b)} \sum \frac{\Gamma(n+b)}{\Gamma(c+n)} (a)_n \frac{z^n}{n!} = \sum_{n=0}^\infty \frac{(a)_n (b)_n}{(c)_n n!} z^n = F(a, b; c; z).$$

When the condition on $\operatorname{Re} c$ and $\operatorname{Re} b$ are both relaxed we have to use the Pochhammer contour

$$F(a, b, c, z) = \frac{\Gamma(c)}{\Gamma(b)\Gamma(c-b)(e^{2\pi ib} - 1)(e^{2\pi i(c-b)} - 1)}$$

$$\times \int_P t^{b-1}(1-t)^{c-b-1}(1-tz)^{-a} dt. \qquad (4.7)$$

For all c except $c = -n$, $n = 0, 1, 2 \ldots$.

From this it follows that $F(a, b; c; z)$ is an entire function of a, b and has simple poles at $c = -n$, $n = 0, 1, 2 \ldots$.

If further

$$\text{Re}(c - a - b) > 0.$$

It follows that

$$F(a, b; c; 1) = \frac{\Gamma(c)\Gamma(c - a - b)}{\Gamma(c - a)\Gamma(c - b)}.$$

It can be verified without difficulty that every one of the finite group of transformations

$$\zeta = z, \ \frac{1}{z}, \ 1 - z, \ \frac{1}{1 - z}, \ \frac{z}{z - 1}, \ \frac{z - 1}{z}$$

leads to an equation whose solution can be expressed as hypergeometric function. It is of interest to note that the region of validity of the corresponding series $|\zeta| < 1$ leads to six regions of the complex plane. Each region is the complement of a second, and has a finite intersection with the other four.

We may now express $F(a, b, c, z)$ as a linear combination of any two independent solutions of $\frac{1}{z}$, $1 - z$, $\frac{z}{z-1}$, etc. A typical one-term relation is the following

$$F(a, b, c, z) = (1 - z)^{-a} F\left(a, c - b, c, \frac{z}{z - 1}\right). \tag{4.9}$$

Setting $t = 1 - \tau$ in Eq. (4.6) we obtain

$$F(a, b, c, z) = \frac{\Gamma(c)}{\Gamma(c - b)\Gamma(b)} \int_0^1 \tau^{c-b-1}(1 - \tau)^{b-1}(1 - z + z\tau)^{-a}$$

$$= (1 - z)^{-a} \frac{\Gamma(c)}{\Gamma(c - b)\Gamma(b)} \int_0^1 \tau^{c-b-1}(1 - \tau)^{b-1}\left(1 - \frac{z\tau}{z - 1}\right)^{-a}$$

$$= (1 - z)^{-a} F\left(a, c - b, c, \frac{z}{z - 1}\right).$$

4.4 Analytic Continuation of the Hypergeometric Series

4.4.1 *The geometric series revisited*

We shall now consider Barnes's theory of analytic continuation of the hypergeometric series. To understand the essential content of Barnes's theory we start from the known problem of continuation of the geometric series:

$$u(z) = 1 + z + z^2 + \cdots = \sum_{n=0}^{\infty} z^n. \tag{4.10}$$

This series converges for $|z| < 1$; to analytically continue it beyond the unit circle, i.e. for $|z| > 1$ we write the nth term of the series as the residue of a meromorphic function

$$\chi(s) = \Gamma(s)\Gamma(1 - s)(-z)^{-s} \tag{4.11}$$

having simple poles at,

$$\left. \begin{array}{l} s = -n \\ s = n + 1 \end{array} \right\} n = 0, 1, 2 \ldots.$$

The residue at the simple pole at $s = -n$ is $\frac{(-)^n}{n!} n!(-z)^n = z^n$.
Thus

$$u(z) = \sum_{n=0}^{\infty} \operatorname{Re} s[\chi(s)]_{s=-n}.$$

If we therefore choose a closed contour C as shown in Fig. 4.1, We have

$$u(z) = \frac{1}{2\pi i} \oint_C \chi(s)ds = \frac{1}{2\pi i} \int_{a-i\infty}^{a+i\infty} \chi(s)ds + \frac{1}{2\pi i} \int_S \chi(s)ds.$$

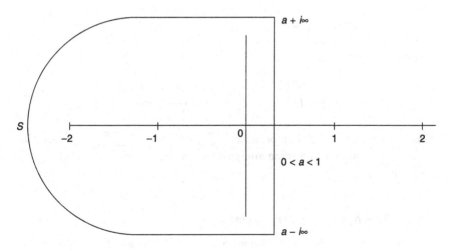

Fig. 4.1 The contour C.

Now on S

$$|\chi(s)| = 2\pi|z|^{-R\cos\theta} \exp[-R|\sin\theta|(\pi - \arg(-z))]$$

Thus $\chi(s) \to 0$ as $R \to \infty$ as long as $|z| < 1$ and $|\arg(-z)| < \pi$. We therefore obtain

$$u(z) = \frac{1}{2\pi i} \int_{a-i\infty}^{a+i\infty} \Gamma(s)\Gamma(1-s)(-z)^{-s} ds \qquad (4.12)$$

This integral has a profound significance because it is defined even for $|z| > 1$ provided $|\arg(-z)| < \pi$. Thus Eq. (4.12) represents an analytic function regular for $|z| > 1$. For $|z| > 1$ we can evaluate the function by closing the contour on the right. The r.h.s. of the line $\mathrm{Re}\, s = a$ contains poles at

$$s = n + 1, \quad n = 0, 1, 2 \ldots$$

and once again the integral for $|z| > 1$ vanishes on the infinite semicircle on the right. Noting that now the contour is in the negative sense we have

$$\frac{1}{2\pi i} \int_{a-i\infty}^{a+i\infty} \chi(s) ds = -\sum_{n=0}^{\infty} \mathrm{Re}\, s[\chi(s)]_{n+1}.$$

Now

$$\mathrm{Re}\, s[\chi(s)]_{n+1} = n! \frac{(-)^{n+1}}{n!}(-z)^{-(n+1)} = z^{-n-1}.$$

Hence

$$\frac{1}{2\pi i} \int_{a-i\infty}^{a+i\infty} \chi(s) ds = -\frac{1}{z} \sum_{n=0}^{\infty} \left(\frac{1}{z}\right)^{n}. \qquad (4.13)$$

This is the expected result because $\sum_{n=0}^{\infty} z^n$ and $-\frac{1}{z}\sum_{n=0}^{\infty}\left(\frac{1}{z}\right)^n$ are the Taylor expansions of the same analytic function $\frac{1}{1-z}$ for $|z| < 1$ and $|z| > 1$ respectively.

4.4.2 *The hypergeometric series*

The above example yields the salient features of Barnes's theory of analytic continuation which is essentially the theory of continuation of power series.

We now apply it to the hypergeometric series

$$F(a, b; c; z) = \sum \frac{\Gamma(a+n)\Gamma(b+n)\Gamma(c)z^n}{\Gamma(a)\Gamma(b)\Gamma(c+n)n!}$$

$$= \sum_{n=0}^{\infty} \frac{(a)_n (b)_n z^n}{(c)_n n!} \qquad (4.14)$$

where $(a)_n = \frac{\Gamma(a+n)}{\Gamma(a)}$, etc.

Following the above procedure we similarly construct a meromorphic function whose sum of residues yields the hypergeometric series. Let

$$\chi(s) = \frac{\Gamma(c)}{\Gamma(a)\Gamma(b)} \frac{\Gamma(a-s)\Gamma(b-s)\Gamma(s)(-z)^{-s}}{\Gamma(c-s)}. \qquad (4.15)$$

We shall assume for simplicity that $\operatorname{Re} a > 0$, $\operatorname{Re} b > 0$. These assumptions can be easily dispensed with. The function $\chi(s)$ is again a meromorphic function having simple poles at

$$s = -n, \ s = a+n, \ s = b+n$$

$$\operatorname{Re} s[\chi(s)]_{s=-n} = \frac{\Gamma(c)}{\Gamma(a)\Gamma(b)} \frac{\Gamma(a+n)\Gamma(b+n)(-)^n(-z)^n}{\Gamma(c+n)n!} = \frac{(a)_n(b)_n z^n}{(c)_n n!}.$$

Thus

$$F(a, b; c; z) = \sum_{n=0}^{\infty} \operatorname{Re} s[\chi(s)]_{s=-n}.$$

Let us now choose the contour C consisting of the line $\operatorname{Re} s = \gamma$, $0 < \gamma < 1$ and an infinite semicircle S on the left. Thus

$$F(a, b, c, z) = \frac{1}{2\pi i} \int_{\gamma-i\infty}^{\gamma+i\infty} \chi(s)ds + \frac{1}{2\pi i} \int_s \chi(s)ds$$

Using Stirling's asymptotic formula we can easily see that $\chi(s)$ goes to zero rapidly on S as long as $|\arg(-z)| < \pi$ and $|z| < 1$. We therefore obtain

$$F(a, b, c, z) = \frac{1}{2\pi i} \frac{\Gamma(c)}{\Gamma(a)\Gamma(b)} \int_{\gamma-i\infty}^{\gamma+i\infty} \frac{\Gamma(a-s)\Gamma(b-s)\Gamma(s)(-z)^{-s}}{\Gamma(c-s)}. \qquad (4.16)$$

Equation (4.16) is meaningful for all $|z|$ as long as $|\arg(-z)| < \pi$ and defines an analytic function for all z. To get the Taylor expansion of this analytic function for $|z| > 1$ we close the contour on the right by an infinite semicircle. $\chi(s)$ once again goes to zero rapidly on this semicircle as long as $|z| > 1$ and $|\arg(-z)| < \pi$. Thus for $|z| > 1$ we have

$$F(a,b,c,z) = -\sum \operatorname{Re} s[\chi(s)]_{s=a+n} - \sum \operatorname{Re} s[\chi(s)]_{s=b+n}.$$

Simple calculation shows

$$-\operatorname{Re} s[\chi(s)]_{s=a+n} = \frac{\Gamma(b-a)\Gamma(c)}{\Gamma(c-a)\Gamma(b)}(-z)^{-a}\frac{(a)_n(1+a-c)_n}{(1+a-b)_n n!}z^{-n}$$

$$-\operatorname{Re} s[\chi(s)]_{s=b+n} = \frac{\Gamma(c)\Gamma(a-b)}{\Gamma(c-b)\Gamma(a)}(-z)^{-b}\frac{(b)_n(1+b-c)_n}{(1+b-a)_n n!}z^{-n}.$$

We therefore obtain

$$F(a,b;c;z) = \frac{\Gamma(c)\Gamma(b-a)}{\Gamma(a)\Gamma(c-a)}(-z)^{-a}F\left(a, 1+a-c, 1+a-b, \frac{1}{z}\right)$$
$$+ \frac{\Gamma(c)\Gamma(a-b)}{\Gamma(a)\Gamma(c-b)}(-z)^{-b}F\left(b, 1+b-c, 1+b-a, \frac{1}{z}\right).$$

$$(4.17)$$

This gives the Taylor expansion of the analytic function $F(a,b;c;z)$ for $|z| > 1$.

Corollary. From Eq. (4.16) setting $c = b$, we obtain

$$(1-z)^{-a} = \frac{1}{\Gamma(a)}\frac{1}{2\pi i}\int_{r-i\infty}^{r+i\infty}\Gamma(a-s)\Gamma(s)(-z)^{-s}ds. \qquad (4.18)$$

4.4.3 *Behavior of the hypergeometric function around* $z = 1$

(i) Barnes's lemma:

$$\frac{1}{2\pi i}\int_{\sigma-i\infty}^{\sigma+i\infty}\Gamma(\alpha-t)\Gamma(\beta-t)\Gamma(\gamma+t)\Gamma(\delta+t)dt$$
$$= \frac{\Gamma(\alpha+\gamma)\Gamma(\alpha+\delta)\Gamma(\beta+\gamma)\Gamma(\beta+\delta)}{\Gamma(\alpha+\beta+\gamma+\delta)}. \qquad (4.19)$$

The constant σ has been chosen such that no pole lies on the path of integration. This assumption is really not necessary because the path can always be deformed appropriately. The simple poles of the integral are shown below in Fig. 4.2:

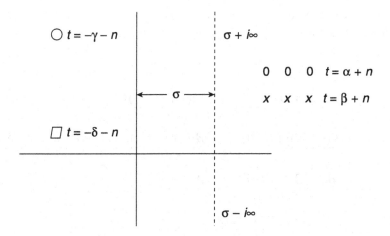

Fig. 4.2

The above integral can be evaluated by closing the contour on the right and calculating the residues of the simple poles of $\Gamma(\alpha - t)\Gamma(\beta - t)$. Thus

$$\frac{1}{2\pi i} \int_{\sigma-i\infty}^{\sigma+i\infty} \Gamma(\alpha - t)\Gamma(\beta - t)\Gamma(\gamma + t)\Gamma(\delta + t)dt$$

$$= -\Gamma(\gamma + \alpha)\Gamma(\delta + \alpha)\Gamma(\beta - \alpha)$$

$$\times \sum_{n=0}^{\infty} \frac{(-)^{n+1}}{n!} \frac{\Gamma(\beta - \alpha - n)\Gamma(\gamma + \alpha + n)\Gamma(\delta + \alpha + n)}{\Gamma(\gamma + \alpha)\Gamma(\delta + \alpha)\Gamma(\beta - \alpha)}$$

$$- \Gamma(\gamma + \beta)\Gamma(\delta + \beta)$$

$$\times \sum_{n=0}^{\infty} \frac{(-)^{n+1}}{n!} \frac{\Gamma(\alpha - \beta - n)\Gamma(\gamma + \beta + n)\Gamma(\delta + \beta + n)}{\Gamma(\gamma + \beta)\Gamma(\delta + \beta)\Gamma(\alpha - \beta)}$$

$$= \Gamma(\gamma + \alpha)\Gamma(\alpha + \delta)\Gamma(\beta - \alpha) \sum_{n=0}^{\infty} \frac{(\gamma + \alpha)_n(\delta + \alpha)_n}{(1 + \alpha - \beta)_n n!}$$

$$+ \Gamma(\gamma + \beta)\Gamma(\delta + \beta)\Gamma(\alpha - \beta) \sum_{n=0}^{\infty} \frac{(\gamma + \beta)_n(\delta + \beta)_n}{(1 + \beta - \alpha)_n n!}$$

$$= \Gamma(\gamma + \alpha)\Gamma(\delta + \alpha)\Gamma(\beta - \alpha)F(\gamma + \alpha, \delta + \alpha, 1 + \alpha - \beta; 1)$$

$$+ \Gamma(\gamma + \beta)\Gamma(\delta + \beta)\Gamma(\alpha - \beta)F(\gamma + \beta, \delta + \beta, 1 + \beta - \alpha; 1).$$

The two-hypergeometric series converge if $\mathrm{Re}(\alpha + \beta + \gamma + \delta - 1) < 0$. Assuming this condition to be satisfied we have

$$\frac{1}{2\pi i}\int_{\sigma-i\infty}^{\sigma+i\infty} \Gamma(\alpha - t)\Gamma(\beta - t)\Gamma(\gamma + t)\Gamma(\delta + t)dt$$

$$= \frac{\Gamma(\gamma + \alpha)\Gamma(\delta + \alpha)\Gamma(\beta - \alpha)\Gamma(1 + \alpha - \beta)\Gamma(1 - \alpha - \beta - \gamma - \delta)}{\Gamma(1 - \beta - \gamma)\Gamma(1 - \beta - \delta)}$$

$$+ \frac{\Gamma(\gamma + \beta)\Gamma(\delta + \beta)\Gamma(\alpha - \beta)\Gamma(1 + \beta - \alpha)\Gamma(1 - \alpha - \beta - \gamma - \delta)}{\Gamma(1 - \alpha - \gamma)\Gamma(1 - \alpha - \delta)}$$

$$= \frac{\Gamma(\alpha + \gamma)\Gamma(\alpha + \delta)\Gamma(\beta + \gamma)\Gamma(\beta + \delta)}{\Gamma(\alpha + \beta + \gamma + \delta)}$$

$$\times \left[\frac{\sin \pi(1 - \beta - \gamma)\sin \pi(1 - \beta - \delta)}{\sin \pi(\beta - \alpha)\sin \pi(1 - \alpha - \beta - \gamma - \delta)} \right.$$

$$\left. + \frac{\sin \pi(1 - \alpha - \gamma)\sin \pi(1 - \alpha - \delta)}{\sin \pi(\alpha - \beta)\sin \pi(1 - \alpha - \beta - \gamma - \delta)} \right]$$

It can be easily verified that the quantity inside the bracket is unity. Thus we obtain Barnes's lemma (4.19) under the condition

$$\mathrm{Re}(\alpha + \beta + \gamma + \delta - 1) < 0.$$

But by the theory of analytic continuation it is true throughout the domain through which both sides of the equation are analytic function of say, α and hence it is true for all values of $\alpha, \beta, \gamma, \delta$ for which none of the poles of $\Gamma(\alpha - t)\Gamma(\beta - t)$ coincide with any of the poles of $\Gamma(\gamma + t)\Gamma(\delta + t)$.

(ii) Behavior around $z = 1$.

We now set in Eq. (4.19) $\alpha = a,\ \beta = b,\ \gamma = -s,\ \delta = c - a - b$. Thus

$$\frac{1}{2\pi i}\int_{\sigma-i\infty}^{\sigma+i\infty} \Gamma(a - t)\Gamma(b - t)\Gamma(-s + t)\Gamma(c - a - b + t)$$

$$= \frac{\Gamma(a - s)\Gamma(c - b)\Gamma(b - s)\Gamma(c - a)}{\Gamma(c - s)}.$$

We therefore obtain

$$\frac{\Gamma(a - s)\Gamma(b - s)}{\Gamma(c - s)} = \frac{1}{2\pi i \Gamma(c - a)\Gamma(c - b)}$$

$$\times \int_{\sigma-i\infty}^{\sigma+i\infty} dt\, \Gamma(a - t)\Gamma(b - t)\Gamma(t - s)\Gamma(c - a - b + t).$$

$$(4.20)$$

Substituting the above in Eq. (4.16) we obtain,

$$F(a, b; c; z) = \frac{\Gamma(c)}{\Gamma(a)\Gamma(b)\Gamma(c-a)\Gamma(c-b)}$$

$$\times \frac{1}{2\pi i} \int_{\sigma-i\infty}^{\sigma+i\infty} \Gamma(a-t)\Gamma(b-t)\Gamma(c-a-b+t)$$

$$\times \left[\frac{1}{2\pi i} \int_{\gamma-i\infty}^{\gamma+i\infty} \Gamma(s)\Gamma(t-s)(-z)^{-s} \right] dt.$$

Now using the Corollary (4.18) we immediately obtain

$$F(a, b; c; z) = \frac{\Gamma(c)}{2\pi i \Gamma(b)\Gamma(c-a)\Gamma(c-b)\Gamma(a)}$$

$$\times \int_{\sigma-i\infty}^{\sigma+i\infty} \Gamma(a-t)\Gamma(b-t)\Gamma(c-a-b+t)\Gamma(t)(1-z)^{-t} dt.$$

For $|1-z| < 1$ and $|\arg(1-z)| < \pi$ this integral can be evaluated by closing the contour by an infinite semicircle on the left, the poles enclosed by this contour are at

$$t = -n, \quad t = a+b-c-n, \quad n = 0, 1, 2 \ldots .$$

Thus

$$F(a, b, c, z)$$

$$= \frac{\Gamma(c)}{\Gamma(a)\Gamma(b)\Gamma(c-a)\Gamma(c-b)}$$

$$\times \left[\sum (-)^n \frac{\Gamma(a+n)\Gamma(b+n)\Gamma(c-a-b-n)}{n!} (1-z)^{(n+c-a-b)} \right.$$

$$\left. + (-)^n \frac{\Gamma(c-a+n)\Gamma(c-b+n)\Gamma(a+b-c-n)}{n!} (1-z)^n \right]$$

$$= \frac{\Gamma(c)\Gamma(c-a-b)}{\Gamma(c-a)\Gamma(c-b)} F(a, b; a+b-c+1; 1-z)$$

$$+ \frac{\Gamma(c)\Gamma(a+b-c)}{\Gamma(a)\Gamma(b)} (1-z)^{c-a-b}$$

$$\times F(c-a, c-b; c-a-b+1; 1-z). \tag{4.21}$$

This formula shows that for arbitrary complex a, b, c the hypergeometric function has a branch point at $z = 1$ and is defined in the cut plane $|\arg(1-z)| < \pi$, i.e. the cut extends from 1 to ∞ as shown in Fig. 4.3.

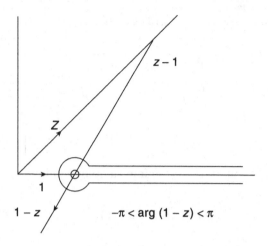

Fig. 4.3

4.5   The Legendre Polynomial and Legendre Function

It is evident that if any one of the numerator parameters a, b is a negative integer the hypergeometric series terminates. The polynomial

$$F\left(-l, l+1; 1; \frac{1}{2}(1-z)\right)$$

where ℓ is a positive integer is known as Legendre polynomial and is denoted as

$$P_l(z) = F\left(-l, l+1; 1; \frac{1}{2}(1-z)\right). \tag{4.22}$$

Expanding the powers of $(1-z)$ in binomial series and using the summation formula

$$F\left(2a, 2b; a+b+\frac{1}{2}; \frac{1}{2}\right) = \frac{\Gamma\left(a+b+\frac{1}{2}\right)\Gamma\left(\frac{1}{2}\right)}{\Gamma\left(a+\frac{1}{2}\right)\Gamma\left(b+\frac{1}{2}\right)} \tag{4.23}$$

the above series can be written in the form

$$P_l(z) = \sum_{r\approx 0}^{m}(-1)^r \frac{(2l-2r)!z^{l-2r}}{2^l r!(1-r)!(l-2r)!} \tag{4.24}$$

where $m = \frac{l}{2}$ or $\frac{l-1}{2}$, whichever is integer. $P_l(z)$ is called Legendre polynomial of degree ℓ.

It is evident that when ℓ is an integer

$$\frac{d^l}{dz^l}(z^2-1)^l = \frac{d^l}{dz^l}\sum \frac{(-1)^r l! z^{2l-2r}}{r!(l-r)!} = \sum_{r=0}^{m} \frac{l!(-)^r}{r!(l-r)!}\frac{(2l-2r)!z^{l-2r}}{(l-2r)!}$$

where $m = \frac{l}{2}$ or $\frac{(l-1)}{2}$, the coefficients of negative powers of z vanishing. From the general formula for $P_\ell(z)$ it follows at once that

$$P_l(z) = \frac{1}{2^l l!} \frac{d^l}{dz^l}(z^2 - 1)^l. \tag{4.25}$$

This result is known as Rodrigues formula.

4.5.1 *Schlafli's integral for* $P_l(z)$

From Cauchy's integral formula we know that

$$f^{(n)}(z) = \frac{n!}{2\pi i} \oint_C \frac{f(t)dt}{(t-z)^{n+1}}.$$

Thus

$$\frac{d^\ell}{dz^\ell}(z^2 - 1)^\ell = \frac{\ell!}{2\pi i} \oint_C \frac{(t^2 - 1)^\ell}{(t - z)^{\ell+1}}$$

$$P_\ell(z) = \frac{1}{\ell! 2^\ell} \frac{d^\ell}{dz^\ell}(z^2 - 1)^\ell$$

$$= \frac{1}{2\pi i} \oint_C \frac{(t^2 - 1)^\ell}{2^\ell (t - z)^{\ell+1}} \tag{4.26}$$

where the contour C encircles the points z once counter clockwise; this is called Schlafli's integral formula for the Legendre polynomials.

We shall now prove the well known orthogonality condition of the Legendre polynomials.

$$\int_{-1}^{1} P_\ell(x)P_n(x)dx = 0 \quad \text{for } \ell \neq n,$$

$$= \frac{2}{2\ell + 1} \quad \text{for } \ell = n. \tag{4.27}$$

From Rodrigue's formula,

$$2^{\ell+n}\ell! n! \int_{-1}^{1} P_\ell(x)P_x(x)dx = \int_{-1}^{1} \left[\frac{d\ell}{dx^\ell}(x^2 - 1)^\ell \right] \left[\frac{d}{dx^n}(x^2 - 1)^n \right]$$

$$= \left\{ \frac{d^{\ell-1}}{dx^{\ell-1}}(x^2 - 1)^\ell \left[\frac{d^n}{dx^n}(x^2 - 1)^n \right] \right\}_{-1}^{1}$$

$$- \int_{-1}^{1} \left\{ \frac{d^{\ell-1}}{dx^{\ell-1}}(x^2 - 1)^\ell \right\} \frac{d^{n+1}}{dx^{n+1}}[(x^2 - 1)^n].$$

Now consider

$$\frac{d^{\ell-1}}{dx^{\ell-1}}(x^2 - 1)^\ell = \frac{(\ell - 1)!}{2\pi i} \oint \frac{(t^2 - 1)^\ell}{(t - x)^\ell}.$$

Thus

$$\left[\frac{d^{\ell-1}}{dx^{\ell-1}}(x^2-1)^\ell\right]_{x=\pm1} = \frac{(\ell-1)!}{2\pi i}\oint\frac{(t^2-1)^\ell}{(t\pm1)^\ell}dt$$

$$= \frac{(\ell-1)!}{2\pi i}\oint dt(t\pm1)^\ell = 0.$$

Continuing this process after we have performed ℓ integrations, we have

$$(-1)^\ell\int(x^2-1)^\ell\left[\frac{d^{\ell+n}}{dx^{\ell+n}}(x^2-1)^n\right].$$

There is no loss of generality if we assume that $\ell > n$. If $\ell > n$, $\ell+n > 2n$ and we have,

$$\therefore\quad\frac{d^{\ell+n}}{dx^{\ell+n}}(x^2-1)^n = 0,$$

identically. Thus for $\ell > n$,

$$\int P_\ell(x)P_n(x)dx = 0\quad\text{Q.e.d.}$$

For $\ell = n$,

$$\frac{d^{2\ell}}{dx^{2\ell}}(x^2-1)^\ell = (2\ell)!$$

whence

$$\int_{-1}^1 P_\ell^2(x)dx = \frac{2}{2\ell+1}.$$

The differential equation satisfied by Legendre polynomials is given by

$$\frac{d}{dx}\left[(1-x^2)\frac{dP_\ell(x)}{dx}\right] + \ell(\ell+1)P_\ell(x) = 0.$$

4.5.2   *The Legendre function*

We define $P_\nu(z)$ for arbitrary ν in terms of Schlafli-type integral as

$$u = \frac{1}{2\pi i2^\nu}\oint_{C_2}\frac{(t-1)^\nu(t+1)^\nu dt}{(t-z)^{\nu+1}}.\tag{4.28}$$

To make the definition precise, we must specify the contour C_2.

The only singularities of the integrand are the branch points at $t = 1$, $t = -1$ and $t = z$. If therefore C_2 be a contour enclosing the points $t = 1$ and $t = z$, but not enclosing the point $t = -1$, then the function $(t^2-1)^\nu$ $(t-z)^{-(\nu+1)}$ will resume its original value after t has described the contour

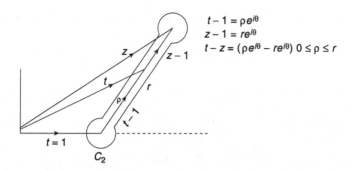

Fig. 4.4

C_2. We therefore choose C_2 to be a "dumbbell" enclosing $t = 1$, $t = z$ as described in Fig. 4.4.

Thus

$$t + 1 = 2 + \rho e^{i\theta} = \left(2 + \frac{\rho}{r}(z - 1)\right) = 2\left[1 - \frac{1}{2}(1 - z)\frac{\rho}{r}\right]$$

$$= 2\left[1 - \frac{1}{2}(1 - z)\tau\right], \quad 0 < \tau < 1$$

$$u(z) = \frac{1}{2\pi i} \oint_{C_2} \left[1 - \frac{1}{2}(1 - z)\tau\right]^{\nu} \tau^{\nu}(\tau - 1)^{-\nu - 1} d\tau$$

$$= \frac{e^{-i\pi(\nu+1)}}{2\pi i} \oint_{C_2} \tau^{\nu}(1 - \tau)^{-\nu - 1}\left[1 - \frac{1}{2}(1 - z)\tau\right]^{\nu} d\tau \qquad (4.29)$$

$$= \frac{e^{-i\pi(\nu+1)}}{2\pi i}\Gamma(-\nu)\Gamma(\nu + 1)(e^{2\pi i\nu} - 1)F\left(-\nu; \nu + 1; 1; \frac{1}{2}(1 - z)\right).$$

Simplifying, we immediately obtain

$$u(z) = F\left(-\nu, \nu + 1; 1; \frac{1}{2}(1 - z)\right) = P_\nu(z). \qquad (4.30)$$

4.5.3 *The Legendre function of second kind*

Another suitable choice of the contour is a figure of eight as in Fig. 4.5, and define the Legendre function of the second kind as:

$$Q_\nu(z) = \frac{1}{4i\sin\nu\pi} \oint_\Sigma \frac{(t^2 - 1)^\nu dt}{(z - t)^{\nu+1}}. \qquad (4.31)$$

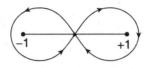

Fig. 4.5

Exercise 1. By squeezing the figure of eight appropriately show that

$$Q_\nu(z) = \frac{1}{2} \int_{-1}^{1} \frac{(1-t^2)^\nu dt}{(z-t)^{\nu+1}}. \tag{4.32}$$

Exercise 2. Show that

$$Q_\nu(z) = \frac{\pi^{\frac{1}{2}} \Gamma(\nu+1)}{\Gamma\left(\nu + \frac{3}{2}\right)(2z)^{\nu+1}} F\left(\frac{1}{2}(\nu+1), \frac{1}{2}\nu + 1; \nu + \frac{3}{2}; \frac{1}{z^2}\right). \tag{4.33}$$

4.6   The Confluent Hypergeometric Function

4.6.1

If the parameter c of the hypergeometric function is not a negative integer, it can be easily verified

$$\lim_{|b|\to\infty} F\left(a, b; c; \frac{z}{b}\right) = \frac{\Gamma(c)}{\Gamma(a)} \sum \frac{\Gamma(a+n)}{\Gamma(c+n)} \frac{z^n}{n!}. \tag{4.34}$$

We shall now introduce the generalized hypergeometric function as

$$_pF_q(a_1, a_2, \ldots a_p; c_1, c_2, \ldots c_q; z) = \sum \frac{(a_1)_n (a_2)_n \cdots (a_p)_n}{(c_1)_n (c_2)_n \cdots (c_q)_n} \frac{z^n}{n!}. \tag{4.35}$$

In this notation, r.h.s. of Eq. (4.34) is $_1F_1(a, c; z)$. Thus

$$\lim_{|b|\to\infty} F\left(a, b; c; \frac{z}{b}\right) = {}_1F_1(a; c; z).$$

It follows that $_1F_1(a, c; z)$ satisfies the differential equation

$$z\omega'' + (c - z)\omega' - a\omega = 0. \tag{4.36}$$

The differential equation has its only singular points a regular singularity at 0 and an irregular singularity at ∞. From the point of view of the limiting process described above, the irregular point at ∞ has been formed by the confluence of the two regular points $z = b$ and $z = \infty$; hence the name confluent hypergeometric function applied to it.

If c is not an integer, a second independent solution valid near the origin is

$$z^{1-c}F(1 + a - c; 2 - c; z).$$

4.6.2 Integral representation

An Eulerian representation for $_1F_1(a, c; z)$ is given by

$$F(a; c; z) = \frac{\Gamma(c)}{\Gamma(a)\Gamma(c-a)} \int_0^1 e^{tz} t^{a-1}(1-t)^{c-a-1} dt \quad \text{Re}\, c > \text{Re}\,(a) > 0.$$
(4.37)

We may remove the restrictions on c and a if we replace the line integral by suitable contour integrals.

The confluent hypergeometric function converges for all finite values of $|z|$ and is therefore an entire function.

Exercise. Show that

$$_1F_1(a; c; z) = e^z {}_1F_1(c - a; c; -z).$$
(4.38)

The Laguerre polynomial $L_n^{(\alpha)}(z)$ for positive integer n is defined as

$$L_n^\alpha(z) = \frac{\Gamma(n + \alpha + 1)}{\Gamma(\alpha + 1)n!} {}_1F_1(-n; \alpha + 1; z).$$
(4.39)

For $\alpha = 0$, it is customary to denote the Laguerre polynomial by $L_n(z)$:

$$L_n(z) = {}_1F_1(-n; 1; z).$$
(4.40)

The Hermite polynomials for odd and even integer can be defined through Laguerre polynomials:

$$H_{2n}(z) = (-1)^n 2^{2n} n! L_n^{-\frac{1}{2}}(z^2)$$
(4.41)

$$H_{2n+1}(z) = (-1)^n 2^{2n+1} n! z L_n^{-\frac{1}{2}}(z^2).$$
(4.42)

4.7 Bessel Function

For every finite z the function

$$\exp\left[\frac{z}{2}\left(t - \frac{1}{t}\right)\right]$$

represents a function of t which is analytic in the t plane from which 0 is omitted. It can therefore be expanded in a Laurent series

$$\exp\left[\frac{z}{2}\left(t - \frac{1}{t}\right)\right] = \sum_{n=-\infty}^{\infty} t^n J_n(z).$$
(4.43)

$J_n(z)$ are called the Bessel function of the first kind, of integral order.

Exercise. Show that

$$J_n(z) = (-1)^n J_{-n}(z), \quad n = 0, 1, 2 \dots. \tag{4.44}$$

Expanding the exponentials $e^{\frac{zt}{2}}$ and $e^{-\frac{z}{2t}}$ and multiplying the resulting series, it follows that

$$J_n(z) = \left(\frac{z}{2}\right)^n \sum_{r=0}^{\infty} \frac{(-1)^r}{(n+r)!r!} \left(\frac{z}{2}\right)^{2r}. \tag{4.45}$$

In the notation of generalized hypergeometric function

$$J_n(z) = \left(\frac{z}{2}\right)^n \frac{1}{n!} {}_0F_1\left(;n+1;\frac{z^2}{4}\right). \tag{4.46}$$

Using the integral formula for the Laurent coefficient, a second expression for $J_n(z)$ is given by,

$$J_n(z) = \frac{1}{2\pi i} \oint_C t^{-n-1} e^{\left[\frac{z}{2}\left(t-\frac{1}{t}\right)\right]} dt \tag{4.47}$$

where C is a closed contour enclosing $t = 0$.

Choosing C to be the unit circle $J_n(z) = \frac{1}{2\pi}\int_{-\pi}^{\pi} \exp(iz\sin\theta - in\theta)d\theta$. Since the imaginary part of the integral is an odd function of θ and the real part is even

$$J_n(z) = \frac{1}{\pi} \int_0^{\pi} \cos(z\sin\theta - n\theta)d\theta. \tag{4.48}$$

The Bessel function defined above is a solution of the ordinary differential equation

$$z^2 w'' + z w' + (z^2 - n^2)w = 0.$$

We now replace n by ν whose ν is an arbitrary complex number and define

$$J_\nu(z) = \frac{z^\nu}{2^{\nu+1}\pi i} \int_{-\infty}^{0+} t^{-\nu-1} \exp\left(t - \frac{z^2}{4t}\right) dt. \tag{4.49}$$

Expanding $e^{-\frac{z^2}{4t}}$ we immediately obtain

$$J_\nu(z) = \frac{z^\nu}{2^{\nu+1}\pi i} \sum_{r=0}^{\infty} \frac{(-1)^r z^{2r}}{2^{2r} r!} \int_{-\infty}^{0+} t^{-\nu-1-r} e^t dt.$$

Using the contour integral representation of the gamma function (Chapter 3), we obtain

$$J_\nu(z) = \left(\frac{z}{2}\right)^\nu \sum \frac{(-1)^r}{\Gamma(\nu+1+r)r!} \left(\frac{z}{2}\right)^{2r}$$

$$= \left(\frac{z}{2}\right)^\nu \frac{1}{\Gamma(\nu+1)} {}_0F_1\left(\nu+1; \frac{z^2}{4}\right) \tag{4.50}$$

the traditional expansion of Eq. (4.45).

Let us now consider the integral

$$J_\nu(z) = \left(\frac{z}{2}\right)^\nu \frac{1}{2\pi i} \int_{-\infty}^{0+} dt\, t^{-\nu-1} \exp\left(t - \frac{z^2}{4t}\right).$$

Since there is a large degree of latitude in deforming a contour, we may substitute

$$t = \frac{1}{2}uz \quad \text{(with } \operatorname{Re} z > 0\text{)}$$

$$= \frac{1}{2\pi i} \int_{-\infty}^{0+} u^{-\nu-1} e^{\frac{z}{2}\left(u - \frac{1}{u}\right)} du.$$

Since we have already seen in Chapter 3 that the integral is independent of the radius ϵ of the circle of indentation, we may choose $\epsilon = 1$. Thus

$$\frac{1}{2\pi i} \int_s u^{-\nu-1} e^{\frac{z}{2}\left(u - \frac{1}{u}\right)} du = \frac{1}{2\pi} \int_{-\pi}^{\pi} e^{-i\nu\theta + iz\sin\theta} d\theta.$$

Similarly the linear part is given by

$$\frac{\sin\pi(\nu+1)}{\pi} \int_1^\infty t^{-\nu-1} e^{-\frac{z}{2}\left(t-\frac{1}{t}\right)} dt$$

$$= \frac{\sin\pi(\nu+1)}{\pi} \int_0^\infty e^{-\nu\tau - z\sin h\tau} dt \quad \text{(Substituting } t = e^\tau\text{)}.$$

We therefore obtain the following integral representation for arbitrary ν

$$J_\nu(z) = \frac{1}{2\pi} \int_{-\pi}^{\pi} e^{-i\nu\theta + iz\sin\theta}$$

$$+ \frac{1}{\pi} \sin\pi(\nu+1) \int_0^\infty e^{-\nu\tau - \sinh\tau} d\tau, |\arg z| < \frac{\pi}{2}. \tag{4.51}$$

4.8   Orthogonal Polynomials

Within an interval (a, b) and a weight function $w(x)$ which is non-negative there, we may associate the scalar product

$$(\phi_1, \phi_2) = \int_a^b \phi_1(x)\phi_2(x)w(x)dx \qquad (4.52)$$

which is defined for all functions ϕ for which $w^{\frac{1}{2}}\phi$ is square integrable in (a, b).

The orthogonal polynomials belonging to the intervals and weight functions which frequently appear in physics are given below:

a	b	$w(x)$	**Name**
-1	1	1	Legendre or Spherical
-1	1	$(1-x)^\alpha(1+x)^\beta$	$P_n^{(\alpha,\beta)}(x)$ Jacobi
$-\infty$	∞	e^{-x^2}	Hermite- $H_n(x)$
0	∞	$x^\alpha e^{-x}$	Laguerre $L_n^\alpha(x)$

All the classical polynomials listed above are solutions of self-adjoint second order differential equation

$$\frac{d}{dx}\left[X w(x)\frac{dy}{dx}\right] + \lambda_n w(x)y = 0.$$

Generating Functions: We define the generating function of an orthogonal polynomial as

$$G(t, x) = \sum N_n t^n P_n(x).$$

By choosing different normalization N_n we get different generating function for the same orthogonal polynomial $P_n(x)$.

(a) Laguerre Polynomial:

$$L_n^0(x) = L_n(x) = {}_1F_1(-n, 1; x).$$

(i) Let us first choose

$$N_n = 1, \quad t < 1$$

$$G(t, x) = \sum t^n \sum \frac{(-n)_r x^r}{(1)_r r!} = \sum t^n \sum (-)^r \frac{n! x^r}{(n-r)! r! r!}$$

$$= (1-t)^{-1} \sum \frac{(-)r(tx/1-t)^r}{r!} = (1-t)^{-1} e^{-\frac{tx}{1-t}}.$$

(ii) We may also choose

$$N_n = \frac{1}{n!}.$$

Then

$$G(t, x) = e^t \sum \frac{(-tx)^r}{r! r!} = e^t J_0(2\sqrt{tx}).$$

(b) Jacobi Polynomials:

$$P_n^{(\alpha, \beta)}(x) = \frac{\Gamma(n + \alpha + 1)}{\Gamma(\alpha + 1)n!} F\left(-n, n + \alpha + \beta + 1; \alpha + 1; \frac{1}{2}(1 - x)\right).$$

We now choose

$$N_n = \frac{\Gamma(n + \alpha + \beta + 1)}{\Gamma(n + \alpha + 1)}; \quad 0 < t < 1.$$

Thus

$$G(t, x)$$

$$= \sum_{n=0}^{\infty} \frac{\Gamma(n + \alpha + \beta + 1)}{\Gamma(\alpha + 1)n!} t^n F\left(-n, n + \alpha + \beta + 1; \alpha + 1; \frac{1}{2}(1 - x)\right)$$

$$= (1 - t)^{-\alpha - \beta - 1} \sum_{r=} \frac{\left[-\frac{2t(1 - x)}{(1 - t)^2}\right]^r}{r!} \frac{\Gamma(\alpha + \beta + 1 + 2r)}{\Gamma(\alpha + 1 + r)}$$

$$= \frac{2^{\alpha + \beta}(1 - t)^{-\alpha - \beta - 1} \Gamma\left(\frac{\alpha + \beta + 1}{2}\right)}{\sqrt{\pi}} \frac{1}{\Gamma(\alpha + 1)} \Gamma\left(\frac{\alpha + \beta + 2}{2}\right)$$

$$\times F\left(\frac{\alpha + \beta + 1}{2}, \frac{\alpha + \beta}{2} + 1; \alpha + 1; -\frac{2t(1 - x)}{(1 - t)^2}\right).$$

The above examples clearly exhibit the non-uniqueness of the generating function.

The hypergeometric function as an analytic function in parameters and variable has been employed in Chapter 9 to tackle the Clebsch–Gordan problem of $SU(1, 1)$ or $SL(2, R)$. The Clebsch–Gordan series appears partly as a line integral in the parameter space corresponding to the contribution of the continuous part of the spectrum and a sum over the residues at simple poles corresponding to the contribution of the discrete part of the spectrum. On the other hand the Clebsch–Gordan coefficients (in the $SO(2)$ basis) appear as coefficients of Taylor or Fourier expansion.

References

[1] E. T. Whittaker and G. N. Watson, *A Course of Modern Analysis.*
[2] E. T. Copson, *Theory of Functions of a Complex Variable.*
[3] L. J. Slater, *Generalized Hypergeometric Functions.*

CHAPTER 5

Bargmann–Segal Spaces of Analytic Functions

5.1

A particular class of spaces of analytic functions turns out to be suitable for many problems in group representation theory and quantum mechanics. These function spaces constitute the so called "Hilbert spaces" whose precise definition is relegated to Chapter 7. The simplest of such Hilbert spaces of analytic functions was introduced by Bargmann for functions of a finite number of complex variables and by Segal for an infinite number of variables. Similar Hilbert space of functions analytic within the unit disc turn out to be the canonical carrier space for a particular class of representations (see Chapter 9) of the three-dimensional Lorentz group. Some of the advantages of these Hilbert spaces are the following.

(a) These provide convenient carrier spaces for many quantum mechanical and group theoretic problems.
(b) One obtains explicit expressions not only for infinitesimal operators of the group but also for their exponentials, the unitary operators of the representation itself.
(c) The infinitesimal generators which are often unbounded may be without difficulty precisely defined as Hilbert space operators.

The concept of the principal vector (or the reproducing kernel) introduced in the next section will turn out to be crucial in Chapter 9 for the construction of the integral kernel of the group ring for a class of representations leading to a simple formulation of the theory of character of infinite dimensional representations. It will also play an important role in Chapter 11 for deriving a class of integral transforms in a Hilbert space of analytic functions. Since the Hilbert space of methods of Bargmann and Segal will play a central role in Part 2 we describe the essential features of their function space in some detail. To avoid inessential complications we present a simple mathematically non-rigorous discussion.

The inner product of two elements ψ_1, ψ_2 in the $L^2(R)$ Hilbert space used for the quantum mechanical description of a system with one degree of freedom is given by

$$(\psi_1, \psi_2) = \int \overline{\psi_1(q)}(\psi_2)dq. \qquad (5.1)$$

Here q is a point of R (unless the domain of integration is explicitly indicated all integrals extend over the whole range of integration variables, i.e. R for q and C for z).

The self adjoint canonical operators q and $p = -i\frac{\partial}{\partial q}$ (with $\hbar = 1$) satisfy the commutation rule

$$[q, p] = i. \qquad (5.2)$$

The well known complex combinations

$$a = \frac{1}{\sqrt{2}}(q + ip), \quad a^\dagger = \frac{1}{\sqrt{2}}(q - ip) \qquad (5.3)$$

which are the harmonic oscillator lowering and raising operators satisfy

$$[a, a^\dagger] = 1. \qquad (5.4)$$

Bargmann and Segal's pioneering work on the Hilbert space of analytic functions is the culmination of Fock's formal operator solution

$$a = \frac{\partial}{\partial z}, \quad a^\dagger = z \qquad (5.5)$$

of the commutation relation Eq. (5.4). These two works constitute a mathematically complete investigation of the function space on which Fock's solution is realized and its connection with the conventional $L^2(R)$ space of functions of quantum mechanics. In order to realize Fock's solution on a suitable Hilbert space, one is led to a space of quantum mechanics. In order to realize Fock's solution on a suitable Hilbert space, one is led to a space of functions that depend on the variable $a^\dagger = z$ but do not explicitly depend on a since $a = \frac{\partial}{\partial z}$. This suggests analytic functions of a single complex variable which we denote by z. Of course in n-dimensions this becomes analytic function of n complex variables $z_1, z_2, \ldots, z_n$.

In addition, the inner product in this Hilbert space (denoted by B(C)) must be determined in such a way that the operators z and $\frac{\partial}{\partial z}$ will be adjoint as required by Eq. (5.3).

This leads to the following two problems:

1. To find a positive real function $\rho(z, \bar{z})$ which defines the inner product in B(C).

$$(f, g) = \int \overline{f(z)} g(z) \rho(z, \bar{z}) d^2 z \tag{5.6}$$

$$z = x + iy \quad d^2 z = dx dy$$

such that z and $\frac{\partial}{\partial z}$ are adjoint.

2. To find an integral kernel A(z, q) such that

$$f(z) = \int A(z, q) \psi(q) dq \tag{5.7}$$

is a unitary mapping of $L^2(R)$ onto B(C) which properly relates the operators a, $a^\dagger$ and z, $\frac{\partial}{\partial z}$ in the two isomorphic Hilbert spaces.

We first proceed to determine ρ by requiring

$$(zf, g) = \left(f, \frac{\partial g}{\partial z} \right) \tag{5.8}$$

which holds for functions that do not grow too fast at infinity. The above equation yields.

$$\int \overline{zf(z)} g(z) \rho(z, \bar{z}) d^2 z = \int \overline{f(z)} \frac{\partial g}{\partial z} \rho(z, \bar{z}) d^2 z. \tag{5.9}$$

Since $f(z)$ is an analytic function we obtain

$$\overline{f(z)} \frac{\partial g}{\partial z} \rho(z, \bar{z}) = \frac{\partial}{\partial z} \left(\overline{f(z)} g(z) \rho(z, \bar{z}) \right) - \overline{f(z)} g(z) \frac{\partial \rho}{\partial z}(z, \bar{z}). \tag{5.10}$$

If we assume that ρ vanishes sufficiently fast at infinity, we have

$$\int \frac{\partial}{\partial z} \left[\overline{f(z)} \, g(z) \, \rho(z, \bar{z}) \right] d^2 z = 0. \tag{5.11}$$

Thus

$$\int \overline{zf(z)} g(z) \rho(z, \bar{z}) d^2 z = - \int \overline{f} g \frac{\partial \rho}{\partial z} d^2 z. \tag{5.12}$$

Since f and g are arbitrary functions we obtain

$$\frac{\partial \rho}{\partial z} = -\bar{z} \rho. \tag{5.13}$$

This immediately leads to

$$\rho = c e^{-|z|^2}. \tag{5.14}$$

Setting the arbitrary constant $C = \frac{1}{\pi}$ we immediately obtain the scalar product as

$$(f, g) = \int \overline{f(z)} g(z) d\mu(z) \tag{5.15}$$

where

$$d\mu(z) = \frac{1}{\pi} e^{-|z|^2} d^2 z. \tag{5.16}$$

The integral kernel A can be determined from the following conditions:

If the integral transform maps ψ into f, it maps $a^\dagger \psi$ into zf and $a\psi$ into $\frac{\partial f}{\partial z}$ where

$$a^\dagger \psi = \frac{1}{\sqrt{2}} \left(q - \frac{\partial}{\partial q} \right) \psi(q) \tag{5.17}$$

$$a\psi = \frac{1}{\sqrt{2}} \left(q + \frac{\partial}{\partial q} \right) \psi(q). \tag{5.18}$$

It is again assumed that ψ is sufficiently smooth and vanishes sufficiently fast at infinity. Since a and $a^\dagger$ are adjoint we have

$$\int A(z, q) \left(a^\dagger \psi \right)(q) dq = \int [a A(z, q)] \psi(q) dq$$

$$= zf(z) = \int z A(z, q) \psi(q) dq. \tag{5.19}$$

The above equation immediately yields

$$\frac{1}{\sqrt{2}} \left(q + \frac{\partial}{\partial q} \right) A(z, q) = z A(z, q). \tag{5.20}$$

Similarly

$$\frac{1}{\sqrt{2}} \left(q - \frac{\partial}{\partial q} \right) A(z, q) = \frac{\partial A}{\partial z}. \tag{5.21}$$

We obtain from the above equations

$$\frac{\partial A(z, q)}{\partial z} = \left(\sqrt{2} z - q \right) A(z, q) \tag{5.22}$$

$$\frac{\partial A(z, q)}{\partial q} = \left(\sqrt{2} z - q \right) A(z, q). \tag{5.23}$$

The integral kernel is therefore given by

$$A(z, q) = \pi^{-\frac{1}{4}} \exp \left[-\frac{1}{2} \left(z^2 + q^2 \right) + \sqrt{2} zq \right] \tag{5.24}$$

where the numerical factor has been chosen for later convenience. The generalization to n-dimensions is obvious,

$$A(z_1, z_2, \ldots, z_n, q_1, q_2, \ldots, q_n)$$
$$= \pi^{-\frac{n}{4}} \exp\left[-\frac{1}{2}\left(z_1^2 + z_2^2 + \cdots + z_n^2 + q_1^2 + q_2^2 + \cdots + q_n^2\right) \right.$$
$$\left. + \sqrt{2}\left(z_1 q_1 + z_2 q_2 + \cdots + z_n q_n\right) \right].$$

We now describe the basic properties of this Hilbert space B(C) and the isomorphic mapping of the quantum mechanical $L^2(R)$ space onto B(C). The elements of B(C) are entire analytic functions $f(z)$ having a finite norm according to the scalar product, Eq. (5.15). Let $f(z)$ be an entire function with the power series

$$f(z) = \sum_{n=0}^{\infty} a_n z^n. \tag{5.25}$$

We show that the powers

$$u_n(z) = \frac{z^n}{\sqrt{n!}}, \quad n = 0, 1, 2, \ldots \tag{5.26}$$

constitute a complete orthonormal set:

$$(u_n, u_m) = \frac{1}{\pi\sqrt{n!\, m!}} \int_0^\infty r^{n+m} e^{-r^2} r\, dr \int_0^{2\pi} e^{i(m-n)\psi} d\psi \tag{5.27}$$
$$= \delta_{mn}.$$

Thus

$$(f, f) = \sum_{n=0}^{\infty} n!\, |a_n|^2 \tag{5.28}$$

i.e. both sides of Eq. (5.28) are infinite or both sides are finite and equal. Every set of coefficients for which the sum converges defines an entire function $f \in B(C)$. By linearity we obtain the inner product of two functions f, g.

$$(f, g) = \sum_{n=0}^{\infty} n!\, \bar{a}_n b_n \tag{5.29}$$

$$g(z) = \sum_{n=0}^{\infty} b_n z^n. \tag{5.30}$$

For any function $f \in$ B(C)

$$(u_n, f) = \sqrt{n!}\, a_n$$

so that Eq. (5.28) expresses the completeness of the system u_n.

We now introduce the principal vectors e_z that are bounded linear functionals in B(C) satisfying

$$f(z) = (e_z, f). \tag{5.31}$$

Thus

$$f(z) = \int \overline{e_z(\xi)} f(\xi) d\mu(\xi)$$
$$= \int K(z, \xi) f(\xi) d\mu(\xi) \tag{5.32}$$

where

$$K(z, \xi) = \overline{e_z(\xi)} \tag{5.33}$$

is the reproducing kernel. Expanding $e_z(\xi)$ in the complete orthonormal set as given by Eq. (5.26) we have

$$e_z(\xi) = \sum a_n u_n(\xi) \tag{5.34}$$

where

$$a_n = (u_n, e_z(\xi)) = \left(\overline{e_z(\xi), u_n(\xi)} \right). \tag{5.35}$$

Using the definition Eq. (5.31) of the principal vector we obtain

$$a_n = \overline{u_n(z)}.$$

Thus

$$e_z(\xi) = \sum_{n=0}^{\infty} a_n \ u_n(\xi) = \sum \overline{u_n(z)} \ u_n(\xi) \tag{5.36}$$

$$\sum_{n=0}^{\infty} \frac{(\bar{z}\xi)^n}{n!} = e^{\bar{z}\xi}. \tag{5.37}$$

Inserting the explicit form of the principal vector we obtain

$$f(z) = \int e^{z\bar{\xi}} f(\xi) d\mu(\xi). \tag{5.38}$$

We now determine the function $\psi_n(q)$ that are mapped into $u_n(z)$ (given by Eq. (5.26)). We start from

$$\int A(z, q) \overline{A(\xi, q)} dq = e^{z\bar{\xi}}. \tag{5.39}$$

Setting $\bar{\xi} = b$ we obtain from this

$$e^{bz} = \int \mathrm{A}(z,q)\mathrm{A}(b,q)dq. \tag{5.40}$$

We now note that

$$u_n(z) = \frac{z^n}{\sqrt{n!}} = \frac{1}{\sqrt{n!}}\left[\frac{\partial^n}{\partial b^n}e^{bz}\right]_{b=0}. \tag{5.41}$$

This immediately yields

$$\psi_n(q) = \frac{1}{\sqrt{n!}}\frac{\partial^n}{\partial b^n}[\mathrm{A}(b,q)]_{b=0}$$
$$= \left[2^n n!\sqrt{\pi}\right]^{-\frac{1}{2}} e^{-\frac{q^2}{2}}\mathrm{H}_n(q) \tag{5.42}$$

where $\mathrm{H}_n(q)$ are the Hermite polynomials. Now from the Taylor expansion

$$\mathrm{A}(z,q) = \sum \frac{z^n}{\sqrt{n!}} \frac{1}{\sqrt{n!}}\left[\frac{\partial^n}{\partial z^n}\mathrm{A}(z,q)\right]_{z=0} \tag{5.43}$$

we immediately obtain

$$\mathrm{A}(z,q) = \sum_{n=0}^{\infty} u_n(z)\,\psi_n(q). \tag{5.44}$$

The above equation suggests $\mathrm{A}^{-1}f = \mathrm{W}f$, where

$$[\mathrm{W}f](q) = \int \overline{\mathrm{A}(z,q)}f(z)d\mu(z). \tag{5.45}$$

It is now sufficient to prove

$$\mathrm{A}[\mathrm{W}f] = f. \tag{5.46}$$

Letting $g = \mathrm{A}[\mathrm{W}f]$ we have

$$g(\xi) = \iint \mathrm{A}(\xi,q)\mathrm{A}(z,q)f(z)dqd\mu(z). \tag{5.47}$$

The integral over q can be evaluated from the formula Eq. (5.39). Thus

$$g(\xi) = \int e^{\xi\bar{z}}f(z)d\mu(z) = f(\xi) \tag{5.48}$$

which follows from the definition of the principal vector. We now collect the most important features of the Bargmann–Segal space B(C)

1. The elements of B(C) are entire analytic functions $f(z)$ having a finite norm according to the scalar product

$$(f, g) = \int \overline{f(z)} g(z) d\mu(z) \tag{5.49}$$

where $d\mu(z)$ is the Gaussian measure

$$d\mu(z) = \frac{e^{-|z|^2}}{\pi} d^2 z, \quad d^2 z = dx\, dy, \quad z = x + iy. \tag{5.50}$$

This scalar product satisfies

$$(zf, g) = \left(f, \frac{\partial g}{\partial z} \right). \tag{5.51}$$

2. A complete orthonormal set in B(C) is given by the powers

$$u_n(z) = \frac{z^n}{\sqrt{n!}}, \quad n = 0, 1, 2 \dots. \tag{5.52}$$

3. The principal vector in B(C) satisfying

$$f(z) = (e_z, f) \tag{5.53}$$

is given by

$$e_z(\xi) = e^{\bar{z}\xi} \tag{5.54}$$

so that

$$f(z) = \int e^{z\bar{\xi}} f(\xi) d\mu(\xi). \tag{5.55}$$

It is seen that the principal vectors play here a role similar to that of the δ functions $[\delta(q - a)]$ in the traditional $L^2(R)$ Hilbert space of quantum mechanics but unlike the δ functions they are elements of the Hilbert space.

4. The isomorphic mapping of B(C) on the conventional $L^2(R)$ space is given by Bargmann's integral transform pair.

$$f(z) = \int A(z, q)\psi(q) dq \tag{5.56}$$

$$\psi(q) = \int \overline{A(z, q)} f(z) d\mu(z) \tag{5.57}$$

where

$$A(z, q) = \pi^{-\frac{1}{4}} \exp\left[-\frac{1}{2}(z^2 + q^2) + \sqrt{2} zq \right]. \tag{5.58}$$

5.2

The connection between the Bargmann–Segal Hilbert space of analytic functions and the coherent states was pointed out by Glauber. As early as 1926 Schrodinger introduced these coherent states to describe non-spreading wave packets in quantum mechanics. Later Von Neumann studied an important subsystem of these coherent states related to the regular cell partition of the phase space for a system with one degree of freedom. This system was used by Von Neumann to analyze the quantum mechanical measurement process. After about three decades interest in coherent states was revived by Klauder, Glauber and Sudarshan who showed that these are the appropriate quantum mechanical states for the description of laser fields containing a large and intrinsically uncertain number of photons.

Since the Glauber coherent states for the electromagnetic field have been the source of inspiration for the development of all types of coherent states in the past three decades, it is worthwhile to emphasize their historical as well as pedagogical importance. The canonical coherent states of Glauber can be constructed from any one of the two mathematical definitions.

1. The coherent states $|\alpha\rangle$ are eigenstates of the annihilation operator a,

$$a|\alpha\rangle = \alpha|\alpha\rangle \tag{5.59}$$

where α is a complex number.

2. The coherent states $|\alpha\rangle$ can be obtained by applying a displacement operator $D(\alpha)$ on vacuum state of the harmonic oscillator

$$|\alpha\rangle = D(\alpha)|0\rangle \tag{5.60}$$

where the displacement operator $D(\alpha)$ is defined as

$$D(\alpha) = \exp\left(\alpha a^\dagger - \bar{\alpha} a\right). \tag{5.61}$$

It can be shown that the same coherent states are obtained from the two mathematical definitions.

Glauber pointed out that the eigenstates of the free field Hamiltonian (i.e. eigenstates of $a^\dagger a$) are poorly suited for the description of laser fields of intense beam of photons. However when the radiation field is in a coherent state the field has properties very similar to that of a classical electromagnetic wave with a certain phase and amplitude. For example, the expectation value of $N = a^\dagger a$ is given by

$$\langle\alpha|N|\alpha\rangle = |\alpha|^2. \tag{5.62}$$

Thus the average number of photons, i.e. intensity in the state $|\alpha\rangle$ is $|\alpha|^2$. If we now calculate the expectation value of the electric field with respect to these states it can be easily verified that it is proportional to

$$\frac{i}{2}\widehat{e}\,[\alpha \exp\{i(k \cdot r - \omega t)\}] - \bar{\alpha} \exp\{-i(k \cdot r - \omega t)\}$$
$$= \widehat{e}|\alpha| \sin(k \cdot r - \omega t + \psi) \tag{5.63}$$

where $\psi = \arg \alpha$ and $\widehat{e}$ is the unit polarization vector. Thus the coherent state can be interpreted as representing a harmonic wave with phase ψ.

Since the coherent states do not form an orthonormal set, they had earlier received little attention as a possible system of basis vectors for the expansion of arbitrary states. However, Glauber showed that such expansions can be carried out conveniently and uniquely and they possess exceedingly useful properties. While orthogonality is a convenient property for a set of basis states it is not a necessary one for the expansibility of an arbitrary vector in terms of such a set, the essential property is that it must be complete. It was Glauber who showed that the coherent states, although non-orthogonal, do form a complete set Expansion of arbitrary states can be developed as a simple generalization of the traditional expansions in terms of complete orthonormal sets.

We start from Glauber's definition of coherent states as

$$a|\alpha\rangle = \alpha|\alpha\rangle, \quad \alpha \in \mathbb{C} \tag{5.64}$$

$$|\alpha\rangle = \sum_{n=0}^{\infty} \langle n|\alpha\rangle |n\rangle \tag{5.65}$$

where $|n\rangle$ are the Fock states

$$a^\dagger a|n\rangle = n|n\rangle. \tag{5.66}$$

It is well known

$$|n\rangle = \frac{(a^\dagger)^n}{\sqrt{n!}}|0\rangle. \tag{5.67}$$

We therefore obtain

$$|\alpha\rangle = \langle 0|\alpha\rangle \sum_{n=0}^{\infty} \frac{\alpha^n}{\sqrt{n!}}|n\rangle. \tag{5.68}$$

Setting $\langle \alpha|\alpha\rangle = 1$ we have

$$|\langle 0|\alpha\rangle|^2 \, e^{|\alpha|^2} = 1. \tag{5.69}$$

Hence ignoring a phase

$$|\alpha\rangle = e^{-|\alpha|^2/2} \sum \frac{\alpha^n}{\sqrt{n!}} |n\rangle. \tag{5.70}$$

The coherent states $|\alpha\rangle$ constitute a non-orthogonal overcomplete, linearly dependent set,

$$\langle\beta|\alpha\rangle = e^{\bar{\beta}\alpha} - \frac{|\alpha|^2}{2} - \frac{|\beta|^2}{2} \tag{5.71}$$

$$\int |\alpha\rangle\langle\alpha|d^2\alpha = \pi d^2\alpha = d(\mathrm{Re}\,\alpha)d(\mathrm{Im}\,\alpha). \tag{5.72}$$

We are now in a position to establish the connection of Glauber coherent states with the Bargmann–Segal space.

Since the occupation number states $|n\rangle$ form a complete set an arbitrary state of an oscillator must possess an expansion.

$$|\psi\rangle = \sum C_n |n\rangle = \sum C_n \frac{(a^\dagger)^n |0\rangle}{\sqrt{n!}}. \tag{5.73}$$

Since $\langle\psi|\psi\rangle = 1$ we have

$$\sum |C_n|^2 = 1. \tag{5.74}$$

Thus the above series must be absolutely convergent and we have

$$\lim_{n\to\infty} \left| \frac{C_{n+1}}{C_n} \right| < 1. \tag{5.75}$$

Let us now consider

$$f(z) = \sum_{n=0}^{\infty} C_n \frac{z^n}{\sqrt{n!}}. \tag{5.76}$$

Denoting the nth term of the power series by u_n, we have

$$\lim_{n\to\infty} \left| \frac{u_{n+1}}{u_n} \right| = \lim_{n\to\infty} \left| \frac{C_{n+1}}{C_n} \right| \frac{|z|}{\sqrt{n}} = 0. \tag{5.77}$$

The radius of convergence of the power series (5.76) is, therefore, infinite and the analytic function represented by it is an entire function.

If we denote the arbitrary state which corresponds to the function $f(z)$ by $|f\rangle$ then we may write

$$|f\rangle = \sum c_n \frac{(a^\dagger)^n}{\sqrt{n!}} |0\rangle \tag{5.78}$$

$$\langle\beta|f\rangle = \langle\beta|0\rangle \sum c_n \frac{\bar{\beta}^n}{\sqrt{n!}} = e^{-|\beta|^2/2} f(\bar{\beta}). \tag{5.79}$$

Thus

$$\langle f|g\rangle = \frac{1}{\pi} \int \langle f|\beta\rangle\langle\beta|g\rangle = \int \overline{f(\bar{\beta})} g(\bar{\beta}) d\mu(\beta) \tag{5.80}$$

so that setting $\bar{\beta} = z$

$$(f,g) = \int \overline{f(z)} g(z) d\mu(z) \tag{5.81}$$

which coincides with the scalar product of Bargmann and Segal. Similarly

$$\langle\beta|f\rangle = \frac{1}{\pi} \int \langle\beta|\alpha\rangle\langle\alpha|f\rangle d^2\alpha. \tag{5.82}$$

Setting $\bar{\beta} = z$ and $\bar{\alpha} = \xi$ we have

$$f(z) = \int e^{z\bar{\xi}} f(\xi) d\mu(\xi) \tag{5.83}$$

which coincides with the principal vector in the Bargmann–Segal space.

Bargmann used the Hilbert space $B(C_2)$ which consists of analytic functions $u(z_1, z_2)$ of two complex variables for the analysis of the three-dimensional rotation group. In a strict mathematical sense the Hilbert space method developed by Bargmann is isomorphic to Schwinger's operator method. Schwinger introduced a construction of the generators of the rotation group in terms of the harmonic oscillator creation and annihilation operators. In addition all other objects to be studied were defined in terms of the boson operators $a, a^\dagger$ including the orthonormal vector basis on which they act. In Bargmann's investigation, however the Hilbert space is *a priori* given as a function space and the standard methods of entire analytic functions can be used for the solution.

Bargmann showed that the subspace $B_j(C)$ invariant under Schwinger's angular momentum operators consists of homogeneous polynomials of degree $2j$, $j = 0, \frac{1}{2}, 1, \ldots$ in z_1, z_2

$$u(z_1, z_2) = [(2j)!]^{\frac{1}{2}} z_2^{2j} f(z), \quad z = \frac{z_1}{z_2} \tag{5.84}$$

where the numerical factor $[(2j)!]^{\frac{1}{2}}$ is introduced for later convenience. To obtain the scalar product in $B_j(C)$ we start from the scalar product in $B(C_2)$.

$$(u,v) = \int \overline{u(z_1,z_2)} v(z_1,z_2) d\mu(z_1) d\mu(z_2) \tag{5.85}$$

and restrict ourselves to functions of the form Eq. (5.84). Replacing the variables z_1, z_2 by $z = \frac{z_1}{z_2}, z_2$ we obtain

$$(u,v) = \int \overline{f(z)} g(z) \mathrm{I}(z,\bar{z}) d^2 z \tag{5.86}$$

where

$$\mathrm{I}(z,\bar{z}) = \frac{1}{\pi^2} [(2j)!(2j')!]^{-\frac{1}{2}} \iint e^{-|z_2|^2(1+|z|^2)} \bar{z}_2^{2j'} z_2^{2j} |z_2|^2. \tag{5.87}$$

The above integral can be evaluated easily and we have

$$\mathrm{I}(z,\bar{z}) = \delta_{jj'} \frac{(2j+1)}{\pi} \left(1+|z|^2\right)^{-2j-2}. \tag{5.88}$$

The scalar product in $B_j(C)$ is therefore given by

$$(f,g) = \int \overline{f(z)} g(z) d\lambda(z) \tag{5.89}$$

where

$$d\lambda(z) = \frac{2j+1}{\pi} (1+|z|^2)^{-2j-2} d^2 z. \tag{5.90}$$

To get the principal vector in $B_j(C)$ we start from the two-dimensional form of Eq. (5.83)

$$u(z_1,z_2) = \iint e^{z_1 \bar{\xi}_1 + z_2 \bar{\xi}_2} u(\xi_1,\xi_2) d\mu(\xi_1) d\mu(\xi_2). \tag{5.91}$$

We now replace the variables z_1, z_2 and ξ_1, ξ_2 by $z = \frac{z_1}{z_2}, z_2$ and $\xi = \frac{\xi_1}{\xi_2}, \xi_2$ respectively and restrict ourselves to functions of form Eq. (5.84). Thus

$$z_2^{2j} f(z) = \int \mathrm{I}(z_2, z, \xi, \bar{\xi}) f(\xi) d^2 \xi \tag{5.92}$$

where

$$\mathrm{I}(z_i, z, \xi, \bar{\xi}) = \frac{1}{\pi^2} \int e^{z_2 \bar{\xi}_2(1+z\bar{\xi}) - |\xi_2|^2(1+|\xi|^2)} 2j \xi_2 |\xi_2|^2 d^2 \xi_2. \tag{5.93}$$

The above integral can be evaluated easily and we have

$$\mathrm{I}(z_2, z, \xi, \bar{\xi}) = z_2^{2j} \frac{2j+1}{\pi} (1+|\xi|^2)^{-2j-2} (1+z\bar{\xi})^{2j}. \tag{5.94}$$

Thus

$$f(z) = \int (1 + z\bar{\xi})^{2j} f(\xi) d\lambda(\xi). \tag{5.95}$$

The principal vector in $B_j(C)$ is therefore given by

$$e_z(\xi) = (1 + \bar{z}\xi)^{2j}. \tag{5.96}$$

The Hilbert space method of Bargmann and Segal has been applied in many problems in Part 2 of this book. This has achieved a significant unification in the theory of integral transforms and substantial simplification in the theory of characters of locally compact groups leading to a simple derivation of the Plancherel formula for the universal covering group of SL(2, R) from which Harish–Chandra's Plancherel formula for SL(2, R) follows as point contributions (see Chapter 9).

Since the above problem as well as those considered later will require explicit use of some aspects of generalized functions we shall consider this concept briefly in the next chapter.

References

[1] V. Bargmann, *Commun. Pure Appl. Math.* **14** (1961) 187; **20** (1967) 1.

[2] I. E. Segal, *Ill. J. Math.* **6** (1962) 500; *Mathematical Problems of Relativistic Physics* (Am. Math. Soc., Providence, R.I. 1963).

[3] J. R. Klauder, *Ann. Phys. (NY)* **11** (1960) 123;
J. R. Klauder and E. C. G. Sudarshan, *Fundamentals of Quantum Optics* (Benjamin, N.Y., 1968) (Dover, 2006);
J. R. Klauder and B. S. Skagerstam, *Coherent States. Applications in Physics and Mathematical Physics* (World Scientific Publishing, Singapore, 1985).

[4] R. J. Glauber, *Phys. Rev. Lett.* **10** (1963) 84; *Phys. Rev.* **130** (1963) 25, 29; **131** (1963) 2766.

[5] E. C. G. Sudarshan, *Phys. Rev. Lett.* **10** (1963) 277.

CHAPTER 6

Elements of the Theory of Generalized Functions

The concept of generalized functions is a convenient link connecting many aspects of analysis, functional analysis and the representation theory of locally compact Lie groups. They have been proved useful particularly in the investigation of the equivalence and irreducibility of representations.[1]

One of the specific questions of the theory of representations where generalized functions can be used conveniently is the so-called Plancherel theorem which gives expansion of $f(g)$ on the Lie group into analogue of the Fourier integral. This problem has been addressed in Chapter 9 and solved completely for the universal covering group of SL(2, R)[2].

The discussion that follows is essentially an adaptation of the relevant parts of Gel'fand and Shilov's treatise on generalized functions (Vol. 1). We define generalized functions (singular as well as regular) and their derivative and introduce special generalized functions like $x_\pm^\lambda$, $(x \pm i0)^\lambda$, $(x_1^2 + x_2^2 + \cdots + x_p^2 - x_{p+1}^2 - x_{p+2}^2 - \cdots - x_{p+q}^2)_+^\lambda$ as analytic functions of λ which will prove to be particularly useful later.

6.1

It is well known that there exists a class of "functions" which cannot be properly accommodated within the traditional framework of function theory. These so-called "singular functions" are now identified with functionals which associates with every "sufficiently good" function some well defined number.

This framework will then not only include classical summable functions but the singular functions as well and hence the name generalized functions has been applied to it. First we must spell out what we precisely mean by a "good function" which will be called test functions spanning, say, a space K. We shall say a sequence $\phi_1(x), \phi_2(x), \ldots, \phi_n(x)$ of test functions converges to zero in K if all these functions vanish outside a certain fixed bounded region and converge uniformly to zero together with their derivatives of any order.

As an example of a test function in K we cite

$$\phi(x,a) = e^{-\frac{a^2}{a^2-x^2}}, \quad -a < x < a; \ a > 0 \tag{6.1}$$
$$= 0 \qquad\qquad |x| \geq a.$$

The sequence $\phi_n = \frac{1}{n}\phi(x,a); \ n = 1, 2, \ldots$ converges to zero in K.

6.2   Generalized Functions

We shall say f is a continuous linear functional on K if there exists some rule according to which we can associate with every $\phi(x)$ in K a real (complex) number (f, ϕ) satisfying the following conditions:

(a) For any two real numbers α_1 and α_2

$$(f, \alpha_1\phi_1 + \alpha_2\phi_2) = \alpha_1(f, \phi_1) + \alpha_2(f, \phi_2).$$

(b) If the sequence $\phi_1, \phi_2, \ldots, \phi_n, \ldots$ converges to zero in K, then the sequence $(f, \phi_1), (f, \phi_2), \ldots, (f, \phi_n) \ldots$ converges to zero (continuity of f).

For example, let $f(x)$ be absolutely integrable in every bounded region of R, i.e. $\left| \int_R f(x)dx \right| < \infty$.

We shall call such functions locally summable. By means of such a function we can associate every $\phi(x)$ in K with,

$$(f, \phi) = \int_R f(x)\phi(x)dx \tag{6.2}$$

where the integral is actually taken only over the bounded region in which $\phi(x)$ fails to vanish. It is easily verified that the conditions (a) and (b) are satisfied for the functional f.

Equation (6.2) represents a very special kind of continuous linear functional on K. Other kinds of functionals are easily shown to exist. For instance the functional which associates with every $\phi(x)$ in K its value at $x = 0$.

$$(\delta, \phi) = \phi(0). \tag{6.3}$$

This also satisfies the desired conditions (a) and (b). However this functional cannot be written in the form (6.2) with any locally summable function $f(x)$.

Indeed let us assume that there exists some locally summable function $f(x)$ such that for every $\phi(x)$ in K we have

$$\int_R f(x)\phi(x)dx = \phi(0).$$

Taking $\phi(x)$ as $\phi(x, a)$ of Eq. (6.1) we have,

$$\int_{-a}^{a} f(x)\phi(x, a)dx = \phi(0, a) = a^{-1}. \tag{6.4}$$

But as $a \to 0$ the integral tends to zero for all locally summable $f(x)$ which contradicts Eq. (6.4).

The functional under consideration will be called delta function which is really a misnomer since delta function is not a function in the classical sense of the word. We shall denote it by $\delta(x)$ so that

$$(\delta(x), \phi(x)) = \phi(0). \tag{6.5}$$

One often deals with "translated" delta function

$$(\delta(x - x_0), \phi(x)) = \phi(x_0). \tag{6.6}$$

We now define a generalized function as any linear continuous functional defined on K. Those functionals which can be given by an equation such as (6.2) shall be called regular and all others, including the delta function, will be called singular.

6.3 Derivative of Generalized Functions

Although not all ordinary functions are differentiable generalized functions have derivatives of all orders which are also generalized functions.

Let $f(x)$ be a continuous function with a continuous first derivative and consider the functional

$$(f', \phi) = \int f'(x)\phi(x)dx.$$

Integrating by parts and recalling that $\phi(x)$ is in K so that outside some interval $[a, b]$ it vanishes we arrive at

$$(f', \phi) = f(x)\phi(x)|_{-\infty}^{\infty} - \int f(x)\phi'(x)dx = (f, -\phi'). \tag{6.7}$$

We shall use this equation to define the derivative of a generalized function.

Let f be any continuous linear functional on K. Then the functional g defined by

$$(g, \phi) = (f, -\phi') \tag{6.8}$$

will be called the derivative of f and be denoted f' (or $\frac{df}{dx}$).

In accordance with our notational convention we may write f' also in the form

$$\int f'(x)\phi(x)dx = -\int f(x)\phi'(x)dx. \tag{6.9}$$

Example. Consider the step function,

$$\theta(x) = 1 \quad x > 0$$
$$= 0 \quad x < 0.$$

Thus

$$(\theta', \phi) = -(\theta(x), \phi'(x)) = -\int_0^\infty \phi'(x)dx = \phi(0). \tag{6.10}$$

Thus $\theta'(x) = \delta(x)$ and equivalently $\theta'(x - h) = \delta(x - h)$.

Example. Derivative of δ-function

$$(\delta'(x - h), \phi(x)) = -\phi'(h)$$
$$(\delta^{(k)}(x - h), \phi(x)) = (-1)^k \phi^{(k)}(h). \tag{6.11}$$

Consider a sequence $f_1, f_2, \ldots, f_n, \ldots$ of generalized functions which converges to the generalized function f. We assert the sequence of derivatives $\frac{\partial f_\nu}{\partial x_j}$ converges to $\frac{\partial f}{\partial x_j}$. This is immediately obvious, since for any $\phi(x)$ in K we have

$$\left(\frac{\partial f_\nu}{\partial x_j}, \phi\right) = \left(f_\nu, -\frac{\partial \phi}{\partial x_j}\right)$$
$$\left(f, -\frac{\partial \phi}{\partial x_j}\right) = \left(\frac{\partial f}{\partial x_j}, \phi\right) \tag{6.12}$$

as asserted.

Similarly, a series of generalized functions $h_1 + h_2 + \cdots + h_n + \cdots$ which converges to the generalized function g can be differentiated term by term. In other words one may write

$$h_1' + h_2' + \cdots + h_n' + \cdots = g'. \tag{6.13}$$

In classical analysis such theorems do not hold, for the derivative of a convergent sequence will not in general converge. Consider for example the sequence $f_n(x) = \frac{\sin nx}{n}$ on the real axis which converges uniformly to zero.

The derivatives $f'_n(x) = \cos nx$ of this sequence fail to converge in the classical sense. But in the sense of generalized function the sequence does converge to zero

$$(f'_n, \phi(x)) = -\frac{1}{n} \int \sin nx \phi'(x) dx \to 0 \quad \text{as } n \to \infty.$$

Note that $\lim_{n \to \infty} \int \cos nx \phi(x) dx = 0$ by Riemann–Lebesgue lemma.

6.4 Delta-Convergent Sequence

There are many ways to construct a sequence of regular functions which converge to the delta function. All that is needed is that the corresponding ordinary functions $f_\nu(x)$ form what we shall call a delta-convergent sequence, which means that they must possess the following two properties.

(a) For any $M > 0$ and for $|a| \leq M$ and $|b| \leq M$, quantities

$$\left| \int_a^b f_\nu(\xi) d\xi \right| \tag{6.14}$$

must be bounded by a constant independent of a, b or ν (in other words depending only on M).

(b) For any fixed non-vanishing a and b, we must have

$$\lim_{\nu \to \infty} \int_a^b f_\nu(\xi) d\xi$$
$$= \begin{cases} 0 & \text{for } a < b < 0 \text{ and } 0 < a < b \\ 1 & a < 0 < b. \end{cases} \tag{6.15}$$

Let $f_\nu(x)$ be such a delta-convergent sequence. Consider also the sequence of primitive functions

$$F_\nu(x) = \int_{-1}^x f_\nu(\xi) d\xi.$$

It follows from the two properties of a delta-convergent sequence that as ν is allowed to increase the $F_\nu(x)$ converges to zero for $x < 0$ and to 1 for $x > 0$. Moreover, these functions are uniformly bounded (in ν) in every interval. This implies that $F_\nu(x)$ converge in the sense of generalized functions to $\theta(x)$, which is equal to zero for $x < 0$ and to 1 for $x > 0$. Then in the sense of generalized functions the sequence $F'_\nu(x) = f_\nu(x)$ converges to $\theta'(x) = \delta(x)$ as asserted.

Example 1. Consider the function

$$f_\epsilon(x) = \frac{1}{\pi} \frac{\epsilon}{x^2 + \epsilon^2} \tag{6.16}$$

$$\left| \int_a^b f_\epsilon(x) dx \right| = \frac{1}{\pi} \left| \lim_{\epsilon \to 0} \int_{\tan^{-1} a/\epsilon}^{\tan^{-1} b/\epsilon} d\left(\tan^{-1} \frac{x}{\epsilon} \right) \right|$$

$$< d\frac{1}{\pi} \left| \int_{-\frac{\pi}{2}}^{\frac{\pi}{2}} d\theta \right| < \frac{1}{\pi} \cdot \pi = 1$$

$$\therefore \quad \lim_{\epsilon \to 0} \left| \int_a^b f_\epsilon(x) dx \right| < 1.$$

Also $\lim_{\epsilon \to 0} \int f_\epsilon(x) dx = \lim_{\epsilon \to 0} \left[\tan^{-1} \frac{b}{\epsilon} - \tan^{-1} \frac{a}{\epsilon} \right]$.
For $a < b < 0$

$$\lim_{\epsilon \to 0} \tan^{-1} \frac{b}{\epsilon} = -\frac{\pi}{2} = \tan^{-1} \frac{a}{\epsilon}$$

$$\therefore \quad \int_a^b f_\epsilon(x) dx = 0 \quad \text{for } a < b < 0$$

$$= 0 \quad \text{for } 0 < a < b.$$

For $a < 0 < b$

$$\lim_{\epsilon \to 0} \tan^{-1} \frac{b}{\epsilon} = \frac{\pi}{2} = -\lim_{\epsilon \to 0} \tan \frac{a}{\epsilon}$$

$$\therefore \quad \int_a^b f_\epsilon(x) dx = \frac{1}{\pi} \times \pi = 1$$

$$\therefore \quad \lim_{\epsilon \to 0} f_\epsilon(x) = \lim_{\epsilon \to 0} \frac{1}{\pi} \frac{\epsilon}{x^2 + \epsilon^2} = \delta(x).$$

Example 2. Consider the function

$$f_\epsilon(x) = \frac{1}{2\sqrt{\pi \epsilon}} e^{-\frac{x^2}{4\epsilon}} \tag{6.17}$$

$$\lim_{\epsilon \to 0} f_\epsilon(x) = \delta(x).$$

Since $f_\epsilon(x) > 0$ so that for any a, b

$$\int_a^b f_\epsilon(x) dx \leq \frac{1}{2\sqrt{\pi \epsilon}} \int_{-\infty}^{\infty} e^{-x^2/4\epsilon} dx = 1.$$

We now substitute

$$\frac{x}{2\sqrt{\epsilon}} = t$$

$$dx = 2\sqrt{\epsilon} \, dt.$$

Thus $\lim_{\epsilon \to 0} \int_a^b f_\epsilon(x)dx = \lim_{\epsilon \to 0} \frac{1}{\sqrt{\pi}} \int_{\frac{a}{2\sqrt{\epsilon}}}^{\frac{b}{2\sqrt{\epsilon}}} e^{-t^2} dt$ for $a < b < 0$

$$\lim_{\epsilon \to 0} \int_a^b f_\epsilon(x)dx = \frac{1}{\sqrt{\pi}} \lim_{\substack{R_1 \to \infty \\ R_2 \to \infty}} \int_{-R_1}^{-R_2} e^{-t^2} dt = 0.$$

Similarly for $0 < a < b$

$$\lim_{\epsilon \to 0} \int_a^b f_\epsilon(x)dx = \frac{1}{\sqrt{\pi}} \lim_{\substack{R_1 \to \infty \\ R_2 \to \infty}} \int_{R_1}^{R_2} e^{-\epsilon^2} dt = 0$$

and

$$\lim_{\epsilon \to 0} \int_a^b f_\epsilon(x)dx = 1 \quad \text{for } a < 0 < b$$

$$\therefore \quad \lim_{\epsilon \to 0} \frac{1}{2\sqrt{\pi\epsilon}} e^{-x^2/4\epsilon} = \delta(x).$$

Example 3. Consider,

$$f_\nu(x) = \frac{1}{\pi} \frac{\sin \nu x}{x}. \tag{6.18}$$

We shall show

$$\lim_{\nu \to \infty} f_\nu(x) = \delta(x).$$

From Chapter 2, it follows that

$$\lim_{\nu \to \infty} \int_{-\infty}^{\infty} f_\nu(x)dx = \frac{1}{\pi} \times \pi.$$

Further for any $b > a > 0$, the integrals

$$\int_a^b f_\nu(x)dx = \frac{1}{\pi} \int_{a\nu}^{b\nu} \frac{\sin y}{y} dy$$

converge to zero as $\nu \to \infty$ for $a < b < 0$ and $b > a > 0$.

Also $\left| \frac{1}{\pi} \int_a^b \frac{\sin \nu x}{x} dx \right|$ is bounded uniformly in a and b for all ν. Thus

$$\lim_{\nu \to \infty} f_\nu(x) = \delta(x).$$

The above result can be formulated in the following way: in the sense of convergence in K'

$$\lim_{\nu \to \infty} \int_{-\nu}^{\nu} e^{ikx} dk = 2\pi\delta(x).$$

It should be pointed out that one can always add a term of the form

$$A \cos \nu x + B \sin \nu x$$

to any delta convergent sequence because

$$f'_\nu(x) = f_\nu(x) + (A \cos \nu x + B \sin \nu x)$$

so that

$$F'_\nu(x) = F_\nu(x) + \frac{A}{\gamma}(A \sin \nu x - B \cos \nu x).$$

In the limit $\nu \to \infty$ the term in the bracket together with its derivatives is zero in the sense of generalized functions,

$$\lim_{\mu \to \infty} \int \sin \nu x \phi(x) = 0$$

$$\lim_{\nu \to \infty} \int \cos \nu x \phi(x) = 0$$

(by Riemann–Lebesgue lemma).

6.5   Regularization of Divergent Integrals

Let $f(x)$ be a function locally summable everywhere except say at $x = 0$. As an example we may cite $f(x) = \frac{1}{x}$ where x is on the real axis. Then in general the integral.

$$\int f(x)\phi(x)dx \tag{6.19}$$

where $\phi(x)$ is in K will diverge. But this integral will converge if $\phi(x)$ vanishes in a neighborhood of zero. It is indeed possible to construct an $f \epsilon K'$ such that for all $\phi(x)$ in K vanishing in the neighborhood of zero the functional has the value given by (6.19). Any such functional f is called regularization of the integral (6.19). For instance for $f(x) = \frac{1}{x}$, we may set

$$(f, \phi) = \int_{-\infty}^{-a} \frac{\phi(x)}{x} + \int_{-a}^{b} \frac{[\phi(x) - \phi(0)]dx}{x} + \int_{b}^{\infty} \frac{\phi(x)}{x} dx \tag{6.20}$$

with any $a, b > 0$.

Let us suppose $f(x)$ is an ordinary function whose derivative $f'(x)$ exists in the usual sense except possibly at isolated points but is not locally summable (e.g. $f(x) = \frac{1}{x}$). For this case

$$(f', \phi) = \int_{-\infty}^{\infty} f'(x)\phi(x)dx \tag{6.21}$$

will in general diverge and thus will not define a functional. In such a case f' will be defined as a regularization of (6.21). In particular the regularization defined by,

$$\int_{-\infty}^{\infty} f'(x)\phi(x)dx = (f', \phi) = -(f, \phi') = -\int_{-\infty}^{\infty} f(x)\phi'(x)dx \tag{6.22}$$

yields a preferred regularization in that it coincides with Cauchy's process of analytic continuation discussed in Chapter 3. This, in a sense, is the natural regularization.

Example 1. Let us find the derivative of the generalized function

$$x_+^\lambda = 0, \quad x \leq 0, \quad -1 < \operatorname{Re}\lambda < 0$$
$$= x^\lambda, \quad x > 0. \tag{6.23}$$

This function is locally summable but its ordinary derivative $\lambda x^{\lambda-1}$ is not and we must regularize the divergent integral

$$\int_0^\infty \lambda x^{\lambda-1}\phi(x)dx. \tag{6.24}$$

To obtain the appropriate regularization we start from,

$$((x_+^\lambda)'\phi) = -(x_+^\lambda, \phi') = -\int_0^\infty x^\lambda \phi'(x)dx. \tag{6.25}$$

Let us integrate it by parts setting

$$\phi'(x)dx = du, \quad x^\lambda = v.$$

So that

$$u = \phi(x) + C$$

and we have

$$((x_+^\lambda)', \phi) = -\lim_{\epsilon \to 0}\left\{x^\lambda[\phi(x) + C]_\epsilon^\infty + \int_\epsilon^\infty \lambda x^{\lambda-1}[\phi(x) + C]dx\right. .$$

The first term goes to zero on setting $C = -\phi(0)$ so that

$$((x_+^\lambda)', \phi) = \int_0^\infty \lambda x^{\lambda-1}[\phi(x) - \phi(0)]dx \tag{6.26}$$

which is then the desired rule for assigning meaning to the integral (6.24).

It can be easily verified that this rule is essentially Cauchy's procedure (see Chapter 3) of analytic continuation of the integral (6.24) outside the region of convergence of the integral.

Let us define

$$((x_+^\lambda)', \phi) = \frac{1}{(e^{2\pi i\lambda} - 1)}\int_\infty^{0+} \lambda x^{\lambda-1}\phi(x)dx \tag{6.27a}$$

$$= \frac{1}{(e^{2\pi i\lambda} - 1)}\int_C \lambda x^{\lambda-1}\phi(x)dx \tag{6.27b}$$

where C (or equivalently $\int_\infty^{0+}$) is the contour, as shown in Fig. 6.1.

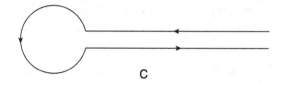

Fig. 6.1

We shall now verify that for $-1 < \text{Re}\,\lambda < 0$ the r.h.s. Eq. (6.27) coincides with Eq. (6.26). We have

$$\frac{1}{(e^{2\pi i\lambda} - 1)} \int_{\infty}^{0+} \lambda x^{\lambda-1} \phi(x) dx$$

$$= \lim_{\epsilon \to 0} \left[\int_{\epsilon}^{\infty} \lambda x^{\lambda-1} \phi(x) dx + \frac{\lambda}{(e^{2\pi i\lambda} - 1)} \int_{s} x^{\lambda-1} \phi(x) dx \right].$$

Now

$$\frac{1}{(e^{2\pi i\lambda} - 1)} \int_{s} \lambda x^{\lambda-1} \phi(x) dx$$

$$= \frac{1}{(e^{2\pi i\lambda} - 1)} \sum_{n=0}^{\infty} \frac{\phi^{(n)}(0)}{n!} \frac{\lambda i \epsilon^{n+\lambda}}{i(n+\lambda)} (e^{2\pi i\lambda} - 1).$$

Since $-1 < \text{Re}\,\lambda < 0,\ n + \text{Re}\,\lambda > 0$ for $n = 1, 2, \ldots .$

Thus $\frac{1}{(e^{2\pi i\lambda}-1)} \int_{s} \lambda x^{\lambda-1} \phi(x) dx = \frac{\lambda \epsilon^{\lambda}}{\lambda} \phi(0) = -\phi(0) \int_{\epsilon}^{\infty} \lambda x^{\lambda-1} dx$

so that

$$\frac{1}{(e^{2\pi i\lambda} - 1)} \int_{\infty}^{0+} \lambda x^{\lambda-1} \phi(x) dx = \int_{0}^{\infty} dx \lambda x^{\lambda-1} [\phi(x) - \phi(0)]. \quad (6.28)$$

Thus Eq. (6.27) is equivalent to Eq. (6.26) which follows from the definition of the derivative of the generalized function x_{+}^{λ}.

6.6 General Theory of Regularization of Functions with Algebraic Singularities

Of the functions with non-summable singularities at isolated points the most important are those with algebraic singularities. These are functions which as x approaches the singular point x_0 increase no faster than some power of $\frac{1}{|x-x_0|}$.

Let us consider a generalized function f depending on a parameter λ running over an open region Λ in the complex λ-plane. Then f_λ is called an analytic function of λ in Λ if (f_λ, ϕ) is an analytic function of λ for ϕ in K.

We discuss the basic features of the analytic continuation method of regularization which closely follows the method of classical analysis explained in Chapter 3.

For instance we start with

$$x_+^\lambda = 0 \quad \text{for } x \leq 0$$
$$= x^\lambda \quad \text{for } x > 0.$$

For Re $\lambda > -1$ this is the regular functional defined by

$$(x_+^\lambda, \phi) = \int_0^\infty x^\lambda \phi(x) dx. \tag{6.29}$$

The above is evidently an analytic function of λ which is regular on the right of the line Re $\lambda = -1$. Let us suppose we extend it to $-2 < \text{Re } \lambda < 0$. We now rewrite the r.h.s. of (6.29) as

$$\int_0^\infty x^\lambda \phi(x) dx = \int_0^1 x^\lambda [\phi(x) - \phi(0)] dx + \int_1^\infty x^\lambda \phi(x) dx + \frac{\phi(0)}{\lambda+1}. \tag{6.30}$$

The first term is defined for $-2 < \text{Re } \lambda < 0$, the second for all λ and the third for $\lambda \neq -1$. Although the l.h.s. of the above identity ceases to exist for $-2 < \text{Re } \lambda < -1$ the r.h.s. is well defined and has a meaning for Re $\lambda > -2$ and $\lambda \neq -1$. Thus the r.h.s. is the analytic continuation of (6.29) outside the region of convergence of the integral (upto Re $\lambda > -2$). We shall see that this is just Cauchy's analytic continuation,

$$\int_0^\infty x^\lambda \phi(x) dx \to \frac{1}{(e^{2\pi i \lambda} - 1)} \int_\infty^{0+} x^\lambda \phi(x) dx. \tag{6.31}$$

To verify this we proceed as in Chapter 3. The r.h.s. of Eq. (6.31) is,

$$\int_\epsilon^1 x^\lambda \phi(x) dx + \int_1^\infty x^\lambda \phi(x) dx + \frac{1}{(e^{2\pi i \lambda} - 1)} \int_s x^\lambda \phi(x) dx$$

$$= \int_\epsilon^1 x^\lambda \phi(x) dx + \int_1^\infty x^\lambda \phi(x) dx + \frac{\phi(0) \epsilon^{\lambda+1}}{\lambda+1}$$

$$= \int_\epsilon^1 x^\lambda [\phi(x) - \phi(0)] dx + \int_1^\infty x^\lambda \phi(x) dx + \frac{\phi(0)}{\lambda+1} \tag{6.32}$$

which is the r.h.s. of Eq. (6.30).

We may proceed similarly and continue x_+^λ into the region $\text{Re}\,\lambda > -n - 1$, $\lambda \neq -1, -2, \ldots, -n$. For $\text{Re}\,\lambda > -n - 1$ we have for the integral over s,

$$\frac{1}{(e^{2\pi i\lambda} - 1)} \int_s x^\lambda \phi(x) dx = -\sum_{r=0}^{n-1} \frac{\phi^{(r)}(0)(1 - \epsilon^{r+\lambda+1})}{r!(r + \lambda + 1)} + \sum \frac{\phi^{(r)}(0)/r!}{r + \lambda + 1}$$

$$= -\sum_{r=0}^{n-1} \frac{\phi^{(r)}(0)}{r!} \int_\epsilon^1 x^{\lambda+r} dx + \sum \frac{\phi^{(r)}(0)/r!}{(r + \lambda + 1)}.$$

Thus

$$(x_+^\lambda, \phi) = \int_1^\infty x^\lambda \phi(x) dx + \int_0^1 x^\lambda \left[\phi(x) - \sum_{r=0}^{n-1} \frac{\phi^{(r)}(0)}{r!} x^r \right] dx$$

$$+ \sum_{r=0}^{n-1} \frac{\phi^{(r)}(0)}{r!(r + \lambda + 1)}. \tag{6.33}$$

This is Gel'fand and Shilov's formula for the functional for values $-n - 1 < \text{Re}\,\lambda < 0$. If in addition $\text{Re}\,\lambda < -n$, then in the strip $-n - 1 < \text{Re}\,\lambda < -n$ the last term of Eq. (6.33) is given by

$$\sum_{r=0}^{n-1} \frac{\phi^r(0)}{r!(r + \lambda + 1)} = -\int_1^\infty dx x^\lambda \sum_{r=0}^{n-1} \frac{\phi^r(0)}{r!} x^r.$$

So that

$$(x_+^\lambda, \phi) = \int_0^\infty x^\lambda \left[\phi(x) - \sum_{r=0}^{n-1} \frac{\phi^{(r)}(0)}{r!} x^r \right] dx$$

$$\text{for } -n - 1 < \text{Re}\,\lambda < -n. \tag{6.34}$$

From the regularized formula (6.33) for the functional x_+^λ it follows that this is a meromorphic function of λ having simple poles at

$$\lambda = -1, -2, \ldots$$

and the residue at the simple pole at $\lambda = -l$, is

$$\frac{\phi^{(l-1)}(o)}{(l - 1)!}, \ l = 1, 2, \ldots.$$

Γ-function discussed in Chapter 3 is a special case of the generalized function for the test function

$$\phi(x) = e^{-x}.$$

Thus for $\text{Re}\,\lambda > -n - 1$

$$\Gamma(\lambda + 1) = \int_0^1 x^\lambda \left[e^{-x} - \sum_{k=0}^{n-1} \frac{(-)^k}{k!} x^k \right] dx$$

$$+ \int_1^\infty x^\lambda e^{-x} dx + \sum_{k=0}^{n-1} \frac{(-)^k}{k!(k+\lambda+1)}. \qquad (6.35)$$

Let us now consider the generalized function

$$(x + i0)^\lambda = x^\lambda \qquad \text{for } x > 0$$
$$= e^{i\pi\lambda} |x|^\lambda \quad \text{for } x < 0.$$

Unlike $x_\pm^\lambda$, $(x \pm i0)^\lambda$ is an entire function, i.e. it has no singularity of finite affix in the complex λ plane. We consider the functional

$$I(\lambda) = \int_{L_\epsilon} (x + i0)^\lambda \phi(x) dx \qquad (6.36)$$

where L_ϵ is the contour shown in Fig. 6.2.

This integral as in Fig. 6.1 independent of ϵ and as an analytic function of λ is an entire function. It can be easily verified that

$$I(\lambda) = \int_1^\infty x^\lambda \left[\phi(x) + e^{i\lambda\pi} \phi(-x) \right]$$

$$+ \int_e^1 x^\lambda \left[\phi(x) + e^{i\lambda\pi} \phi(-x) \right] dx + \int_s \phi(x) x^\lambda dx.$$

On s

$$x = \epsilon e^{i\theta}.$$

Thus

$$\int_s \phi(x) x^\lambda dx = \sum_{n=0}^\infty \frac{\phi^{(n)}(0) \epsilon^{n+\lambda+1}}{n!} \left[1 - e^{i(n+\lambda+1)\pi} \right].$$

Further

$$\int_e^1 x^\lambda \left[\phi(x) + e^{i\lambda\pi} \phi(-x) \right] = \sum \frac{\phi^{(n)}(0)(1 - \epsilon^{n+\lambda+1})}{n!(n+\lambda+1)} \left[1 - e^{i(n+\lambda+1)\pi} \right].$$

Fig. 6.2

Thus

$$
\begin{aligned}
I(\lambda) &= \int_1^\infty x^\lambda [\phi(x) + e^{i\lambda\pi}\phi(-x)]dx + \sum \frac{\phi^{(n)}(0)[1 - e^{i(n+\lambda+1)\pi}]}{n!(n+\lambda+1)} \\
&= \int_1^\infty x^\lambda \left[\phi(x) + e^{i\lambda\pi}\phi(-x)\right] dx - 2i \sum_{n=0}^\infty \frac{\phi^{(n)}(0)e^{i(n+\lambda+1)\frac{\pi}{2}}}{n!} \\
&\quad \times \frac{\sin(n+\lambda+1)\frac{\pi}{2}}{(n+\lambda+1)}.
\end{aligned}
\tag{6.37}
$$

Since the sum over n is an entire function of λ it follows that $I(\lambda)$ is an entire function of λ.

6.7 Regularization in a Finite Interval

We now consider the functional defined over a finite interval, for instance

$$
(x_{0 \le x \le b}^\lambda, \ \phi) = \int_0^b x^\lambda \phi(x)dx.
\tag{6.38}
$$

We have chosen the interval $0 \le x \le b$ as being particularly typical in that it has a single point of possible divergence which is, further, one of the end points ($x = 0$ or $x = b$).

The interval $a \le x \le b$ without any singular points is of no interest and if the singular points lies within the interval then the integral can be broken up into two parts each of which has the singular point at an end.

We consider, therefore, the integral (6.38) which converges for $\operatorname{Re}\lambda > -1$. Writing

$$
\begin{aligned}
\int_0^b & x^\lambda \phi(x)dx \\
&= \int_0^b x^\lambda \left[\phi(x) - \phi(0) - x\phi'(0) - \frac{x^2}{2!}\phi''(0) - \cdots - \frac{x^n - 1}{(n-1)!}\phi^{(n-1)}(0)\right] dx \\
&\quad + \left[\phi(0)\frac{b^{\lambda+1}}{\lambda+1} + \phi'(0)\frac{b^{\lambda+2}}{\lambda+2} + \cdots + \frac{\phi^{(n-1)}(0)}{(n-1)!}\frac{b^{\lambda+n}}{\lambda+n}\right]
\end{aligned}
\tag{6.39}
$$

this gives the analytic continuation for $\operatorname{Re}\lambda > -n - 1$.

This obviously can be continued into the entire λ plane except for the points $\lambda = -1, -2, -3\ldots$, where it has simple poles.

We now consider the regularization of the integral at the upper limit we consider

$$((b-x)_{0 \leq x \leq b}^{\lambda}, \phi(x)) = \int_0^b (b-x)^{\lambda} \phi(x) dx$$

which converges at the upper limit $\operatorname{Re} \lambda > -1$. To get the corresponding Gelfand–Shilov formula for regularization it is convenient to start from the contour integral

$$I(\lambda) = \frac{1}{(1 - e^{2\pi i \lambda})} \int_0^{b+} (b-x)^{\lambda} \phi(x) dx \tag{6.40}$$

where the symbol $\int_0^{b+}$ indicates a contour that starts from the origin along the real axis encircles the point $x = b$ once counterclockwise and returns to the origin along the real axis.

$$I(\lambda) = \int_0^{b-\epsilon} (b-x)^{\lambda} \phi(x) dx + \frac{1}{(1 - e^{2\pi i \lambda})} \int_s (b-x)^{\lambda} \phi(x) dx$$

$$= \int_0^c (b-x)^{\lambda} \phi(x) dx + \int_c^{b-\epsilon} (b-x)^{\lambda} \phi(x) dx + \sum \frac{\phi^{(r)}(b)}{r!} \frac{(-)^r \epsilon^{r+\lambda+1}}{(r+\lambda+1)}.$$

Further

$$\int_c^{b-\epsilon} (b-x)^{\lambda} \phi(x) dx = \sum \frac{\phi^{(r)}(b)(-)^r}{r!(r+\lambda+1)} \left[(b-c)^{r+\lambda+1} - \epsilon^{r+\lambda+1} \right].$$

Thus

$$I(\lambda) = \int_0^c (b-x)^{\lambda} \phi(x) dx + \sum \frac{(-)^r \phi^{(r)}(b)(b-c)^{r+\lambda+1}}{r!(r+\lambda+1)} \tag{6.41}$$

which is as usual independent of ϵ.

We now set $c = b - \epsilon$.

So that

$$I(\lambda) = \int_0^{b-\epsilon} (b-x)^{\lambda} \phi(x) dx + \sum_{r=0}^{\infty} \frac{(-)^r \phi^{(r)}(b) \epsilon^{r+\lambda+1}}{r!(r+\lambda+1)}.$$

For $-n - 1 < \operatorname{Re} \lambda < 0$, we have (proceeding to the limit $\epsilon \to 0$, $(I(\lambda) =$ independent of ϵ)

$$I(\lambda) = \int_0^{b-\epsilon} (b-x)^{\lambda} \phi(x) dx + \sum_{r=0}^{n-1} \frac{(-1)^r}{r!} \frac{\phi^{(r)}(b) \epsilon^{r+\lambda+1}}{(r+\lambda+1)}.$$

Noting that

$$\frac{\epsilon^{r+\lambda+1}}{r+\lambda+1} = \frac{b^{r+\lambda+1}}{r+\lambda+1} - \int_0^{b-\epsilon} (b-x)^{r+\lambda} dx.$$

We have

$$I(\lambda) = \int_0^b (b-x)^\lambda \left[\phi(x) - \sum_{r=0}^{n-1} \frac{(-)^r}{r!} \phi^{(r)}(b)(b-x)^r \right]$$

$$+ \sum_{r=0}^{n-1} \frac{(-)^r}{r!} \frac{\phi^{(r)}(b)b^{r+\lambda+1}}{r+\lambda+1}. \tag{6.42}$$

This is Gelfand and Shilov's formula for the functional:

$$((b-x)^\lambda_{0\le x\le b}, \phi(x)) = \int_0^b (b-x)^\lambda \left[\phi(x) - \sum_{r=0}^{n-1} \frac{(-)^r}{r!} \phi^{(r)}(b)(b-x)^r \right]$$

$$+ \sum_{r=0}^{n-1} \frac{(-1)^r}{r!} \frac{\phi^{(r)}(b)}{(r+\lambda+1)} b^{r+\lambda+1} \tag{6.43}$$

for $-n-1 < \operatorname{Re}\lambda < 0$.

Let us now consider a function $f(x)$ of the form,

$$f(x) = x^\lambda p(x), \quad 0 < x \le b$$
$$= 0 \qquad\qquad x < 0, \ x > b \tag{6.44}$$

where $p(x)$ is an infinitely differentiable function. We shall consider the regularization of $f(x)$ to be defined by

$$(\operatorname{Reg} f(x), \ \phi(x)) = (x^\lambda_{0\le a\le b}, p(x)\phi(x)). \tag{6.45}$$

Let us now consider a function $f(x)$ with algebraic singularities at a and b so that in the neighborhood of a we "may write"

$$f(x) = (x-a)^\lambda p_a(x) \tag{6.46a}$$

and in the neighborhood of b

$$f(x) = (b-x)^\mu p_b(x) \tag{6.46b}$$

where $p_a(x)$ and $p_b(x)$ are infinitely differentiable functions in the intervals $a \le x < b$ and $a < x \le b$ respectively. We shall define the regularization of this integral through the contour integral

$$(\operatorname{Reg} f(x), \phi(x)) = -\frac{1}{(1-e^{2\pi i\lambda})(1-e^{2\pi i\mu})} \int_p f(x)\phi(x)dx \tag{6.47}$$

where P stands for Pochhammer's double circuit contour (see Fig. 3.12).

As an application of this contour integral form for regularization we shall derive Gelfand and Shilov's formula for the beta function.

$$B(\lambda, \mu) = -\frac{1}{(1 - e^{2\pi i\lambda})(1 - e^{2\pi i\mu})} \int_P x^{\lambda-1}(1 - x)^{\mu-1}dx.$$

Proceeding as usual and simplifying we obtain

$$B(\lambda, \mu) = \int_0^{\frac{1}{2}} x^{\lambda-1}\left[(1 - x)^{\mu-1} - \sum_{r=0}^{k-1}\frac{(1 - \mu)_r}{r!}x^r\right]dx$$

$$+ \int_{\frac{1}{2}}^1 (1 - x)^{\mu-1}\left[x^{\lambda-1}\sum_{r=0}^{p-1}\frac{(1 - \lambda)_r}{r!}(1 - x)^r\right]dx$$

$$+ \sum_{r=0}^{p-1}\frac{(1 - \lambda)_r}{r!}\frac{2}{r + \mu} + \sum_{r=0}^{k-1}\frac{(1 - \mu)_r}{r!}\frac{2^{-\lambda-r}}{\lambda + r} \qquad (6.48)$$

valid for $\operatorname{Re}\lambda > -k$, $\operatorname{Re}\mu > -p$.

6.8 The Generalized Function r^λ

Let $r = \sqrt{x_1^2 + x_2^2 + \cdots + x_n^2}$ and consider the functional defined by

$$(r^\lambda, \phi) = \int_{\mathbf{R}_n} r^\lambda\phi(x)d^n x. \qquad (6.49)$$

The integral converges in the usual sense for $\operatorname{Re}\lambda > -n$. For $\operatorname{Re}\lambda \leq -n$ it is to be understood in the sense of its regularization.

Let us go over to spherical coordinates writing it in the form

$$(r^\lambda, \phi) = \int_0^\infty r^\lambda\left\{\int_\Omega \phi(r\omega)d\omega\right\}r^{n-1}dr \qquad (6.50)$$

where $d\omega$ is the hypersurface element on the unit sphere (solid angle). The integral appearing in the integral (6.50) can be written as

$$\int_\Omega \phi(r\omega)d\omega = \Omega_n S_\phi(r). \qquad (6.51)$$

Thus

$$(r^\lambda, \phi) = \Omega_n\int_0^\infty r^{\lambda+n-1}S_\phi(r)dr \qquad (6.52)$$

Since $\phi(x) \in K$, $S_\phi(r)$ vanishes outside some $r = C$. Further

$$\Omega_n S_\phi(r) = \int_\Omega \left[\phi(0) + \sum_{j=}x_j\left(\frac{\partial\phi(0)}{\partial x_j}\right) + \sum x_i x_j \cdot \frac{\partial\phi(0)}{\partial x_i\partial x_j} + \cdots\right]d\omega.$$

$$(6.53)$$

Clearly every term with an odd number of factors of x_j fails to contribute to the integral. Thus

$$S_\phi(r) = \phi(0) + a_1 r^2 + a_2 r^4 + \cdots + . \qquad (6.54)$$

So $S_\phi(r)$ has only derivatives of even order.

Then the functional (6.52) represents the application of $\Omega_n \, x_+^\mu$ (with $\mu = \lambda + n - 1$) to $S_\phi(x)$. Hence from the previous results it now follows that the functional (6.52) is a meromorphic function of λ having simple poles at

$$\lambda = -n, \quad -n - 1, \dots$$

and the residue at $\lambda = -n - m + 1$ is

$$\frac{S_\phi^{(m-1)}(0)}{(m-1)!}. \qquad (6.55)$$

But since the derivatives of odd order vanish at $x = 0$ the poles corresponding to even values of m do not in fact exist.

This leaves us with simple poles at

$$\lambda = -n, \quad -n - 2, \quad -n - 4.$$

Accordingly the residue of (r^λ, ϕ) at $\lambda = -n - 2k$ for non-negative integral k is given by

$$\Omega_n \frac{(S^{(2k)}(x), S_\phi(x))}{(2k)!} = \Omega_n \frac{S^{(2k)}(0)}{(2k)!}.$$

In particular at $\lambda = -n$ the functional as an analytic function of λ has a simple pole whose residue is

$$\Omega_n S_\phi(0) = \Omega_n \phi(0). \qquad (6.56)$$

6.9   The Generalized Function P_+^λ

Let us now consider

$$P(x) = x_1^2 + x_2^2 + \cdots + x_p^2 - x_{p+1}^2 - x_{p+2}^2 - \cdots x_{p+q}^2$$

with $p + q = n$. We define the generalized function P_+^λ, where λ is a complex number, by

$$(P_+^\lambda, \ \phi) = \int_{P>0} P^\lambda(x) \phi(x) d^n x$$

where $x = (x_1, \dots, x_n)$ and $d^n x = dx_1 \cdots dx_n$.

For $\mathrm{Re}\,\lambda \geq 0$ this integral converges in usual sense. For $\mathrm{Re}\,\lambda < 0$ this is to be understood in the sense of its regularization (analytic continuation).

Let us proceed to find the singularities of the functional (P_+^λ, ϕ). We write

$$x_1 = r\omega_1, \quad x_2 = r\omega_2, \ldots, x_p = r\omega_p \tag{6.57}$$

$$x_{p+1} = s\omega_{p+1}, \ldots, x_{p+q} = s\omega_{p+q} \tag{6.58}$$

where

$$r = \sqrt{x_1^2 + x_2^2 + \cdots + x_p^2}$$

$$s = \sqrt{x_{p+1}^2 + x_{p+2}^2 + \cdots + x_{p+q}^2}.$$

Thus we have

$$(P_+^\lambda, \phi) = \int_{p>0} (r^2 - s^2)^\lambda \phi r^{p-1} s^{q-1} dr \, ds \, d\Omega^p \times d\Omega^q. \tag{6.59a}$$

Now we write

$$\psi(r,s) = \int \phi d\Omega^{(p)} d\Omega^{(q)}. \tag{6.59b}$$

So that

$$(P_+^\lambda, \phi) = \int (r^2 - s^2)^\lambda r^{p-1} s^{q-1} \psi(r,s) dr \, ds. \tag{6.60}$$

Since $\phi(x)$ is in K, $\psi(r,s)$ as defined in Eq. (6.59b) is an infinitely differentiable function of r^2 and s^2 with bounded support. We make the change of variables

$$u = r^2 \quad v = s^2$$

$$\psi(r,s) = \psi_1(u,v)$$

so that

$$(P_+^\lambda, \phi) = \frac{1}{4} \int_{u-0}^\infty \int_{v=0}^u (u-v)^\lambda \psi_1(u,v) u^{\frac{1}{2}(p-2)} v^{\frac{1}{2}(q-2)} \times du \, dv.$$

Finally we write

$$v = ut \quad dv = u \, dt$$

so that

$$= \frac{1}{4} \int_{u=0}^\infty du \int u^\lambda (1-t)^\lambda \psi(u,tu) \times u^{\frac{1}{2}(p-2)} (ut)^{\frac{1}{2}(q-2)} u \cdot dt$$

$$= \frac{1}{4} \int_0^\infty du u^{\lambda+\frac{1}{2}(p+q)-1} \int_0^1 (1-t)^\lambda t^{\frac{1}{2}(q-2)} \psi_1(u,tu) dt. \tag{6.61}$$

To cover the whole λ-plane we replace the above integral by its analytic continuation

$$(\mathrm{P}^\lambda_+, \phi) = \frac{1}{[e^{2i\pi(\lambda+\frac{n}{2})} - 1]} \int_\infty^{0+} \phi(\lambda, u) u^{\lambda + \frac{n}{2} - 1} du \qquad (6.62)$$

where

$$\phi(\lambda, u) = \frac{1}{(e^{2\pi i\lambda} - 1)} \int_0^{1+} (1 - t)^\lambda t^{\frac{1}{2}(q-2)} \psi_1(u, tu) dt. \qquad (6.63)$$

The function $\phi(\lambda, u)$ is regular for all λ except at

$$\lambda = -1, -2, -3, \ldots$$

where it has simple poles.

On the other hand, even at regular points of $\phi(\lambda, u)$ the integral (6.62) has poles at

$$\lambda = -\frac{n}{2} - l, l = 0, 1, 2, \ldots.$$

The residue at the simple pole at $-\frac{n}{2} - l$ is given by,

$$[\mathrm{Res}(P^\lambda_+, \phi)]_{-\frac{n}{2}-l} = \frac{1}{l!} \left\{ \frac{\partial^l}{\partial u^l} \left[\phi\left(-\frac{n}{2} - l, u\right) \right] \right\}_{u=0}.$$

There will be poles at half-integers

$$\lambda = -\frac{n}{2}, \quad -\frac{n}{2} - 1, \ldots$$

only if n is odd.

The problems discussed here are the bare minimum for use in later chapters. Their usefulness and efficacy can be seen in due course.

References

[1] G. W. Mackay, *Ann. Math.* **55** (1952) 101; **58** (1953) 193;
    F. Y. Mautner, *Proc. Nat. Acad. Sci. USA* **39** (1953).
[2] D. Basu, arXiv 0710.2224V3 [hep-th].
[3] I. M. Gelfand and G. E. Shilov, *Generalized Functions* (Vol. 1) Chapter 1.

PART II

Applications to Group Representation Theory

CHAPTER 7

Lie Groups and Their Representations

The scope and applicability of the representation theory of Lie groups keep on growing seemingly endless. It now covers a vast area starting from particle classification to quantum optics and problems in quantum gravity. The representation theory of compact groups studied in this chapter and in the next are an essential prerequisite for the study of infinite dimensional representations of locally compact groups considered in Chapters 9–11.

7.1 Abstract Groups

Although the study of group theory originated with finite groups, the theory of Lie groups has achieved greater importance than that of finite groups. We start with the definition of abstract groups.

(a) A group G is a set in which an operation is defined which associates with every ordered pair (a, b) of elements in G a third element c denoted by

$$c = ab \quad \text{(in general } ba \neq ab\text{)}.$$

(b) The multiplication is associative, that is

$$(ab)c = a(bc).$$

(c) There is one element, the identity e, with the property that for all $a \in G$

$$ae = ea = a.$$

(d) For each $a \in G$ there is an element $a^{-1} \in G$ such that

$$a^{-1}a = aa^{-1} = e.$$

7.2 Transformation Groups

By a transformation of a set M we mean a one to one mapping of the set onto itself. We shall denote the image of an element x of M under a mapping

g by gx. Let G be some group. We say G is a transformation group of the set M if with each element $g \in$ G one can associate a transformation $x \rightarrow gx$ in M where the following conditions are satisfied.

(a) With identity element e of G we associate the identity transformation of the set M:

$$ex = x.$$

(b) For any two elements g_1 and g_2 of G we have

$$(g_1 g_2)x = g_1(g_2 x).$$

As examples of transformation groups we have the group S_n of all permutations of n elements, the group of rotations in 2, 3 or more dimensions, the group of all Lorentz transformations, etc.

7.3 Lie Groups of Transformations

A group is said to be continuous if some generalized definition of continuity or nearness is imposed upon the elements of the group. We shall restrict ourselves to the simpler case where the elements of the group manifold can be labelled by a finite set of continuously varying parameters. For example the set of transformations

$$x' = ax + b, \quad a \neq 0 \tag{7.1}$$

forms a group. The two (real) parameters a and b vary continuously from $-\infty$ to $+\infty$ and we say that the group is a two-parameter continuous group. Every group element can be symbolically denoted by $T(a, b)$

$$\mathrm{T}(a, b)x = x' = ax + b. \tag{7.2}$$

The law of composition can be obtained as follows

$$\begin{aligned}
\mathrm{T}(a_1, b_1)\mathrm{T}(a_2, b_2)x &= \mathrm{T}(a_1, b_1)x' = \mathrm{T}(a_1, b_1)(a_2 x + b_2) \\
&= a_1(a_2 x + b_2) + b_1 \\
&= a_1 a_2 x + (a_1 b_2 + b_1)
\end{aligned} \tag{7.3}$$

so that

$$\mathrm{T}(a_1, b_1)\mathrm{T}(a_2, b_2) = \mathrm{T}(a_3, b_3) \tag{7.4}$$

$$a_3 = a_1 a_2, \quad b_3 = a_1 b_2 + b_1, \quad e = \mathrm{T}(1, 0), \quad \mathrm{T}^{-1}(a, b) = \mathrm{T}\left(\frac{1}{a}, -\frac{b}{a}\right).$$

In general an r parameter continuous group is a group of transformations

$$x_i' = f(x_1, x_2, \ldots, x_n; a_1, a_2, \ldots, a_r)$$

or symbolically

$$x' = f(x; a). \tag{7.5}$$

The transformation must satisfy all the group requirements. Thus given a transformation of the form (7.5) we can find a parameter set $\bar{a}$ such that

$$x'' = f(x', \bar{a}) = x \quad \text{(inverse)}. \tag{7.6}$$

If we perform in succession two transformations of the set

$$\begin{aligned} x' &= f(x; a) \\ x'' &= f(x'; b) \end{aligned} \tag{7.7}$$

we require the resulting transformation be also a number of the set

$$x'' = f(x; c). \tag{7.8}$$

The parameter c must be functions of the parameters (a, b), i.e.

$$c = \phi(a, b). \tag{7.9}$$

We assume that the functions ϕ are analytic functions of a, b and $\bar{a}$ are analytic functions of a. The groups of transformation under these conditions will be called Lie groups. A particular Lie group will naturally admit many different parametrizations. Generally the parameters will be chosen such that $a = 0$ will correspond to the identity e.

The theory of representations has particularly simple results for a certain class of groups, namely, the compact groups. A Lie group is defined to be compact if there is a parametrization in which the group is covered by a **finite** number of **bounded** parameter domains.

If there is no parametrization in which the parameters vary over a closed bounded interval then the group will be called non-compact.

The groups treated in this text will all be "locally compact" in the sense that these will have a discrete finite number of parameters. The transformation groups appearing in modern gauge theory, for example, are not locally compact and the representation theory of such groups are still at their infancy.

The locally compact groups considered here will all be subgroups of general linear groups $\mathrm{GL}(n, c)$ of $n \times n$ complex matrices. Most Lie groups can be constructed in this way as groups of matrices.

7.4   Group Integration and Invariant Measure

Consider a finite group G consisting of the elements $l, g, h, \ldots$. Let $F(g)$ be a function on G. Then

$$\sum_g F(g) = \sum_g F(hg) = \sum_g F(gh) \qquad (7.10)$$

because multiplication on the left or right just reshuffles the elements keeping the sum unchanged. By a similar analogy we define the invariant measure of continuous groups by:

$$I = \int f(g) d\mu_l(g) = \int f(g_0 g) d\mu_l(g) \qquad (7.11)$$

where the subscript l stands for the left invariant measure:

$$d\mu_l(g) = \omega_l(g) d\alpha_1 \ldots d\alpha_p \qquad (7.12)$$

where $\alpha_1, \alpha_2, \ldots, \alpha_p$ are the group parameters (in a particular parametrization); $\omega_l(g)$ is a suitable positive definite weight function.

Similarly the right invariant measure is defined by

$$\int f(g) d\mu_r(g) = \int f(gg_0) d\mu_r(g) \qquad (7.13)$$

where

$$d\mu_r(g) = \omega_r(g) d\alpha_1 \ldots d\alpha_p. \qquad (7.14)$$

Example. Consider the group of diagonal matrices

$$\Lambda = \begin{pmatrix} \lambda^{-1} & 0 \\ 0 & \lambda \end{pmatrix} \qquad (7.15)$$

where $\lambda = \sigma + i\tau$.

This is a two-parameter Lie group having real parameters σ, τ.

Then we write

$$d\mu_l(\lambda) = \omega_l(\lambda) d\sigma d\tau. \qquad (7.16)$$

We start from the obvious equality

$$\int f(\Lambda) d\mu_l(\lambda) = \int f(\Lambda') d\mu_l(\lambda') = \int f(\Lambda') \omega_l(\lambda') d\sigma' d\tau'. \qquad (7.17)$$

If we write

$$\Lambda' = \Lambda_0 \Lambda \qquad (7.18)$$

we have

$$\lambda' = \lambda_0 \lambda$$

$$\sigma' + i\tau' = (\sigma_0\sigma - \tau_0\tau) + i(\sigma_0\tau + \tau_0\sigma)$$

so that

$$d\sigma'd\tau' = \left|\frac{\partial(\sigma',\tau')}{\partial(\sigma,\tau)}\right| d\sigma d\tau$$

$$= |\lambda_0|^2 d\sigma d\tau = \frac{|\lambda'|^2}{|\lambda|^2} d\sigma d\tau. \tag{7.19}$$

Thus

$$\int f(\Lambda)d\mu_l(\lambda) = \int f(\Lambda_0\Lambda)\omega_l(\lambda')\frac{|\lambda'|^2}{|\lambda|^2} d\sigma d\tau. \tag{7.20}$$

But from the definition of left invariance we have

$$\int f(\Lambda)d\mu_l(\lambda) = \int f(\Lambda_0\Lambda)d\mu_l(\lambda) = \int f(\Lambda_0\Lambda)\omega_l(\lambda)d\sigma d\tau. \tag{7.21}$$

Hence from Eqs. (7.20) and (7.21) we have, in view of the arbitrariness of the function $f(\Lambda)$,

$$\omega_l(\lambda) = \omega_l(\lambda')\frac{|\lambda'|^2}{|\lambda|^2} \tag{7.22}$$

$$|\lambda|^2\omega_l(\lambda) = |\lambda'|^2\omega_l(\lambda').$$

This equality can be valid if both sides are a constant

$$\omega_l(\lambda) = \frac{c}{|\lambda|^2} = \frac{1}{|\lambda|^2} \quad \text{(setting } c = 1)$$

$$d\mu_l(\lambda) = \frac{d\sigma d\tau}{|\lambda|^2}. \tag{7.23}$$

An important theorem.

Let us consider the Jacobian

$$J = \left|\frac{\partial(u_1, v_1,\ u_2, v_2, \ldots, u_n, v_n)}{\partial(x_1, y_1,\ x_2, y_2, \ldots, x_n, y_n)}\right| \tag{7.24}$$

and let u_k, v_k be the real and imaginary parts of an **analytic** function $\omega_k(z_k)$ of the complex variable $z_k = x_k + iy_k$. Then

$$J = \left|\frac{\partial(\omega_1, \omega_2, \ldots, \omega_n)}{\partial(z_1, z_2, \ldots, z_n)}\right|^2 \tag{7.25}$$

Eq. (7.19) is a special case of this theorem for $n = 1$. For proof of this theorem see Naimark (Appendix).

Exercise. Show that the definitions (7.11) and (7.13) imply

$$d\mu_l(g) = d\mu_l(g_0 g)$$

$$d\mu_r(g) = d\mu_r(g g_0).$$

From the definition of the invariant measures it follows that its existence requires the Jacobian to factorize in a particular way which is not always apparently ensured. Even for the three dimensional rotation group the Jacobian calculation from the definition turns out to be practically intractable. These difficulties are greatly simplified by using the concept of differential invariants. We introduce the left and right invariant differentials as

$$d\omega^l(g) = g^{-1} dg, \quad d\omega^r(g) = dg \cdot g^{-1} \qquad (7.26)$$

where dg stands (for matrix groups) for matrix of the differentials

$$dg = \|dg_{pq}\|.$$

For a fixed g_0 it therefore follows

$$d(g_0 g) = g_0 dg$$

so that

$$\begin{aligned} d\omega^l(g_0 g) &= (g_0 g)^{-1} g_0 dg = g^{-1} g_0^{-1} g_0 dg \\ &= g^{-1} dg = d\omega^l(g). \end{aligned} \qquad (7.27)$$

Similarly

$$d\omega^r(g g_0) = d\omega^r(g). \qquad (7.28)$$

Thus the differentials $d\omega^l$ and $d\omega^r$ are invariant under the left and right translation. Let us now consider a p-parameter Lie group; then out of all $d\omega$ there are only p independent differentials $d\omega$. Then we define the left and right invariant measure as

$$d\mu_l(g) = \prod_{p=1}^{p} d\omega_p^l \quad d\mu_r(g) = \prod_{p=1}^{p} d\omega_p^r. \qquad (7.29)$$

7.5 Invariant Measure for the Three-Dimensional Rotation Group

The three-dimensional rotation consists of 3×3 real orthogonal matrices a satisfying

$$\begin{aligned} \tilde{a} a &= I \\ \tilde{a} &= a^{-1} \end{aligned} \qquad (7.30)$$

where $\tilde{a}$ denotes transpose of the matrix a. From this it immediately follows that $(\det a)^2 = 1$, $\det a = \pm 1$. Orthogonal transformation for which $\det a = -1$ are called improper orthogonal transformations.

Examples.

$$
1. \begin{pmatrix} -1 & 0 & 0 \\ 0 & -1 & 0 \\ 0 & 0 & -1 \end{pmatrix} \quad 2. \begin{pmatrix} 0 & 0 & 1 \\ 0 & 1 & 0 \\ 1 & 0 & 0 \end{pmatrix} \quad 3. \begin{pmatrix} 1 & 0 & 0 \\ 0 & 0 & 1 \\ 0 & 1 & 0 \end{pmatrix}
$$

On the other hand the matrices satisfying $\det a = 1$, are all continuously connected to the identity and are called proper orthogonal transformations.

The proper orthogonal group will be denoted by SO(3). [S stands for unit determinant, O for rotation and 3 for three dimension].

The condition

$$\tilde{a} = a^{-1} \tag{7.31}$$

has the following consequence. Since $\det a = 1$, $a^{-1} = \|A_{ji}\|$ where A_{ji} is the cofactor of a_{ji} and $\|\tilde{a}_{ij}\| = \|a_{ji}\|$ we have

$$a_{ji} = A_{ji}. \tag{7.32}$$

Thus for a proper rotation matrix every element coincides with its cofactor. Let us now introduce the left invariant differential,

$$d\omega = a^{-1}da = \tilde{a}da$$
$$d\omega_{ij} = \sum a_{ki}da_{kj}. \tag{7.33}$$

Hence

$$d\omega_{ji} = \sum a_{kj}da_{ki}.$$

Now $\tilde{a}a = I$ demands

$$\sum_k a_{ki}a_{kj} = \delta_{ij} \tag{7.34}$$

$$\therefore \quad \sum a_{ki}da_{kj}^{\dagger} \sum a_{kj}da_{ki} = 0.$$
$$d\omega_{ij} + d\omega_{ji} = 0$$

Thus $d\omega_{ij}$ is fully antisymmetric in i and j

$$d\omega_{ij} = -d\omega_{ji}. \tag{7.35}$$

Hence $d\omega$ can be written as

$$d\omega = \begin{pmatrix} 0 & -d\omega_{21} & -d\omega_{31} \\ d\omega_{21} & 0 & -d\omega_{32} \\ d\omega_{31} & d\omega_{32} & 0 \end{pmatrix}. \tag{7.36}$$

Hence the left invariant measure is given by

$$d\mu_l(g) = c\,d\omega_{21}\,d\omega_{32}\,d\omega_{31}$$

where c is an arbitrary constant

$$d\omega = \tilde{a}da = \begin{pmatrix} a_{11} & a_{21} & a_{31} \\ a_{12} & a_{22} & a_{32} \\ a_{13} & a_{23} & a_{33} \end{pmatrix} \begin{pmatrix} da_{11} & da_{12} & da_{13} \\ da_{21} & da_{22} & da_{23} \\ da_{31} & da_{32} & da_{33} \end{pmatrix}. \tag{7.37}$$

Out of the various da_{ij} only three are independent. Take them da_{21}, da_{31}, da_{32}.

$$d\omega_{11} = 0 = a_{11}da_{11} + a_{21}da_{21} + a_{31}da_{31} \tag{7.38}$$

which expresses da_{11} as a function of da_{21}, da_{31}. Now

$$\begin{aligned} d\omega_{21} &= a_{12}da_{11} + a_{22}da_{21} + a_{32}da_{31} \\ &= \frac{(a_{11}a_{22} - a_{12}a_{21})}{a_{11}}da_{21} + \frac{(a_{11}a_{32} - a_{12}a_{31})}{a_{11}}da_{31} \\ &= \frac{A_{33}}{a_{11}}da_{21} - \frac{A_{23}}{a_{11}}da_{31} \end{aligned}$$

where A_{33} and A_{23} are the cofactors of a_{33} and a_{23}. Since the cofactors coincide with the elements

$$d\omega_{21} = \frac{a_{33}}{a_{11}}da_{21} - \frac{a_{23}}{a_{11}}da_{31}. \tag{7.39}$$

Similarly

$$d\omega_{31} = -\frac{a_{32}}{a_{11}}da_{21} + \frac{a_{22}}{a_{11}}da_{31}. \tag{7.40}$$

Calculation of $d\omega_{32}$ will be a little more tedious than the above.

$$d\omega_{32} = a_{13}da_{12} + a_{23}da_{22} + a_{33}da_{32}. \tag{7.41}$$

Since da_{12} and da_{22} are not independent arbitrary increments the main problem is now to express da_{12} and da_{22} in terms of $da_{21}, da_{31}, da_{32}$. Let us start from

$$\sum a_{ki}a_{kj} = \sum a_{ik}a_{jk} = \delta_{ij}.$$

Setting $i = j = 2$

$$a_{12}^2 + a_{22}^2 + a_{32}^2 = 1$$

$$a_{12}da_{12} = -(a_{22}da_{22} + a_{32}da_{32}). \qquad (7.42)$$

Thus

$$dw_{32} = -\frac{a_{13}}{a_{12}}(a_{22}da_{22} + a_{32}da_{32}) + a_{23}da_{22} + a_{33}da_{32}$$

$$= da_{22}\left(a_{23} - \frac{a_{13}}{a_{12}}a_{22}\right) + da_{32}\left(a_{33} - \frac{a_{13}}{a_{12}}a_{32}\right).$$

Identifying the r.h.s. with the appropriate cofactors as explained above we have

$$dw_{32} = \frac{a_{31}}{a_{12}}da_{22} - \frac{a_{21}}{a_{12}}da_{32}. \qquad (7.43)$$

Similarly setting $i = 1$, $j = 2$ in Eq. (7.34)

$$a_{11}a_{12} + a_{21}a_{22} + a_{31}a_{32} = 0$$

which yields

$$a_{21}da_{22} + a_{22}da_{21} = -a_{12}da_{11} - a_{11}da_{12} - a_{31}da_{32} - a_{32}da_{31}.$$

Finally using Eqs. (7.38) and (7.42) we have

$$a_{21}da_{22} + a_{22}da_{21} = \frac{a_{12}}{a_{11}}(a_{21}da_{21} + a_{31}da_{31})$$

$$+ \frac{a_{11}}{a_{12}}(a_{22}da_{22} + a_{32}da_{32}) - a_{31}da_{32} - a_{32}da_{31}.$$

Transferring da_{22} dependent term appearing on the r.h.s. to the left and identifying the appropriate cofactors

$$-\frac{a_{33}}{a_{12}}da_{22} = -\frac{a_{33}}{a_{11}}da_{21} + \frac{a_{23}}{a_{11}}da_{31} - \frac{a_{23}}{a_{12}}da_{32}$$

from which da_{22} can be expressed in terms of the independent increments $da_{21}, da_{31}, da_{32}$. Substituting this in Eq. (7.43) we have

$$dw_{32} = \frac{a_{31}}{a_{11}}da_{21} - \frac{a_{31}a_{23}}{a_{11}a_{33}}da_{31} + \frac{da_{32}}{a_{33}}. \qquad (7.44)$$

We therefore obtain from Eqs. (7.39), (7.40) and (7.44)

$$d\mu_l(g) = c\left|\frac{\partial(\omega_{21}, \omega_{31}, \omega_{32})}{\partial(a_{21}, a_{31}, a_{32})}\right| da_{21}da_{31}da_{32}.$$

Since the Jacobian

$$\frac{\partial(\omega_{21}, \omega_{31}, \omega_{32})}{\partial(a_{21}, a_{31}, a_{32})} = \begin{vmatrix} \dfrac{a_{33}}{a_{11}} & -\dfrac{a_{32}}{a_{11}} & \dfrac{a_{31}}{a_{11}} \\[2mm] -\dfrac{a_{23}}{a_{11}} & \dfrac{a_{22}}{a_{11}} & -\dfrac{a_{31}a_{23}}{a_{11}a_{33}} \\[2mm] 0 & 0 & \dfrac{1}{a_{33}} \end{vmatrix} = \frac{1}{a_{11}a_{33}}.$$

Thus

$$d\mu_l(g) = c\frac{da_{21}\,da_{31}\,da_{32}}{|a_{11}a_{33}|}. \tag{7.45a}$$

For compact groups c is chosen such that

$$\int d\mu_l(g) = 1.$$

The most popular parametrization of $SO(3)$ is the Euler angle parametrization

$$a(\phi_2, \theta, \phi_1)$$

$$= \begin{pmatrix} \begin{matrix} \cos\phi_1\cos\phi_2 \\ -\cos\theta\sin\phi_1\sin\phi_2 \end{matrix} & \begin{matrix} -\cos\phi_1\sin\phi_2 \\ -\cos\theta\sin\phi_1\cos\phi_2 \end{matrix} & \sin\phi_1\sin\theta \\[4mm] \begin{matrix} \sin\phi_1\cos\phi_2 \\ +\cos\theta\cos\phi_1\sin\phi_2 \end{matrix} & \begin{matrix} -\sin\phi_1\sin\phi_2 \\ +\cos\theta\cos\phi_1\cos\phi_2 \end{matrix} & -\cos\phi_1\sin\theta \\[4mm] \sin\phi_2\sin\theta & \cos\phi_2\sin\theta & \cos\theta \end{pmatrix}. \tag{7.45b}$$

The invariant measure can now be easily expressed in terms of Euler angles:

$$d\mu_l(g) = \frac{c}{|a_{11}a_{33}|}da_{21}\,da_{31}\,da_{32} = \frac{c}{|a_{11}a_{33}|}\left|\frac{\partial(a_{31}, a_{32}, a_{21})}{\partial(\phi_2, \theta, \phi_1)}\right|d\phi_1\,d\phi_2\,d\theta$$

$$= \frac{c}{|a_{11}a_{33}|}\begin{Vmatrix} \cos\phi_2\sin\theta & -\sin\phi_2\sin\theta & \dfrac{\partial a_{21}}{\partial\phi_2} \\[3mm] \sin\phi_2\cos\theta & \cos\phi_2\cos\theta & \dfrac{\partial a_{21}}{\partial\theta} \\[3mm] 0 & 0 & a_{11} \end{Vmatrix}d\phi_1\,d\phi_2\,d\theta$$

$$= c\sin\theta\,d\theta\,d\phi_1\,d\phi_2.$$

If we choose c such that the group volume is normalized to unity

$$c = \frac{1}{8\pi^2}.$$

Thus

$$d\mu_l(g) = \frac{1}{8\pi^2} \sin\theta d\theta d\phi_1 d\phi_2. \qquad (7.46a)$$

Another important parametrization is the axis and angle of rotation,

$$R(\hat{n}, \theta) = R(a_2, a_3, \theta); \quad \hat{n}^2 = a_1^2 + a_2^2 + a_3^2 = 1$$

$$\begin{pmatrix} a_1^2(1 - \cos\theta) + 1 & \begin{array}{c} a_1 a_2(1 - \cos\theta) \\ -a_3 \sin\theta \end{array} & \begin{array}{c} a_1 a_3(1 - \cos\theta) \\ +a_2 \sin\theta \end{array} \\ \begin{array}{c} a_1 a_2(1 - \cos\theta) \\ +a_3 \sin\theta \end{array} & a_2^2(1 - \cos\theta) + 1 & \begin{array}{c} a_2 a_3(1 - \cos\theta) \\ -a_1 \sin\theta \end{array} \\ \begin{array}{c} a_1 a_3(1 - \cos\theta) \\ -a_2 \sin\theta \end{array} & \begin{array}{c} a_2 a_3(1 - \cos\theta) \\ +a_1 \sin\theta \end{array} & a_3^2(1 - \cos\theta) + 1 \end{pmatrix}.$$

$$(7.46b)$$

Determine the invariant measure and normalize the same.

We shall now show that the left and right invariant measure of the rotation group are identical:

$$d\omega = g^{-1} dg; \quad d\omega' = dg \cdot g^{-1} = g d\omega g^{-1}.$$

Hence the eigenvalues of the matrices $d\omega$ and $d\omega'$ are identical. This immediately yields

$$d\omega_{21}^2 + d\omega_{31}^2 + d\omega_{32}^2 = d\omega_{21}'^2 + d\omega_{31}'^2 + d\omega_{32}'^2.$$

Thus $d\omega$ and $d\omega'$ are connected by rotation so that

$$d\omega_{21}' d\omega_{31}' d\omega_{32}' = d\omega_{21} d\omega_{31} d\omega_{32}$$

i.e. $d\mu_r(g) = d\mu_l(g)$.

In general for all simple unimodular Lie groups $d\mu_r(g) = d\mu_l(g)$.

Exercise. Show that for all unimodular groups

$$d\mu_r(g^{-1}) = d\mu_l(g)$$

$$d\mu_l(g^{-1}) = d\mu_r(g).$$

Hence we have

$$d\mu_r(g) = d\mu_l(g) = d\mu(g^{-1}).$$

Thus

$$\int f(g) d\mu(g) = \int f(g^{-1}) d\mu(g).$$

7.6   Representations of Lie Groups

7.6.1   *Linear vector spaces*

We shall consider the representation of the group in linear vector spaces. A linear vector space is an ordered set of complex numbers for which we can define certain operations like the following. Let L be a linear vector space. Then if $x, y \in L$, $\alpha x \in L$, $x + y = y + x \in L$. The multiplication and addition must satisfy the condition

$$(\alpha + \beta)x = \alpha x + \beta x$$

$$(\alpha \beta)x = \alpha(\beta x)$$

$$\alpha(x + y) = \alpha x + \alpha y.$$

The space will contain a null vector such that $x + 0 = x$.

7.6.2   *Unitary space and Hilbert space*

We associate with each pair of vectors x, y a complex number (x, y). The complex number is called the scalar product of x and y and is required to satisfy the conditions

$$(x, y) = \overline{(y, x)}$$

$$(x, \alpha y) = \alpha(x, y) \tag{7.47}$$

$$(x_1 + x_2, y) = (x_1, y) + (x_2, y)$$

$$(x, x) \geq 0$$

$$(x, x) = 0 \quad \text{only } x = 0.$$

The linear space of all n-tuples of complex numbers becomes a unitary space if we define the scalar product of two elements u, v as the complex number

$$(u, v) = \sum_{i=1}^{n} \overline{u_i} v_i.$$

Then

$$(u, u) = \sum |u_i|^2 = \|u\|^2.$$

Let us consider a convergent sequence

$$\lim_{n \to \infty} \|u_n - u\| \to 0$$

$$\lim_{n \to \infty} u_n = u.$$

A sequence $\{u_n\}$ is said to be a Cauchy sequence if

$$\lim_{\substack{n\to\infty \\ m\to\infty}} |u_n - u_m| \to 0.$$

Let us now identify the sequence $\{u^k\}$ as the sequence of vectors in an n-dimensional space, i.e. a sequence of n-tuples in their vector space

$$\{u^k\} = \{u_1^k, u_2^k, \ldots, u_n^k\}.$$

We say that this is a Cauchy sequence if

$$\lim_{\substack{k\to\infty \\ m\to\infty}} \left| u_i^{(k)} - u_i^{(m)} \right| = 0 \quad \text{for} \quad 1 \le i \le n.$$

Similarly the sequence is said to converge to a limit, $u \equiv (u_1, u_2, \ldots, u_n)$ if

$$\lim_{m\to\infty} \left| u_i^{(m)} - u_i \right| = 0.$$

We are now ready to define a Hilbert space. Let us consider a unitary space L. If every Cauchy sequence of elements belonging to L has a limit which also belongs to L, the space is said to be complete. A complete unitary space is called a Hilbert space.

Example. Unitary spaces, like vector spaces of n-tuples are Hilbert spaces.

Space of rational numbers is not complete because e.g. successive approximation to $\sqrt{3}$

$$1.7, \ 1.73, \ 1.732$$

is a Cauchy sequence whose limit $(\sqrt{3})$ is not a rational number.

7.6.3 *Linear operators and their matrix representation*

An operator L acting on a vector x of a Hilbert space H is called linear if

$$L(x_1 + x_2) = Lx_1 + Lx_2.$$

The outward appearance of linear operators will vary from space to space. But all linear operators can be regarded as matrices. This is a famous theorem due to Wigner. Proof of this theorem is quite simple and we refer to Wigner's book.

We illustrate this theorem by an example. Let us consider the operators z and $\frac{\partial}{\partial z}$ defined in the space of entire analytic function. A possible matrix representation of these operators is obtained by considering the Taylor expansion.

$$f(z) = \sum_{n=0}^{\infty} a_n u_n(z); \quad u_n(z) = \frac{z^n}{\sqrt{n!}}.$$

$f(z)$ may be represented by the infinite dimensional column matrix,

$$f = \begin{pmatrix} a_0 \\ a_1 \\ a_2 \\ \vdots \end{pmatrix} \tag{7.48}$$

$$g(z) = zf(z) = \sum a_n \frac{z^{n+1}}{\sqrt{n!}}$$

$$= \sum a^n \sqrt{n+1} \frac{z^{n+1}}{\sqrt{(n+1)!}}$$

$$= 0 \cdot \frac{z^0}{\sqrt{0!}} + a_0\sqrt{1} + \frac{z^1}{\sqrt{1!}} + a_1\sqrt{2}\frac{z^2}{\sqrt{2!}} + \cdots$$

$$g = \begin{pmatrix} 0 \\ \sqrt{1}a_0 \\ \sqrt{2}a_1 \\ \vdots \end{pmatrix}$$

$$f \xrightarrow{z} g = \begin{pmatrix} 0 \\ \sqrt{1}a_0 \\ \sqrt{2}a_1 \\ \vdots \end{pmatrix} = \begin{pmatrix} 0 & 0 & 0 & \cdots \\ \sqrt{1} & 0 & 0 & \cdots \\ 0 & \sqrt{2} & 0 & \cdots \\ 0 & 0 & \sqrt{3} & \cdots \end{pmatrix} \begin{pmatrix} a_0 \\ a_1 \\ a_2 \\ \vdots \end{pmatrix}.$$

Thus

$$z \to \begin{pmatrix} 0 & 0 & 0 & \cdots \\ \sqrt{1} & 0 & 0 & \cdots \\ 0 & \sqrt{2} & 0 & \cdots \\ \vdots & \vdots & \vdots & \end{pmatrix}.$$

In a similar manner we calculate the matrix of the operator $\frac{\partial}{\partial z}$

$$g(z) = \frac{\partial f}{\partial z} = \sum a_n \sqrt{n} \frac{z^{n-1}}{\sqrt{(n-1)!}} = \sum a_{n+1} \sqrt{n+1} \times \frac{z^n}{\sqrt{n!}}$$

$$g = \begin{pmatrix} a_1\sqrt{1} \\ a_2\sqrt{2} \\ a_3\sqrt{3} \\ \vdots \end{pmatrix} = \begin{pmatrix} 0 & \sqrt{1} & 0 & 0 & \cdots \\ 0 & 0 & \sqrt{2} & 0 & \cdots \\ 0 & 0 & 0 & \sqrt{3} & \cdots \\ \vdots & \vdots & \vdots & \vdots & \end{pmatrix} \begin{pmatrix} a_0 \\ a_1 \\ a_2 \\ a_3 \\ \vdots \end{pmatrix}$$

$$\therefore \quad \frac{\partial}{\partial z} \rightarrow \begin{pmatrix} 0 & \sqrt{1} & 0 & 0 & \cdots \\ 0 & 0 & \sqrt{2} & 0 & \cdots \\ 0 & 0 & 0 & \sqrt{3} & \cdots \\ \vdots & \vdots & \vdots & \vdots & \end{pmatrix}.$$

7.6.4 General definition of matrix representations of linear operators

Let A be a linear operator, let us suppose that its action on the basis vectors e_i is known. Let us assume that the space is finite dimensional

$$A e_j = \sum t_{ij}(A) e_i. \tag{7.49}$$

Of course the matrix representation depends upon the choice of basis. Let us first assume that the basis is orthonormal. Since the action of the operator A on the bases is known the matrix can be easily obtained

$$t_{ij}(A) = (e_i, A e_j). \tag{7.50}$$

7.6.5 Nonorthogonal bases

Let us consider the metric

$$(e_i, e_j) = g_{ij} = \overline{(e_j, e_i)} = \overline{g_{ji}} \tag{7.51}$$

$g = g^\dagger$, g is a Hermitian matrix.

We define (when $\det g \neq 0$)

$$g^{ij} = (g^{-1})_{ij}$$

$$\sum_{j=1}^{n} g^{ij} g_{jk} = \delta_k^i. \tag{7.52}$$

Since g^{-1} is also Hermitian we have

$$\overline{g^{ij}} = g^{ji}.$$

Let us now introduce the dual

$$e^k = \sum_{p=1}^{n} g^{pk} e_p. \tag{7.53}$$

Then

$$(e^k, e_i) = \left(\sum g^{pk} e_p, e_i\right) = \sum g^{kp} g_{pi} = \delta_i^k.$$

Thus if we define

$$A e_i = \sum t_i^j(A) e_j$$

then

$$t_i^j(A) = (e^j, A e_i)$$

so that

$$A e_i = \sum (e^j, A e_i) e_j. \tag{7.54}$$

7.6.6   *Transformation of the basis*

Let B be a non-singular operator mapping the space L onto itself and

$$f_i = B e_i.$$

Then in the basis f_i the matrix representation has the form

$$b^{-1} t(A) b \tag{7.55}$$

where b is the matrix representation of the operator B in the basis e_i,

$$b_{ji} = (e_j, B e_i) \tag{7.56}$$

assuming the basis e_i to be orthonormal.

7.6.7   *Some basic properties of the matrix representation of operators in an orthonormal basis*

If A is a nonsingular operator so that

$$A A^{-1} = I$$

$$\sum (e_i, A e_j)(e_j, A^{-1} e_k) = \delta_{ik}$$

$$\sum_j t_{ij}(A) t_{jk}(A^{-1}) = \delta_{ik}$$

which in matrix notation implies

$$t(A)t(A^{-1}) = I$$
$$t(A^{-1}) = t^{-1}(A). \tag{7.57}$$

Thus matrix representation of the inverse operator A^{-1} is equal to the inverse of the representation matrix.

For any operator A in a space L we can define another operator $A^\dagger$ by

$$(Af, g) = (f, A^\dagger g). \tag{7.58}$$

$A^\dagger$ will be called the Hermitian conjugate or adjoint of the operator A.

Let us now consider an orthonormal basis $e_1, e_2, \ldots, e_n$. Then

$$(e_j, A^\dagger e_i) = (Ae_j, e_i) = \overline{(e_i, Ae_j)}$$
$$t_{ji}(A^\dagger) = \overline{t_{ij}(A)}. \tag{7.59}$$

The r.h.s. is the transpose conjugate of the left, i.e.

$$t(A^\dagger) = \overline{\tilde{t}(A)} = t^\dagger(A) \tag{7.60}$$

An operator is self adjoint or Hermitian if it is identical with its adjoint:

$$A^\dagger = A \tag{7.61}$$

This yields

$$t(A^\dagger) = t(A). \tag{7.62}$$

Using Eq. (7.58)

$$t^\dagger(A) = t(A). \tag{7.63}$$

Thus a Hermitian operator is represented by a Hermitian matrix in an orthonormal basis.

An operator is said to be unitary if

$$(Af, Af') = (f, f')$$
$$\text{i.e.} \quad (f, A^\dagger Af') = (f, f')$$
$$\text{so that} \quad A^\dagger A = I$$
$$A^\dagger = A^{-1}. \tag{7.64}$$

Hence in matrix notation

$$t(A^\dagger) = t(A^{-1}) = t^{-1}(A).$$

Thus using Eq. (7.58)

$$t^\dagger(A) = t^{-1}(A). \tag{7.65}$$

Thus in an orthonormal basis the matrix representation of a unitary operator is a unitary matrix.

7.7 Representation of Groups

A representation of the group G is a mapping $g \to T_g$ which associates to every element $g \in G$ a linear operator T_g acting on a Hilbert space H such that the group multiplication is preserved, i.e.

$$T_{g_1} T_{g_2} = T_{g_1 g_2}. \tag{7.66}$$

From this it follows that

$$T_e = I, \ T_{g^{-1}} = (T_g)^{-1}$$

$$g_1 = g, \ g_2 = e$$

$$T_g T_e = T_{ge} = T_g; \ T_e = I.$$

Setting $g_1 = g, \ g_2 = g^{-1}$

$$T_g T_{g^{-1}} = T_{gg^{-1}} = T_e = I.$$

Thus $T_{g^{-1}} = (T_g)^{-1}$.

7.7.1 *The matrix representation*

Let us consider the space H of representation T_g to be finite dimensional. Let us consider an orthonormal basis $e_1, e_2, \ldots, e_n$. Decomposing $T_g e_j$ into basis elements we get

$$T_g e_j = \sum_{i=1}^{n} t_{ij}(g) e_i = \sum_{i=1}^{n} (e_i, T_g e_j) e_i. \tag{7.67}$$

On the other hand for a nonorthogonal basis,

$$T_g e_j = \sum_{i=1}^{n} (e^i, T_g e_j) e_i \tag{7.68}$$

where e^i is the dual of e_i. Henceforth we shall consider orthonormal bases only. For a change of orthonormal basis $e \to f$ the representation matrix changes into

$$t(g) \to a^{-1} t(g) a$$

where the symbols have been explained before.

7.7.2 Equivalent representation

Let T_g be a representation of a group G in the space L_1 and let A be a linear mapping of L_1 into the space L_2 with a continuous inverse A^{-1}. Then the linear operator

$$V_g = AT_g A^{-1}$$

determines a representation of G in the space L_2. Indeed we have

$$V_{g_1} V_{g_2} = AT_{g_1} A^{-1} AT_{g_2} A^{-1} = AT_{g_1} T_{g_2} A^{-1} = AT_{g_1 g_2} A^{-1} = V_{g_1 g_2}.$$
$$(7.69)$$

In this way one can, starting with a given representation T_g construct as many new representations of the same group as desired. These representations are regarded as equivalent to T_g.

Henceforth we shall not distinguish between equivalent representation. Now we shall prove an important theorem.

Theorem. *If representations T_g and V_g are equivalent then for a suitable choice of basis they are given by identical matrices.*

Proof. Let us choose basis $\{e_k\}$ in the space L_1 of T_g. In the space L_2 of V_g one must choose the basis $\{Ae_k\}$.

We have

$$V_g Ae_k = AT_g A^{-1} Ae_k = AT_g e_k = A \sum t_{jk}(g)e_j = \sum t_{jk}(g)Ae_j.$$
$$(7.70)$$

Thus the matrix of $V_g = \|t_{jk}(g)\| = $ matrix of T_g.

This theorem is quite remarkable in that it exhibits the matrix representation to be independent of the carrier space of the representation. For this reason in many older textbooks, the representation matrices are called "invariant matrices". This nomenclature however is misleading because the matrices are dependent on the basis. It is the character which is the real invariant of the group.

7.7.3 Unitary representation

A representation T_g of a group in a space L is called unitary relative to the scalar product (x, y) if the operator T_g leaves the scalar product (x, y)

invariant, i.e. if for all vectors x and y of L and all elements $g \in$ G one has

$$(T_g x, T_g y) = (x, y)$$

$$T_g^\dagger T_g = I, \ T_g^\dagger = T_g^{-1} = T_{g^{-1}}. \tag{7.71}$$

In an orthonormal basis the matrix representation of the unitary operator T_g is a unitary matrix:

$$[t^{-1}(g)]_{ij} = [t^\dagger(g)]_{ij} = \overline{t_{ji}}(g).$$

Unitary representations play a very important role in the theory of group representations. For instance it will be shown later that any finite dimensional representation of a compact Lie group can be made unitary relative to a certain scalar product.

7.7.4 *Basic properties of representations of compact groups*

The representation theory of compact groups constitutes the starting point of the more complex representation theory of non-compact groups which possess infinite dimensional irreducible unitary representations.

We first show that in the case of compact groups we can restrict ourselves without loss of generality to an analysis of unitary representations only.

Theorem 1. *Let $g \to T_g$ be an arbitrary representation of a compact group G in H. Then there exists in H a scalar product relative to which the representation is unitary.*

Proof. Let us introduce in H a scalar product $\langle u, v \rangle$.

$\|u\|' = c$, a finite positive constant, and introduce a new scalar product

$$(u, v) = \int \langle T_g u, T_g v \rangle d\mu(g). \tag{7.92}$$

Setting $u = v$

$$\|u\|^2 = (u, u) = \int \|T_g u\|'^2 d\mu(g). \tag{7.93}$$

Since the operators are bounded

$$\|T_g u\|'^2 \leq c$$

$$\|u\|^2 = \int \|T_g u\|'^2 d\mu(g)$$

$$\leq c \int d\mu(g) = c$$

$$\|u\|^2 \leq c \quad \text{(bounded)}$$

$$(T_{g'}u, T_{g'}v) = \int \langle T_g T_{g'}u, T_g T_{g'}v \rangle d\mu(g)$$

$$= \int \langle T_{gg'}u, T_{gg'}v \rangle d\mu(g)$$

$$= \int \langle T_g u, T_g v \rangle d\mu(g)$$

$$= (u, v). \tag{7.94}$$

Hence $T_{g'}$ is a unitary representation. □

7.8 Invariant Subspaces and Irreducible Representation

Irreducible representations may be regarded as the alphabets of representation theory. Every representation can be written in terms of these representations. Therefore, in a sense, the task of classifying all representations reduces to that of classifying all irreducible representations.

We start by analyzing the concept of an invariant subspace. Let us consider a 3×3 triangular matrix of the form

$$\begin{pmatrix} a & b & e \\ c & d & f \\ 0 & 0 & g \end{pmatrix}. \tag{7.72}$$

If we consider vectors which are in the two-dimensional subspace of the first two components, i.e. consider the subspace of vectors in the x_1-x_2 plane

$$\begin{pmatrix} x_1 \\ x_2 \\ 0 \end{pmatrix}. \tag{7.73}$$

These vectors undergo the transformation

$$\begin{pmatrix} a & b & e \\ c & d & f \\ 0 & 0 & g \end{pmatrix} \begin{pmatrix} x_1 \\ x_2 \\ 0 \end{pmatrix} = \begin{pmatrix} ax_1 + bx_2 \\ cx_1 + dx_2 \\ 0 \end{pmatrix}. \tag{7.74}$$

which is again a vector in the x_1-x_2 plane. Thus the above matrix maps a vector in the x_1-x_2 plane to another vector in the same plane. In other words the two-dimensional subspace is invariant under all transformations of the form (7.72). On the other hand, vectors of the form $\begin{pmatrix} 0 \\ 0 \\ x_3 \end{pmatrix}$ are transformed into

$$\begin{pmatrix} a & b & e \\ c & d & f \\ 0 & 0 & g \end{pmatrix} \begin{pmatrix} 0 \\ 0 \\ x_3 \end{pmatrix} = \begin{pmatrix} ex_3 \\ fx_3 \\ gx_3 \end{pmatrix}. \tag{7.75}$$

So the complementary one-dimensional subspace is not invariant. If, in addition $e = f = 0$, i.e. the representation is fully reducible then, of course, the one-dimensional subspace will also be invariant.

At the start the matrices of the representation $t(g)$ may not have the simple form (7.72); but if we can find a basis transformation which brings all matrices of the representation to this form we say the representation is reducible. In general if we can find a basis in which all matrices $t(g)$ of an n-dimensional representation can be brought simultaneously to the form

$$t(g) = \begin{pmatrix} t^{(1)}(g) & \vdots & a(g) \\ \cdots\cdots\cdots & \vdots & \cdots\cdots\cdots \\ 0 & \vdots & t^{(2)}(g) \end{pmatrix}. \tag{7.76}$$

where $t^{(1)}(g)$ are $m \times m$ matrices, $t^{(2)}(g)$ are $(n-m) \times (n-m)$ matrices, $a(g)$ is a rectangular matrix with m rows and $(n-m)$ columns and 0 denotes a null matrix with $(n-m)$ rows and m columns we say that the representation is reducible. Clearly

$$t(g_1 g_2) = \begin{pmatrix} t^{(1)}(g_1 g_2) & \vdots & a(g_1 g_2) \\ \cdots\cdots\cdots & \vdots & \cdots\cdots\cdots \\ 0 & \vdots & t^{(2)}(g_1 g_2) \end{pmatrix}. \tag{7.77}$$

Since $t(g_1 g_2)$ is a representation

$$t(g_1 g_2) = t(g_1)t(g_2)$$

$$= \begin{pmatrix} t^{(1)}(g_1) & \vdots & a(g_1) \\ \cdots\cdots\cdots & \vdots & \cdots\cdots\cdots \\ 0 & \vdots & t^{(2)}(g_1) \end{pmatrix} \begin{pmatrix} t^{(1)}(g_2) & \vdots & a(g_2) \\ \cdots\cdots\cdots & \vdots & \cdots\cdots\cdots \\ 0 & \vdots & t^{(2)}(g_2) \end{pmatrix}$$

$$= \begin{pmatrix} t^{(1)}(g_1)t^{(1)}(g_2) & \vdots & t^{(1)}(g_1)a(g_2) + a(g_1)t^{(2)}(g_2) \\ \cdots\cdots\cdots\cdots\cdots & \vdots & \cdots\cdots\cdots\cdots\cdots\cdots\cdots\cdots\cdots\cdots \\ 0 & \vdots & t^{(2)}(g_1)t^{(2)}(g_2) \end{pmatrix}.$$

Hence

$$\begin{aligned} t^{(1)}(g_1 g_2) &= t^{(1)}(g_1)t^{(1)}(g_2) \\ t^{(2)}(g_1 g_2) &= t^{(2)}(g_1)t^{(2)}(g_2). \end{aligned} \qquad (7.78)$$

Hence $t^{(1)}(g)$ provides us with an m-dimensional representation and similarly $t^{(2)}(g)$ with $(n-m)$-dimensional representation.

We can now continue the process as follows; we transform the bases in the m-dimensional representation such that all matrices $t^{(1)}(g)$ is brought into the block triangular form (7.76) and so on. Clearly the process comes to an end and then all the matrices $t(g)$ of the representation are expressed in the form,

$$t(g) = \begin{pmatrix} t^{(1)}(g) & \vdots & & a^{(1)}(g) & & \\ \cdots\cdots\cdots & \vdots & & & & \\ 0 & \vdots & t^{(2)}(g) & \vdots & & a^{(2)}(g) & \\ & \vdots & \cdots\cdots\cdots & \vdots & & & \\ 0 & \vdots & 0 & \vdots & t^{(3)}(g) & \vdots & a^{(3)}(g) \\ & & & \vdots & \cdots\cdots\cdots & & \\ & & & & & \ddots & \\ & & & & & & t^{(k)}(g) \end{pmatrix}$$

where k sets of matrices $t^{(1)}(g) \cdots t^{(k)}(g)$ are irreducible (matrix) representation of dimension $m_i (n = \sum_{i=1}^{k} m_i)$.

We are now in a position to give the most general definition of irreducible representation. Let us suppose that we have set up a representation in an n-dimensional space L_n. A subspace L_m $((m < n)$-dimensional) of the representation will be called invariant if $x \in L_m$ implies that for all $g \in G$ we have $T_g x \in L_m$. In other words all operators of the representation T_g transform vectors of the subspace L_m into vectors of the same subspace.

For every representation there are at least two invariant subspaces, namely, the null subspace and the whole space L_n. These invariant subspaces are called trivial. If a representation possesses only trivial invariant subspaces it is called irreducible. A representation with non-trivial invariant subspaces is reducible. If for example the representation space of T_g

decomposes into k non-trivial invariant subspaces we shall say that the representation decomposes into k-irreducible representations.

Generally the irreducible representations may be of finite or infinite dimensions and they are characterized or labelled by a set of real and/or complex numbers. For example the irreducible (unitary) representations of the special unitary group in two dimensions is characterized by a single positive integer or half-integer.

Later on we shall see that the reducible representations of a compact group is in general fully reducible, i.e. they are of the block diagonal form.

7.9   Schur's Lemma 1

If $g \to T_g^{(p)}$ is an irreducible representation of a group G and if a bounded operator P commutes with all operators $T_g^{(p)}$ of the representation,

$$[P, T_g^{(p)}] = 0, \quad \forall g \in G \tag{7.79}$$

then P must be a scalar, i.e.

$$P = \lambda I. \tag{7.80}$$

We prove the above theorem for finite dimensional representations in which case T_g^p and P can be replaced by their matrix representations $t^{(p)}(g)$ and H respectively.

We first show that every Hermitian matrix H which commutes with all the matrices of the representation $t^p(g)$,

$$Ht^{(p)}(g) = t^{(p)}(g)H, \quad \forall g \in G \tag{7.81}$$

is a multiple of the identity.

Let U be a unitary transformation which diagonalizes H

$$U^{-1}HU = H_d \tag{7.82}$$

$$UH_dU^{-1} = H \tag{7.83}$$

where H_d is a diagonal matrix with diagonal elements, say, λ_i which are the eigenvalues of H.

Let us pick up a certain eigenvalue which is repeated, say k times $1 \le k \le n$. Since the ordering of the eigenvalues can be chosen in any way we please we choose the k equal eigenvalues in the first k positions of H_d so that

$$\lambda_1 = \lambda_2 = \lambda_3 = \cdots = \lambda_k \neq \lambda_\mu, \quad k+1 \le \mu \le n.$$

Now

$$Ht^{(p)}(g) = t^{(p)}(g)H.$$

Using Eq. (81) we have

$$UH_d U^{-1} t^{(p)}(g) = t^p(g) U H_d U^{-1}.$$

Multiplying both sides by U^{-1} on the left and U on the right we have

$$H_d U^{-1} t^{(p)}(g) U = U^{-1} T^{(p)}(g) U H_d$$

$$H_d t^{'(p)}(g) = t^{'(p)}(g) H_d$$

where $t^{'(p)}(g)$ are the matrices of a representation equivalent to $t^{(p)}(g)$.
Taking the $j\mu^{\text{th}}$ element from both the sides

$$\sum_m (H_d)_{jm} t^{'(p)}_{m\mu} = \sum_m t^{'(p)}_{jm} (H_d)_{m\mu}$$

$$(\lambda_j - \lambda_\mu) t^{'(p)}_{j\mu} = 0. \tag{7.84}$$

But according to our ordering of eigenvalues

$$\lambda_j \neq \lambda_\mu \quad 1 \leq j \leq k, \ k+1 \leq \mu \leq n$$

$$t^{'(p)}_{j\mu}(g) = 0 \quad 1 \leq j \leq k, \ k+1 \leq \mu \leq n$$

$$t^{'(p)}_{1,k+1} = 0 = t^{'(p)}_{1,k+2} = \cdots = t^{'(p)}_{1n}$$

$$t^{'(p)}_{2,k+2} = 0 = t^{'(p)}_{2,k+2} = \cdots = t^{'(p)}_{2n}$$

$$\vdots$$

$$t^{'(p)}_{k,k+1} = 0 = t^{'(p)}_{k,k+2} = \cdots = t^{'(p)}_{kn}.$$

Also $\quad \lambda_j \neq \lambda_\mu \quad k+1 \leq j \leq n, \ 1 \leq \mu \leq k$

$$\therefore \quad t^{'(p)}_{j\mu} = 0 \quad \text{for} \quad k+1 \leq j \leq n, \ 1 \leq \mu \leq k$$

$$t^{'(p)}_{k+1,1} = 0 =, \ t^{'(p)}_{k+2,1} = 0, \ldots, t^{'(p)}_{n1} = 0$$

$$t^{'(p)}_{k+2,2} = t^{'(p)}_{k+2,2} = \cdots = t^{'(p)}_{n2}$$

$$\vdots$$

$$t^{'(p)}_{k+1,k} = t^{'(p)}_{k+2,k} = \cdots = t^{'(p)}_{nk} = 0.$$

Hence the matrix $t'^{(p)}(g)$ is of the form

1st k rows $t'^p(g)$

$$= \text{next}(n-k) \text{ rows} \begin{array}{cc} & \begin{array}{cc} \text{1st } k & \text{next } (n-k) \\ \text{columns} & \text{columns} \end{array} \\ \begin{pmatrix} t''^{(p)}(g) & \vdots & 0 \\ \cdots\cdots\cdots\cdots & \vdots & \cdots\cdots\cdots \\ 0 & \vdots & t'''^{(p)}(g) \end{pmatrix} \end{array}. \quad (7.85)$$

which shows that $t'^{(p)}(g)$ and hence its equivalent representation $t^{(p)}(g)$ is reducible. But by assumption $t^{(p)}(g)$ is irreducible which is possible only if $k = n$, i.e. if and only if all the eigenvalues of H are the same

$$H_d = \lambda I$$

$$H = U H_d U^{-1} = \lambda I.$$

Now every operator P can be written as

$$P = H_1 + i H_2$$

where H_1 and H_2 are Hermitian operators.

Let us suppose H_1 and H_2 are Hermitian operators such that

$$[H_1, T_g^p] = 0 = [H_2, T_g^p].$$

We have just proved

$$H_1 = \lambda_1 I, \quad H_2 = \lambda_2 I$$

$$P = (\lambda_1 + i\lambda_2)I = \lambda I.$$

The importance of this theorem lies in the fact that if no operator other than a constant operator (scalar) commutes with all the operators of a representation then the representation is irreducible.

7.10 Schur's Lemma 2

Consider two irreducible representations of the same group $t^{(p_1)}(g), t^{(p_2)}(g)$ of dimensions l_1 and l_2. If there exists a matrix M with l_2 rows and l_1 columns such that

$$M t^{(p_1)}(g) = t^{(p_2)}(g) M$$

for $l_1 \neq l_2$, the matrix M is a null matrix; for $l_1 = l_2$, M is either a null matrix or a matrix with non-vanishing determinant. In the latter case M has an inverse and the two irreducible representations are equivalent.

Before we proceed to prove this theorem we shall state an important theorem (sec. (7.7.4)) proved before.

"Every representation of a compact Lie group can be made unitary relative to a certain scalar product."

We shall assume at the outset that the representation is already in the unitary form. We further assume $l_1 \leq l_2$. The matrix M is a rectangular matrix of the form,

$$M = \begin{pmatrix} M_{11} & M_{12} & \cdots & M_{1l_1} \\ M_{21} & M_{22} & \cdots & M_{2l_1} \\ \vdots & & & \\ M_{l_21} & M_{l_22} & \cdots & M_{l_2l_1} \end{pmatrix}.$$

It is given

$$Mt^{(p_1)}(g) = t^{(p_2)}(g)M$$
$$t^{(p_1)\dagger}(g)M^\dagger = M^\dagger t^{(p_2)\dagger}(g)$$
$$t^{(p_1)}(g^{-1})M^\dagger = M^\dagger t^{(p_2)}(g^{-1}) \quad \text{for} \quad \forall g \in G.$$

Thus

$$t^{(p_1)}(g)M^\dagger = M^\dagger t^{(p_2)}(g).$$

Multiplying both sides on the left by M

$$Mt^{(p_1)}(g)M^\dagger = MM^\dagger t^{(p_2)}(g)$$
$$t^{(p_2)}(g)MM^\dagger = MM^\dagger t^{(p_2)}(g). \tag{7.86}$$

Since $t^{(p_2)}$ is an irreducible representation,

$$MM^\dagger = cI. \tag{7.87}$$

If the dimension of the irreducible representations are the same, $l_1 = l_2$, there are two possibilities

(i) $c = 0$
(ii) $c \neq 0$.

For $c \neq 0$, $\det M \neq 0$, M has an inverse and the two irreducible representations are equivalent

(ii) $c = 0$, $MM^\dagger = 0$

$$\sum M_{ik}\overline{M_{ik}} = 0$$
$$\sum |M_{ik}|^2 = 0$$

which is possible only if

$$M_{ik} = 0. \tag{7.88}$$

On the other hand for $l_1 < l_2$, we introduce a square matrix ($l_2 \times l_2$ matrix)

$$M' = \begin{pmatrix} M_{11} & M_{12} & \cdots & M_{1l_1} & 0 & \cdots & 0 \\ M_{21} & M_{22} & \cdots & M_{2l_1} & 0 & \cdots & 0 \\ M_{l_21} & M_{l_22} & \cdots & M_{l_2l_1} & 0 & \cdots & 0 \end{pmatrix} \overset{l_2-l_1 \text{columns}}{} \tag{7.89}$$

$$M' = \begin{pmatrix} M & 0 \end{pmatrix}; \quad M'^{\dagger} = \begin{pmatrix} M^{\dagger} \\ 0 \end{pmatrix}. \tag{7.90}$$

Thus

$$M'M'^{\dagger} = \begin{pmatrix} M & 0 \end{pmatrix} \begin{pmatrix} M^{\dagger} \\ 0 \end{pmatrix} = MM^{\dagger}$$

$$MM^{\dagger} = cI$$

$$M'M'^{\dagger} = cI.$$

Now, $\det M' = 0$ trivially

$$c^{l_2} = \det M' \det M'^{\dagger} = 0$$

$$c = 0$$

$$MM^{\dagger} = 0.$$

As before

$$M_{ik} = 0, \ M = 0. \tag{7.91}$$

7.11   Weyl's Lemma

Let $g \to T_g$ be a unitary representation of the group G and let u be any fixed vector in the carrier space H. Then the Weyl operator for all $v \in H$ defined by the formula

$$K_u v = \int (T_g u, v) T_g u d\mu(g) \tag{7.95}$$

satisfies

$$K_u T_g = T_g K_u. \tag{7.96}$$

The nature of the operator K_u can be better understood by using Dirac notation,

$$K_u = \int T_g |u\rangle \langle u| T_g^\dagger d\mu(g).$$

From the definition

$$T_{g'} K_u v = \int (T_g u, v) T_{g'g} u \, d\mu(g)$$

$$= \int (T_{g'g}^{-1} u, v) T_g u \, d\mu(g)$$

$$= \int (T_{g'}^\dagger T_g u, v) T_g u \, d\mu(g)$$

$$= \int (T_g u, T_{g'} v) T_g u \, d\mu(g)$$

$$= K_u T_{g'} v.$$

Since v is an arbitrary vector in H

$$T_{g'} K_u = K_u T_{g'}.$$

We shall now prove the following theorem:

Every irreducible unitary representation of a compact group is finite dimensional.

Proof. Since K_u commutes with all $T_{g'}$ by Schur's lemma

$$K_u = \alpha(u) I \tag{7.97}$$

$$K_u v = \int (T_g u, v) T_g u \, d\mu(g) = \alpha(u) v.$$

Hence

$$(v, K_u v) = \int (T_g u, v)(v, T_g u) d\mu(g) = \alpha(u) \|v\|^2 \tag{7.98}$$

or,

$$\int |(T_g u, v)|^2 d\mu(g) = \alpha(u) \|v\|^2.$$

Interchanging the role of u and v we have,

$$\int |(T_g v, u)|^2 d\mu(g) = \alpha(v) \|u\|^2$$

$$\int |(u, T_g v)|^2 d\mu(g) = \alpha(v) \|u\|^2 \tag{7.99}$$

$$\int |\mathrm{T}_g^\dagger u, v)|^2 d\mu(g) = \alpha(v)\|u\|^2$$

$$\int |(\mathrm{T}_{g^{-1}}u, v)|^2 d\mu(g) = \alpha(v)\|u\|^2.$$

Using the invariance of the measure under inversion

$$\int f(g)d\mu(g) = \int f(g^{-1})d\mu(g)$$

$$\int |(\mathrm{T}_g u, v)|^2 d\mu(g) = \alpha(v)\|u\|^2. \tag{7.100}$$

Hence

$$\alpha(u)\|v\|^2 = \alpha(v)\|u\|^2$$

$$\frac{\alpha(u)}{\|u\|^2} = \frac{\alpha(v)}{\|v\|^2} = c$$

$$\alpha(u) = c\|u\|^2 = c, \quad \|u\|^2 = 1.$$

Setting $v = u$ we have

$$\int |(\mathrm{T}_g u, v)|^2 d\mu(g) = c. \tag{7.101}$$

We now prove the essential part of the theorem. Let $e_p, p = 1, 2, 3, \ldots, n$ be any orthonormal vectors in H.

Setting $u = e_k$ and $v = e_1$

$$\int |(\mathrm{T}_g e_k, e_1)|^2 d\mu(g) = c$$

$$c = \int |(\mathrm{T}_g e_k, e_1)|^2 d\mu(g)$$

$$nc = \sum \int |t_{1k}(g)|^2 d\mu(g)$$

$$= \int \sum |t_{1k}(g)|^2 d\mu(g). \tag{7.102}$$

Since the rows and columns of a unitary matrix are normalized we have,

$$\sum_{k=1}^n |t_{1k}(g)|^2 \leq 1$$

$$nc \leq \int d\mu(g) = 1$$

$$\therefore \quad nc \leq 1. \tag{7.103}$$

Hence the dimension of the space cannot exceed $\frac{1}{c}$ and hence is finite. In fact if N is the dimension of the representation

$$N = \frac{1}{c}. \qquad \square$$

Wigner's orthogonality theorem

The matrices t^p of a unitary irreducible representation of a compact group constitute an orthogonal set w.r.t. the invariant measure

$$\int t^p_{km}(g)\overline{t^q_{jn}(g)}d\mu(g) = \frac{\delta_{pq}\delta_{kj}\delta_{mn}}{N}. \qquad (7.104)$$

Proof. Let us construct for $p = q$ a matrix

$$X = \int t^p(g)Zt^p(g^{-1})d\mu(g) \qquad (7.105)$$

where Z is a numerical matrix independent of the group parameters to be chosen later.

$$t^p(a)Xt^p(a^{-1}) = \int t^p(ag)Zt^p(g^{-1}a^{-1})d\mu(g)$$

$$= \int t^p(ag)Zt^p((ag)^{-1})d\mu(g)$$

$$= \int t^p(g)Zt^p(g^{-1})d\mu(g) \quad \text{(left translation } g \to a^{-1}g)$$

$$= X.$$

Thus

$$t^p(a)X = Xt^p(a), \quad \forall a \in G. \qquad (7.106)$$

Since $t^p(a)$ is the matrix of an irreducible representation by Schur's lemma,

$$X = \lambda I$$

$$\text{Tr}X = \lambda \text{Tr}I = N\lambda$$

$$\lambda = \frac{\text{Tr}X}{N}. \qquad (7.107)$$

Since t^p is a finite-dimensional matrix

$$\text{Tr}X = \int \text{Tr}[t^p(g)Zt^p(g^{-1})]d\mu(g)$$

$$= \int \text{Tr}[t^p(g^{-1})t^p(g)Z]d\mu(g)$$

$$= \text{Tr}Z \int d\mu(g) = \text{Tr}Z. \qquad (7.108)$$

Hence finally

$$\int t^p(g)Zt^p(g^{-1})d\mu(g) = \frac{\operatorname{Tr}Z}{N}I.$$

Taking kj^{th} matrix element from both sides

$$\sum_{rs} \int t^p_{kr}(g)Z_{rs}t^p_{sj}(g^{-1})d\mu(g) = \frac{\operatorname{Tr}Z}{N}\delta_{kj}. \tag{7.109}$$

Since t^p is a unitary matrix,

$$t^p(g^{-1}) = t^{p^{-1}}(g) = t^{p\dagger}(g)$$

$$t^p_{sj}(g^{-1}) = \overline{t^p_{js}(g)}.$$

Hence

$$\sum_{rs} \int t^p_{kr}(g)Z_{rs}\overline{t^{(p)}_{js}}(g)d\mu(g) = \frac{\operatorname{Tr}Z}{N}\delta_{kj}.$$

Let us choose the numerical matrix Z such that the mn^{th} matrix element is 1 and all other elements zero:

$$Z_{rs} = \delta_{rm}\delta_{ns}$$

so that

$$\operatorname{Tr}Z = \sum_r Z_{rr} = \sum \delta_{rm}\delta_{nr} = \delta_{nm} = \delta_{mn}.$$

Hence

$$\int t^p_{km}(g)\overline{t^p_{jn}(g)}d\mu(g) = \frac{\delta_{mn}\delta_{kj}}{N}. \tag{7.110}$$

We shall now prove that

$$\int t^p_{km}(g)\overline{t^q_{jn}(g)}d\mu(g) = 0 \quad \text{for} \quad p \neq q.$$

Once again we introduce

$$X = \int t^p(g)Zt^q(g^{-1})d\mu(g)$$

where $p \neq q$ and Z is a rectangular matrix with appropriate number of rows and columns. As before

$$t^p(a)X = Xt^q(a).$$

We now recall Schur's second lemma.

If $t^p(a)$ and $t^q(a)$ are matrices of irreducible representations labelled by p and q and X is a matrix such that

$$t^p(a)X = Xt^q(a)$$

then either X has an inverse or X = 0. Since here $p \neq q$, the irreducible representations are inequivalent and hence

$$X = 0$$

$$\therefore \quad \int t^p(g)Zt^q(g^{-1})d\mu(g) = 0$$

choosing Z as before we have

$$\int t^p_{km}(g)\overline{t^q_{jn}(g)}d\mu(g) = 0.$$

Hence combining these results

$$\int t^p_{km}(g)t^q_{jn}(g)d\mu(g) = \frac{\delta_{pq}\delta_{kj}\delta_{mn}}{N}. \qquad \square$$

7.12 Peter–Weyl Theorem

Let $\{T^s\}$ be the set of all irreducible inequivalent unitary representations of G. The functions

$$\sqrt{N_s}t^s_{jk}(g) \qquad (7.111)$$

where N_s is the dimension of T^s_g and $t^s_{jk}(g)$ are the matrix elements of T^s_g (in an orthonormal basis) form a complete orthonormal system in $L^2(G)$.

Let $L^\perp$ be the orthogonal complement to L consisting of the above mentioned functions. L is obviously invariant under $T^R_{g_0}$: $T^R_{g_0} \rightarrow$ right translation.

$$T^R_{g_0}t^s_{jk}(g) = t^s_{jk}(gg_0)$$

$$= \sum_r t^s_{jr}(g)t^s_{rk}(g_0) \qquad (7.112)$$

$$t^s_{jk}(g) = e^s_k$$

$$T^R_{g_0}e^s_k = \sum t^s_{rk}(g_0)e^s_r.$$

Thus the subrepresentation $^{H^s}T^R_{g_0}$ is irreducible and equivalent to T^s.

Let $v \in L^\perp$ and set

$$u(g) = \int [T^R_g v(g_0)]\overline{v(g_0)}d\mu(g_0).$$

Now

$$\int u(g)\overline{t_{jk}^s(g)}d\mu(g) = \int d\mu(g)[T_g^R v(g_0)]\overline{v(g_0)} \times \overline{t_{jk}^s(g)}d\mu(g_0)$$

$$= \int d\mu(g_0)\overline{v(g_0)} \int v(g_0 g)\overline{t_{jk}^s(g)}d\mu(g)$$

$$= \int d\mu(g_0)\overline{v(g_0)} \int v(g)\overline{t_{jk}^s(g_0^{-1}g)}d\mu(g)$$

$$= \sum \int \overline{v(g_0)}\overline{t_{rj}^s}(g_0)d\mu(g_0) \int v(g)\overline{t_{rk}^s(g)}d\mu(g) = 0.$$

Hence $u(g) \in L^\perp$.

Moreover

$$u(e) = \|v\|^2 > 0.$$

Let us now set

$$\omega(g) = u(g) + \overline{u(g^{-1})} \qquad (7.113)$$

and consider the operator

$$A\psi(g) = \int \omega(gg_0^{-1})\psi(g_0)d\mu(g_0).$$

We shall now prove that A is a self-adjoint compact operator.

From the definition of $\omega(g)$ we have

$$\overline{\omega(gg_0^{-1})} = \omega(g_0 g^{-1}) \qquad (7.114)$$

$$(A\phi, \psi) = (A\phi(g), \psi(g)) = \int \overline{\omega(gg_0^{-1})}\,\overline{\phi(g_0)} \times \psi(g)d\mu(g_0)d\mu(g)$$

$$= \int \omega(g_0 g^{-1})\overline{\phi(g_0)}\psi(g)d\mu(g)d\mu(g_0)$$

$$= (\phi(g_0), A\psi(g_0)) = (\phi, A\psi).$$

Hence A is a self adjoint operator. The boundedness A follows from the fact that the group volume is finite.

Let λ be an eigenvalue (degenerate) of A belonging to the eigenfunction $\psi_\lambda(g)$ so that

$$A\psi_\lambda(g) = \lambda\psi_\lambda(g).$$

Now

$$\int \psi_\lambda(g)\overline{t_{jk}^s(g)}d\mu(g) = \frac{1}{\lambda}\int A\psi_\lambda(g)\overline{t_{jk}^s(g)}d\mu(g)$$

$$= \frac{1}{\lambda}\int \omega(gg_0^{-1})\psi_\lambda(g_0)d\mu(g_0)\overline{t_{jk}^s(g)}d\mu(g_0)d\mu(g)$$

$$= \frac{1}{\lambda}\int \omega(g)\overline{t_{jk}^s(gg_0)}\psi_\lambda(g_0)d\mu(g)d\mu(g_0)$$

$$= \frac{1}{\lambda}\sum_r \int \omega(g)\overline{t_{jr}^s(g)}d\mu(g)\int \overline{t_{rk}^s(g_0)}\psi_\lambda(g_0)d\mu(g_0).$$

Now

$$\int \omega(g)\overline{t_{jr}^s(g)}d\mu(g) = 0 \tag{7.115}$$

because

$$\omega(g) = u(g) + \overline{u(g^{-1})} \quad \text{and} \quad u(g) \in L^\perp$$

and

$$\int u(g)\overline{t_{jr}^s(g)}d\mu(g) = 0.$$

Thus finally we have

$$\int \psi_\lambda(g)t_{jk}^s(g)d\mu(g) = 0.$$

Hence $\psi_\lambda \in L^\perp$.

Now the irreducible subspace $H^s(\lambda)$ is invariant under T^R because A is T^R invariant.

$$(T_{g_0}^R A\psi)(g) = A\psi(gg_0)$$

$$= \int \omega(gg_0g'^{-1})\psi(g')d\mu(g')$$

$$= \int \omega(gg'^{-1})\psi(g'g_0)d\mu(g')$$

$$= \omega(gg')T_{g_0}^R\psi(g')d\mu(g')$$

$$= \int \omega(gg'^{-1})\phi_{g_0}(g')d\mu(g') \quad [\phi_{g_0}(g') = T_{g_0}^R\psi(g')]$$

$$= A\phi_{g_0}(g) = AT_{g_0}^R\psi(g).$$

Thus

$$AT_{g_0}^R = T_{g_0}^R A$$

i.e.

$$AT_{g_0}^R A^{-1} = T_{g_0}^R.$$

Consequently

$$AT_{g_0}^R \psi_\lambda = T_{g_0}^R A\psi_\lambda = \lambda T_{g_0}^R \psi_\lambda.$$

$T_{g_0}^R \psi_\lambda$ is also in eigenfunction of A belonging to the eigenvalue λ. Hence $H(\lambda)$ is invariant under $T_{g_0}^R$.

Let $e_k^s(g)$ be a basis of an irreducible subspace $H^s(\lambda)$.

$$T_{g_0}^R e_k^s(g) = e_k^s(gg_0) = \sum t_{jk}^s(g_0)e_j^s(g).$$

Thus

$$e_k^s(g_0) = e_j^s(e)t_{jk}^s(g_0)$$

$$\therefore \quad e_k^s(g_0) \in L$$

$$\therefore \quad H^s(\lambda) = \{0\} \quad L^\perp = 0.$$

We shall now put the completeness condition in a form which will be useful later

$$u(g) = \sum c_{jk}^s t_{jk}^s(g)$$

$$c_{jk}^s = N_s \int u(g)\overline{t_{jk}^s(g)}d\mu(g)$$

$$\int \overline{u(g)}h(g)d\mu(g) = \sum N_s \overline{c_{jk}^s}b_{jk}^s$$

$$= \sum_s N_s \sum_{j,k} c_{kj}^{\dagger s}b_{jk}^s$$

$$= \sum_s N_s \text{Tr}(c^{\dagger s}b^s).$$

This is the form of the Plancherel formula which will be derived in a completely invariant way later for $(1,1)$.   (7.116)

7.13   The Clebsch–Gordan Problem and the Tensor Operators

Let T^{μ_1} and T^{μ_2} be two irreducible unitary representations of G in Hilbert spaces H^{μ_1} and H^{μ_2} respectively. Let $\phi_{\nu_1}^{\mu_1}$ and $\phi_{\nu_2}^{\mu_2}$ be the orthonormal basis vectors in H^{μ_1} and H^{μ_2} respectively. Here $\mu_1(\mu_2)$ stands for all the representation labels and $\nu_1(\nu_2)$ stands for all the state labels. In the tensor product space $H = H^{\mu_1} \otimes H^{\mu_2}$ we can construct two sets of basis vectors. The first consists of the Kronecker product of the original basis vectors,

$$\phi^{\mu_1,\mu_2}_{\nu_1,\nu_2} = \phi^{\mu_1}_{\nu_1}\phi^{\mu_2}_{\nu_2} \qquad (7.117)$$

while the second contains the basis vectors

$$\psi_\gamma\begin{pmatrix} \mu_1 & \mu_2 & \mu \\ & & \nu \end{pmatrix} \qquad (7.118)$$

which span an irreducible carrier space H^μ_γ contained in the tensor product space

$$H^{\mu_1} \otimes H^{\mu_2} = \sum \oplus H^\mu_\gamma$$

where γ stands for the multiplicity, i.e. number of times the representation T^μ occurs in the decomposition. Consequently H^μ_γ, $\gamma = 1, 2, \ldots, \gamma$ is the number of mutually orthogonal subspaces supporting the representation T^μ.

The Clebsch–Gordon coefficients (in the ν-basis) are the matrix elements of the unitary operator connecting the above two bases,

$$\psi_\gamma\begin{pmatrix} \mu_1 & \mu_2 & \mu \\ & & \nu \end{pmatrix} = \sum_{\nu_1,\nu_2} \overline{C_\gamma\begin{pmatrix} \mu_1 & \mu_2 & \mu \\ \nu_1 & \nu_2 & \nu \end{pmatrix}}\phi^{\mu_1}_{\nu_1}\phi^{\mu_2}_{\nu_2}. \qquad (7.119)$$

Inverting the above relation we have,

$$\phi^{\mu_1}_{\nu_1}\phi^{\mu_2}_{\nu_2} = \sum_{\mu\nu\gamma} C_\gamma\begin{pmatrix} \mu_1 & \mu_2 & \mu \\ \nu_1 & \nu_2 & \nu \end{pmatrix}\psi_\gamma\begin{pmatrix} \mu_1 & \mu_2 & \mu \\ & & \nu \end{pmatrix}. \qquad (7.120)$$

If we know the range of sum over μ and γ we know the corresponding representations and their multiplicity. Hence the above relation may be regarded as the **CG series**.

Using the above equations the CG series can be expressed in terms of the representation matrices:

$$t^{\mu_1}_{\lambda_1\nu_1}(g)t^{\mu_2}_{\lambda_2\nu_2}(g) = \sum_{\mu\nu\lambda\gamma} C_\gamma\begin{pmatrix} \mu_1,\mu_2,\mu \\ \nu_1,\nu_2,\nu \end{pmatrix}\overline{C_\gamma\begin{pmatrix} \mu_1 & \mu_2 & \mu \\ \lambda_1 & \lambda_2 & \lambda \end{pmatrix}}t^\mu_{\lambda\nu}(g). \qquad (7.121)$$

The tensor operator (transforming according to the μ-irreducible representation of G) is defined by

$$T_g T^\mu_\nu T^{-1}_g = \sum_\sigma t^\mu_{\sigma\lambda}(g)T^\mu_\sigma \qquad (7.122)$$

where $t^{\mu}_{\sigma\lambda}(g)$ are the representation matrices in an appropriate orthonormal basis.

The Wigner–Eckart theorem of the compact group G can now be stated as:

$$\left\langle \begin{matrix} \mu \\ \lambda \end{matrix} \left| T^{\mu_2}_{\lambda_2} \right| \begin{matrix} \mu_1 \\ \lambda_1 \end{matrix} \right\rangle = \sum_r C_r \begin{pmatrix} \mu_1 & \mu_2 & \mu \\ \lambda_1 & \lambda_2 & \lambda \end{pmatrix} \langle \mu \| T^{\mu_2} \| \mu_1 \rangle_\gamma \qquad (7.123)$$

where $\langle \mu \| T^{\mu_2} \| \mu_1 \rangle_\gamma$ are the reduced matrix elements (dependent on μ_1, μ_2, μ and γ alone).

Proof. We start from

$$T_g T^{\mu_2}_{\lambda_2} T^{-1}_g = \sum_{\sigma_2} t^{\mu_2}_{\sigma_2 \lambda_2}(g) T^{\mu_2}_{\sigma_2}. \qquad (7.124)$$

Multiplying both sides on the left by T^{-1}_g and on the right by T_g and using $T^{-1}_g = T_{g^{-1}}$ we have

$$T^{\mu_2}_{\lambda_2} = \sum_{\sigma_2} t^{\mu_2}_{\sigma_2 \lambda_2}(g) T_{g^{-1}} T^{\mu_2}_{\sigma_2} T_g$$

so that

$$\left\langle \begin{matrix} \mu \\ \lambda \end{matrix} \left| T^{\mu_2}_{\lambda_2} \right| \begin{matrix} \mu_1 \\ \lambda_1 \end{matrix} \right\rangle = \sum t^{\mu_2}_{\sigma_2 \lambda_2}(g) \left\langle \begin{matrix} \mu \\ \lambda \end{matrix} \left| T^\dagger_g T^{\mu_2}_{\sigma_2} T_g \right| \begin{matrix} \mu_1 \\ \lambda_1 \end{matrix} \right\rangle$$

$$= \sum t^{\mu_2}_{\sigma_2 \lambda_2}(g) \left\langle \begin{matrix} \mu \\ \lambda \end{matrix} \left| T^\dagger_g \right| \begin{matrix} \mu \\ \sigma \end{matrix} \right\rangle \left\langle \begin{matrix} \mu \\ \sigma \end{matrix} \left| T^{\mu_2}_{\sigma_2} \right| \begin{matrix} \mu_1 \\ \sigma_1 \end{matrix} \right\rangle$$

$$\times \left\langle \begin{matrix} \mu_1 \\ \sigma \end{matrix} \left| T_g \right| \begin{matrix} \mu_1 \\ \lambda_1 \end{matrix} \right\rangle$$

$$= \sum \overline{t^{\mu}_{\sigma\lambda}(g)} t^{\mu_1}_{\sigma_1 \lambda_1}(g) t^{\mu_2}_{\sigma_2 \lambda_2}(g) \left\langle \begin{matrix} \mu \\ \sigma \end{matrix} \left| T^{\mu_2}_{\sigma_2} \right| \begin{matrix} \mu_1 \\ \sigma_1 \end{matrix} \right\rangle$$

$$= \sum C_\gamma \begin{pmatrix} \mu_1 & \mu_2 & \tau \\ \lambda_1 & \lambda_2 & \rho \end{pmatrix} \overline{C_\gamma \begin{pmatrix} \mu_1 & \mu_2 & \tau \\ \sigma_1 & \sigma_2 & \delta \end{pmatrix}} t^\tau_{\delta\rho}(g) \overline{t^{\mu}_{\sigma\lambda}(g)}$$

$$\times \left\langle \begin{matrix} \mu \\ \sigma \end{matrix} \left| T^{\mu_2}_{\sigma_2} \right| \begin{matrix} \mu_1 \\ \sigma_1 \end{matrix} \right\rangle. \qquad (7.125)$$

Integrating both sides with respect to the normalized invariant measure and using the orthogonality theorem,

$$\int t^{\tau}_{\delta\rho}(g)\overline{t^{\mu}_{\sigma\lambda}(g)}d\mu(g) = \frac{\delta_{\tau\mu}\delta_{\delta\sigma}\delta_{\rho\lambda}}{N(\mu)}$$

we have

$$\left\langle \begin{matrix} \mu \\ \lambda \end{matrix} \middle| T^{\mu_2}_{\lambda_2} \middle| \begin{matrix} \mu_1 \\ \lambda_1 \end{matrix} \right\rangle = \sum C_{\gamma} \begin{pmatrix} \mu_1, \mu_2, \mu \\ \lambda_1, \lambda_2, \lambda \end{pmatrix} \sum_{\sigma_1\sigma_2\sigma} \frac{1}{N_{\mu}}$$

$$\times \overline{C_{\gamma} \begin{pmatrix} \mu_1, \mu_2, \mu \\ \sigma_1, \sigma_2, \sigma \end{pmatrix}} \left\langle \begin{matrix} \mu \\ \sigma \end{matrix} \middle| T^{\mu_2}_{\sigma_2} \middle| \begin{matrix} \mu_1 \\ \sigma_1 \end{matrix} \right\rangle .$$

Since $\sigma_1, \sigma_2, \sigma$ is summed over

$$\frac{1}{N_{\mu}} \sum_{\sigma_1\sigma_2\sigma} \overline{C_{\gamma} \begin{pmatrix} \mu_1 & \mu_2 & \mu \\ \sigma_1 & \sigma_2 & \sigma \end{pmatrix}} \left\langle \begin{matrix} \mu \\ \sigma \end{matrix} \middle| T^{\mu_2}_{\sigma_2} \middle| \begin{matrix} \mu_1 \\ \sigma_1 \end{matrix} \right\rangle = \langle \mu \| T^{\mu_2} \| \mu_1 \rangle_{\gamma}$$

is a function of μ_1, μ_2, μ and γ only.
Hence

$$\left\langle \begin{matrix} \mu \\ \lambda \end{matrix} \middle| T^{\mu_2}_{\lambda_2} \middle| \begin{matrix} \mu_1 \\ \lambda_1 \end{matrix} \right\rangle = \sum_{\gamma} C_{\gamma} \begin{pmatrix} \mu_1 & \mu_2 & \mu \\ \lambda_1 & \lambda_2 & \lambda \end{pmatrix} \langle \mu \| T^{\mu_2} \| \mu_1 \rangle_{\gamma} \qquad (7.126)$$

which is the Wigner–Eckart theorem. $\qquad\qquad\square$

7.14 Infinitesimal Operators (Generators) of a Representation

Let us suppose we are given a unitary representation of an r-parameter Lie group G.

$$T(a) = T(a_1 \cdots a_r) = T_g(a_1 \cdots a_r) = T(0) + \sum_{k=1}^{r} a_k X_k + \cdots$$

$$= I + \sum a_k X_k + \cdots \qquad (7.127)$$

where

$$X_k = \frac{\partial T(a)}{\partial a_k}\bigg|_{a=0} \qquad (7.128)$$

are called the generators (infinitesimal operators) of the group.

T$_0$ explore neighborhood of the identity we only need retain terms linear in a_k

$$T(a) = \left(1 + \sum_{k=1}^{r} a_k X_k\right). \tag{7.129}$$

Let us consider an arbitrary vector

$$(a_1, a_2, \ldots, a_r) \tag{7.130}$$

and consider two transformations about this vector

$$g(ta_1, ta_2, \ldots, ta_r) \tag{7.131}$$

and

$$g(sa_1, sa_2, \ldots, sa_r). \tag{7.132}$$

The product of these two transformations is obviously a transformation about the same vector given by the parameters

$$((t+s)a_1, (t+s)a_2, \ldots, (t+s)a_r).$$

Thus

$$T((s+t)a_1, (s+t)a_2, \ldots, (s+t)a_r) = T(sa_1, sa_2, \ldots, sa_r)T(ta_1, ta_2, \ldots, ta_r). \tag{7.133}$$

Differentiating both sides with respect to s and setting $s = 0$, we have

$$\frac{d}{dt}T(ta_1, ta_2, \ldots, ta_r) = \left.\frac{d}{ds}T(sa_1, sa_2, \ldots, sa_r)\right|_{s=0} \cdot T(ta_1, ta_2, \ldots, ta_r).$$

Setting

$$X(t) = T(ta_1, ta_2, \ldots, ta_r) \tag{7.134}$$

we have

$$\frac{dX(t)}{dt} = HX(t) \tag{7.135}$$

where

$$H = \frac{d}{ds}T(sa_1, sa_2, \ldots, sa_r)$$

$$= \frac{d}{ds}\left[I + s\sum_{k=1}^{r} a_k X_k\right]$$

$$= \sum_{k=1}^{r} a_k X_k.$$

Thus

$$X(t) = e^{t \sum_{k=1}^{r} a_k X_k}. \tag{7.136}$$

Hence setting $t = 1$

$$
\begin{aligned}
T_g(a_1 \cdots a_r) &= T(a_1, a_2, \ldots, a_r) \\
&= X(1) = e^{\sum_{k=1}^{r} a_k X_k}.
\end{aligned}
$$

Let us consider two one-parameter subgroups

$$
\begin{aligned}
R(a_\rho)R(a_\sigma) &= e^{\epsilon X_\rho} e^{\epsilon X_\sigma} = R(a^{1\tau} X_\tau) \\
&= e^{a_{\rho\sigma}'^{\tau} X_\tau}. \tag{7.137}
\end{aligned}
$$

Similarly

$$R(a_\sigma)R(a_\rho) = e^{a''^{\tau} X_\tau}. \tag{7.138}$$

Thus setting $\epsilon \to 0$ and retaining terms of the order of ϵ^2

$$\epsilon^2 \cdot [X_\rho, X_\sigma] = (a_{\rho\sigma}'^{\tau} - a_{\sigma\rho}''^{\tau}) X_\tau.$$

Now

$$
\begin{aligned}
a_{\rho\sigma}'^{\tau} &= \epsilon(\delta_\sigma^\tau + \delta_\rho^\tau) + \epsilon^2 a_{\rho\sigma}^\tau + \cdots \\
a_{\sigma\rho}''^{\tau} &= \epsilon(\delta_\sigma^\tau + \delta_\rho^\tau) + \epsilon^2 b_{\rho\sigma}^\tau + \cdots
\end{aligned}
$$

so that

$$[X_\rho, X_\sigma] = c_{\rho\sigma}^\tau X_\tau \quad [c_{\rho\sigma}^\tau = a_{\rho\sigma}^\tau - b_{\rho\sigma}^\tau]. \tag{7.139}$$

$c_{\rho\sigma}^\tau$ are called the structure constants.

The adjoint representation of a Lie group (or a Lie algebra) is an r-dimensional irreducible representation defined by

$$\overset{\circ}{X}_\tau X_\rho = [X_\tau, X_\rho] = c_{\tau\rho}^\sigma X_\sigma.$$

The matrix representation can be written as,

$$[\overset{\circ}{X}_\tau]_\rho^\sigma = c_{\tau\rho}^\sigma. \tag{7.140}$$

The r infinitesimal operators X_k, $k = 1, 2, \ldots, r$ can be considered as a basis in an r-dimensional space L_r. The scalar product of any two vectors

A and B in L_r is required to be independent of the choice of the basis $\{X_k\}$, i.e. it must be defined in terms of a matrix invariant

$$g_{\alpha\beta} = (X_\alpha, X_\beta) = \text{tr}(\overset{\circ}{X}_\alpha \overset{\circ}{X}_\beta) = (\overset{\circ}{X}_\alpha)^\sigma_\rho (\overset{\circ}{X}_\beta)^\rho_\sigma.$$

Thus

$$g_{\alpha\beta} = c^\sigma_{\alpha\rho} c^\rho_{\beta\sigma}. \tag{7.141}$$

If we restrict ourselves to semisimple groups, we have,

$$X^\alpha = g^{\alpha\beta} X_\beta$$
$$X_\alpha = g_{\alpha\beta} X^\beta$$

where

$$g^{\alpha\beta} = (g^{-1})_{\alpha\beta}$$

g^{-1} exists, because for semi-simple groups

$$\det g \neq 0.$$

The Invariants and Casimir Operators of Lie Groups

If I is an operator built out of the r infinitesimal operators and it is invariant under the group G,

$$I'(X) = T(a)I(X)T(a)^{-1} = I$$

i.e.

$$[T(a), I] = 0. \tag{7.142}$$

With the help of the metric tensor we have

$$C^{(2)} = C^{\beta_2}_{\alpha_1\beta_1} C^{\beta_1}_{\alpha_2\beta_2} X^{\alpha_1} X^{\alpha_2}$$
$$= g_{\alpha_1\alpha_2} X^{\alpha_1} X^{\alpha_2} = X_{\alpha_1} X^{\alpha_1}$$
$$C^{(3)} = C^{\beta_3}_{\alpha_1\beta_1} C^{\beta_1}_{\alpha_2\beta_2} C^{\beta_2}_{\alpha_3\beta_3} X^{\alpha_1} X^{\alpha_2} X^{\alpha_3}.$$

It can be shown that for a semisimple group of rank l there are l invariants: $C^{(2)}, C^{(3)}, \ldots, C^{l+1}$.

Generally the irreducible unitary representations are characterized by real and/or complex numbers which are essentially the eigenvalues of the appropriate invariant operators of the group. For example, representations of the special unitary group in two dimensions is characterized by the eigenvalues of the quadratic Casimir operator

$$C^{(2)} = J_1^2 + J_2^2 + J_3^2 = J(J+1)$$

where the eigenvalues of J in a unitary irreducible representation is a positive integer or half-integer.

References

[1] E. P. Wigner, *Group Theory and Its Applications to Quantum Mechanics of Atomic Spectra* (Academic, New York, 1964).

[2] A. O. Barut and R. Raczka, *Theory of Group Representations and Applications* (PWN Polish Scientific Publishers).

[3] J. D. Talman, *Special Functions: A Group Theoretic Approach* (Benjamin, New York, 1968).

[4] E. Loebl (ed.), *Group Theory and Its Applications* (Academic, New York, 1975) [Article by L. Ó Raifeartaigh].

CHAPTER 8

The Three-Dimensional Rotation Group and SU(2) and Elements of SU(3)

It is well known that the special unitary group in two dimensions $SU(2)$[1], is the quantal angular momentum group because it alone (as distinct from $SO(3)$) can account for the half integral angular momenta. In elementary particle physics the unitary group $SU(3)$ is the symmetry group or the approximate symmetry group of the Hamiltonian, the $SU(2)$ subgroup being identified with charge independence.

Gell-Mann and Neéman[2] assigned octet and the decuplet representations to the baryons while pseudoscalar mesons were fitted in the octet representation. The states of the particle are labeled by irrep labels of $SU(3)$ and its subgroups and the diagonal generators, namely, $p, q\text{II}_3 Y$.

We give in this chapter a closed formula (Majumdar's formula[3]) from which $\text{II}_3 Y$ can be calculated for every given (p, q)[3].

We have seen that the 3×3 real orthogonal matrix a with $\det a = 1$, may be regarded as representing the rotation of a vector in three dimensions. It can be expressed in terms of Euler angles (see Eq. (7.45b)) or in terms of axis and angle of rotation (see Eq. (7.46b)) etc. The inverse of the corresponding matrices can be obtained by simply transposing them. For example the transpose of the matrix (7.45a) is equivalent to the replacement $\phi_1 \to \pi - \phi_2, \phi_2 \to \pi - \phi_1$. Hence we may write $a^{-1}(\phi_2; \theta, \phi_1) = a(\pi - \phi_1, \theta, \pi - \phi_2)$.

The invariant measure, as shown in Chapter 7, can be written as

$$d\mu(g) = c \frac{da_{21} da_{31} da_{32}}{|a_{11} a_{33}|}.$$

Setting the group volume to be 1 the normalized invariant measure can be easily expressed in terms of Euler angles

$$d\mu(g) = \frac{1}{8\pi^2} \sin\theta \, d\theta \, d\phi_1 \, d\phi_2.$$

8.1   Connection Between the Rotation Group and the Group SU(2)

We consider the group of special unitary matrices in two dimensions,

$$u = \begin{pmatrix} \alpha & \beta \\ -\bar{\beta} & \bar{\alpha} \end{pmatrix} \tag{8.1a}$$

$$|\alpha|^2 + |\beta|^2 = 1. \tag{8.1b}$$

To establish its connection with rotation group we introduce the traceless Hermitian matrix

$$X = \begin{pmatrix} x_3 & x_1 - ix_2 \\ x_1 + ix_2 & -x_3 \end{pmatrix} \tag{8.2a}$$

and define

$$X' = uXu^{-1} = uXu^\dagger \tag{8.2b}$$

where $u\epsilon$SU(2). Since u is unitary it follows that X' is also traceless and Hermitian. Thus

$$X' = \begin{pmatrix} x_3' & x_1' - ix_2' \\ x_1' + ix_2' & -x_3' \end{pmatrix}. \tag{8.3}$$

Now $\det X = \det X'$, so that

$$x_1^2 + x_2^2 + x_3^2 = x_1'^2 + x_2'^2 + x_3'^2.$$

Thus the transformation (8.2b) is a rotation. From (8.2) it easily follows

$$\begin{pmatrix} x_1' - ix_2' \\ x_1' + ix_2' \\ x_3' \end{pmatrix} = \begin{pmatrix} \alpha^2 & -\beta^2 & -2\alpha\beta \\ -\bar{\beta}^2 & \bar{\alpha}^2 & -2\bar{\alpha}\bar{\beta} \\ \alpha\bar{\beta} & \bar{\alpha}\beta & |\alpha|^2 - |\beta|^2 \end{pmatrix} \begin{pmatrix} x_1 - ix_2 \\ x_1 + ix_2 \\ x_3 \end{pmatrix} \tag{8.4a}$$

i.e.

$$x' = T^{-1}UTx \tag{8.4b}$$

where

$$T = \begin{pmatrix} 1 & -i & 0 \\ 1 & i & 0 \\ 0 & 0 & 1 \end{pmatrix}, \; U = \begin{pmatrix} \alpha^2 & -\beta^2 & -2\alpha\beta \\ -\bar{\beta}^2 & \bar{\alpha}^2 & -2\bar{\alpha}\bar{\beta} \\ \alpha\bar{\beta} & \bar{\alpha}\beta & |\alpha|^2 - |\beta|^2 \end{pmatrix}, \; x = \begin{pmatrix} x_1 \\ x_2 \\ x_3 \end{pmatrix}.$$

This yields the rotation matrix in terms of the parameters of the SU(2) group.

$$R(u) = \begin{pmatrix} \frac{1}{2}(\alpha^2 + \bar{\alpha}^2 - \beta^2 - \bar{\beta}^2) & \frac{i}{2}(\bar{\alpha}^2 - \alpha^2 - \bar{\beta}^2 + \beta^2) & -(\alpha\beta + \bar{\alpha}\bar{\beta}) \\ \frac{i}{2}(\alpha^2 - \beta^2 + \bar{\beta}^2 - \bar{\alpha}^2) & \frac{1}{2}(\alpha^2 + \bar{\alpha}^2 + \beta^2 + \bar{\beta}^2) & -i(\alpha\beta - \bar{\alpha}\bar{\beta}) \\ \alpha\bar{\beta} + \bar{\alpha}\beta & -i(\alpha\bar{\beta} - \bar{\alpha}\beta) & |\alpha|^2 - |\beta|^2 \end{pmatrix}.$$

$$(8.5)$$

We can easily verify that the SU(2) matrix (one-parameter subgroup)

$$\begin{pmatrix} e^{i\phi/2} & 0 \\ 0 & e^{-i\phi/2} \end{pmatrix} \longrightarrow \begin{pmatrix} \cos\phi & \sin\phi & 0 \\ -\sin\phi & \cos\phi & 0 \\ 0 & 0 & 1 \end{pmatrix} \qquad (8.6)$$

corresponds to rotation about the 3rd (Z)-axis, $R_z(\phi)$, the one-parameter subgroup

$$\begin{pmatrix} \cos\frac{\phi}{2} & \sin\frac{\phi}{2} \\ -\sin\frac{\phi}{2} & \cos\frac{\phi}{2} \end{pmatrix} \longrightarrow \begin{pmatrix} \cos\phi & 0 & -\sin\phi \\ 0 & 1 & 0 \\ \sin\phi & 0 & \cos\phi \end{pmatrix} \qquad (8.7)$$

corresponds to rotation about second (Y) axis, $R_y(\phi)$ and the one parameter subgroup

$$\begin{pmatrix} \cos\frac{\phi}{2} & i\sin\frac{\phi}{2} \\ i\sin\frac{\phi}{2} & \cos\frac{\phi}{2} \end{pmatrix} \longrightarrow \begin{pmatrix} 1 & 0 & 0 \\ 0 & \cos\phi & \sin\phi \\ 0 & -\sin\phi & \cos\phi \end{pmatrix} \qquad (8.8)$$

corresponds to rotation about 1st (X) axis, $R_x(\phi)$. Thus we have shown that to every SU(2) matrix there corresponds a rotation in three-dimensional space. Conversely to each rotation there corresponds two such matrices differing only in sign. The two parameter domains (and consequently the group volume) are of course different. In the Euler angle decomposition

$$u(\phi_1, \theta, \phi_2) = \begin{pmatrix} e^{i\phi_1/2} & 0 \\ 0 & e^{-i\phi_1/2} \end{pmatrix} \begin{pmatrix} \cos\frac{\theta}{2} & i\sin\frac{\theta}{2} \\ i\sin\frac{\theta}{2} & \cos\frac{\theta}{2} \end{pmatrix} \begin{pmatrix} e^{i\phi_2/2} & 0 \\ 0 & e^{-i\phi_2/2} \end{pmatrix}$$

$$= \begin{pmatrix} \cos\frac{\theta}{2} e^{i(\phi_1+\phi_2)/2} & i\sin\frac{\theta}{2} e^{+i(\phi_1-\phi_2)/2} \\ i\sin\frac{\theta}{2} e^{-i(\phi_1-\phi_2)/2} & \cos\frac{\theta}{2} e^{-i(\phi_1+\phi_2)/2} \end{pmatrix} \qquad (8.9)$$

$$0 \le \phi_1 \le 4\pi, \quad 0 \le \phi_2 \le 2\pi, \quad 0 \le \theta \le \pi.$$

The group volume is $16\pi^2$, and the normalized invariant measure is given by

$$d\mu(u) = \frac{1}{16\pi^2} \sin\theta d\theta d\phi_1 \, d\phi_2. \tag{8.10}$$

We have just seen

$$R_x(\phi) = \begin{pmatrix} 1 & 0 & 0 \\ 0 & \cos\phi & \sin\phi \\ 0 & -\sin\phi & \cos\phi \end{pmatrix}$$

$$R_y(\phi) = \begin{pmatrix} \cos\phi & 0 & -\sin\phi \\ 0 & 1 & 0 \\ \sin\phi & 0 & \cos\phi \end{pmatrix}$$

$$R_z(\phi) = \begin{pmatrix} \cos\phi & \sin\phi & 0 \\ -\sin\phi & \cos\phi & 0 \\ 0 & 0 & 1 \end{pmatrix}.$$

Thus if we write

$$R(\hat{a}_0\phi) = e^{(\underline{a}\cdot\underline{A})\phi} = e^{\phi_1 A_1 + \phi_2 A_2 + \phi_3 A_3} \tag{8.11}$$

according to Chapter 7,

$$A_1 = \left(\frac{dR_x}{d\phi}\right)_{\phi=0} = \begin{pmatrix} 0 & 0 & 0 \\ 0 & 0 & 1 \\ 0 & -1 & 0 \end{pmatrix} \tag{8.12a}$$

$$A_2 = \left(\frac{dR_y}{d\phi}\right)_{\phi=0} = \begin{pmatrix} 0 & 0 & -1 \\ 0 & 0 & 0 \\ 1 & 0 & 0 \end{pmatrix} \tag{8.12b}$$

$$A_3 = \begin{pmatrix} 0 & 1 & 0 \\ -1 & 0 & 0 \\ 0 & 0 & 0 \end{pmatrix}. \tag{8.12c}$$

Explicit calculation yields

$$[A_i, A_j] = -\epsilon_{ijk} A_k. \tag{8.13}$$

In a unitary representation

$$A_l = iJ_l; \quad l = 1, 2, 3. \tag{8.14}$$

Thus

$$[J_l, J_m] = i\epsilon_{lmn} J_n. \tag{8.15}$$

8.2   Determination of the Irreducible Representation of the Rotation Group and SU(2)

Since every SU(2) matrix corresponds to a rotation it is sufficient to find all irreducible unitary representations of SU(2) which satisfies the same commutation relation (8.15) (local isomorphism). To do this we introduce

$$J_{\pm} = J_1 \pm iJ_2, \ J_3.$$

The invariant operator (Casimir operator) is given by

$$Q = \underline{J}^2 = J_1^2 + J_2^2 + J_3^2$$

$$= \frac{1}{2}(J_+J_- + J_-J_+) + J_3^2 \tag{8.16}$$

$$[Q, \ J_+] = [Q, \ J_-] = [Q, \ J_3] = 0.$$

Q is obviously a bounded positive Hermitian operator.

We now choose a basis in which SO(2) subgroup (about the Z-axis) is reduced, i.e.

$$\begin{aligned} J_3 \ \ f_m^q &= mf_m^q \\ Q \ \ f_m^q &= qf_m^q. \end{aligned} \tag{8.17}$$

Henceforth we shall suppress the superscript q.

Now

$$[J_{\pm}, \ J_3] = \mp J_{\pm}$$

so that

$$J_+(J_3 \ f_m) = J_3J_+f_m - J_+f_m$$

i.e.

$$J_3(J_+f_m) = (m+1)J_+f_m. \tag{8.18}$$

Hence J_+f_m is an eigenvector J_3 with the eigenvalue $m+1$:

$$J_+f_m = \beta_{m+1}f_{m+1}. \tag{8.19}$$

Similarly J_-f_m is an eigenvector of J_3 with eigenvalue $m-1$:

$$J_-f_m = \alpha_m f_{m-1}. \tag{8.20}$$

By successively applying J_+ and J_- we can construct the ladder,

$$\ldots, m+2, \ m+1, \ m, \ m-1, \ m-2, \ldots.$$

Since the irreducible unitary representations of SU(2) must be finite dimensional (Chapter 7) the ladder must be truncated, i.e. m must have the highest and lowest value. Let $m_{max} = j$, $m_{min} = m_0$. Then $\beta_{j+1} = 0 = \alpha_{m_0}$. The existence of the highest and lowest value of m follows also from the fact that $q - m^2 > 0$.[a]

Further using the unitarity condition

$$(f_m, \ J_+ f_{m-1}) = (J_- f_m, \ f_{m-1}). \tag{8.21}$$

We have

$$\alpha_m = \bar{\beta}_m \tag{8.22}$$

so that

$$\alpha_{j+1} = \bar{\beta}_{j+1} = 0. \tag{8.23}$$

Substituting (8.22) in

$$Q f_m = \left[\frac{1}{2}(J_+ J_- + J_- J_+) + J_3^2 \right] f_m \tag{8.24}$$

and

$$\frac{1}{2}(J_+ J_- - J_- J_+) f_m = m f_m \tag{8.25}$$

we have

$$\frac{1}{2}(|\alpha_m|^2 + |\alpha_{m+1}|^2) = q - m^2 \tag{8.26a}$$

$$\frac{1}{2}(|\alpha_m|^2 - |\alpha_{m+1}|^2) = m \tag{8.26b}$$

so that

$$|\alpha_m|^2 = q - m^2 + m; \quad |\alpha_{m+1}|^2 = q - m^2 - m. \tag{8.27}$$

Setting $m = j$ in Eq. (8.27) we have

$$\alpha_{j+1} = 0$$

$$q = j^2 + j = j(j+1).$$

Also

$$\alpha_{m_0} = 0 = q - m_0^2 + m_0$$

$$m_0(m_0 - 1) - j(j+1) = 0$$

---

[a]$(J_1^2 + J_2^2)f_m = (q - m^2)f_m$. As the operator $J_1^2 + J_2^2$ is a positive Hermitian operator, $(q - m^2) > 0$, i.e. $-\sqrt{q} < m < \sqrt{q}$.

$$(m_0^2 - j^2 - m_0 - j) = 0$$

$$(m_0 + j)(m_0 - j - 1) = 0.$$

Since

$$m_{\max} = j \quad \therefore \ m_0 \neq j + 1$$

$$m_0 = -j.$$

Hence ladder of m-values is,

$$j, \ j - 1, \ j - 2, \ldots, -j$$

$$\therefore \quad j - n = -j$$

$$n = 2j \quad \therefore \ j = \frac{n}{2}. \tag{8.28}$$

Hence j is an integer or half-integer and the representation is $(2j + 1)$–dimensional.

Ignoring a phase we have

$$\alpha_{m+1} = \sqrt{q - m^2 - m}$$

$$= \sqrt{(j - m)(j + m + 1)}$$

$$\alpha_m = \sqrt{(j + m)(j - m + 1)}.$$

Hence finally we have

$$J_{\pm} f_m = \sqrt{(j \mp m)(j \pm m + 1)} f_{m \pm 1} \quad j = 0, \frac{1}{2}, 1, \frac{3}{2}, \ldots. \tag{8.29}$$

Objects transforming according to

$j = 0$ representation is called scalar,

$j = \frac{1}{2}$ representation is called a spinor,

$j = 1$ representation is called a vector, etc.

Hence every irreducible unitary representation of SU(2) is characterized or labeled by a single non-negative integer or half-integer j and the representation is $(2j + 1)$ dimensional.

8.3   The Finite Element of the Group

The representations of SU(2) will be realized in the Bargman–Segal space $B(C_2)$ which consists of entire analytic function $\phi(z_1, z_2)$ where z_1, z_2 are spinors transforming according to the fundamental representation of SU(2):

$$(z_1', z_2') = (z_1, z_2)u. \tag{8.30}$$

The action of the finite element of the group in $B(C_2)$ is given by

$$T_u\phi(z_1, z_2) = \phi(\alpha z_1 - \bar{\beta} z_2, \beta z_1 + \bar{\alpha} z_2). \tag{8.31}$$

To decompose $B(C_2)$ into the direct sum of the subspaces $B_j(C)$ invariant under SU(2) we introduce the infinitesimal operators of SU(2) as given by Eqs. (8.6)–(8.8). Thus

$$J_3 = \frac{1}{2}\left(z_1\frac{\partial}{\partial z_1} - z_2\frac{\partial}{\partial z_2}\right) \tag{8.32a}$$

$$J_2 = -\frac{i}{2}\left(z_1\frac{\partial}{\partial z_2} - z_2\frac{\partial}{\partial z_1}\right) \tag{8.32b}$$

$$J_1 = \frac{1}{2}\left(z_1\frac{\partial}{\partial z_2} + z_2\frac{\partial}{\partial z_1}\right). \tag{8.32c}$$

These are identical with Schwinger's angular momentum operators realized in $B(C_2)$.

Explicit calculation yields

$$Q = J_1^2 + J_2^2 + J_3^2 = J(J+1) \tag{8.34}$$

where

$$J = \frac{1}{2}\left(z_1\frac{\partial}{\partial z_1} + z_2\frac{\partial}{\partial z_2}\right). \tag{8.35}$$

Since J commutes with J_1, J_2, J_3 in an irreducible representation it can be replaced by j, $j = 0, \frac{1}{2}, 1, \ldots$.

The subspace $B_j(C)$ is, therefore, the space of homogeneous polynomials of degree $2j$ in z_1, z_2.

$$\phi(z_1, z_2) = z_2^{2j} f(z), \quad z = \frac{z_1}{z_2}. \tag{8.36}$$

The finite element of the group in $B_j(C)$ can be obtained easily as follows.

$$T_u\phi(z_1, z_2) = \phi(z_1', z_2')$$
$$= \phi(\alpha z_1 - \bar{\beta} z_2, \beta z_1 + \bar{\alpha} z_2)$$

$$T_u z_2^{2j} f(z) = (\beta z_1 + \bar{\alpha} z_2)^{2j} f\left(\frac{\alpha z_1 - \bar{\beta} z_2}{\beta z_1 + \bar{\alpha} z_2}\right)$$

$$= z_2^{2j} (\beta z + \bar{\alpha})^{2j} f\left(\frac{\alpha z - \bar{\beta}}{\beta z + \bar{\alpha}}\right)$$

$$(z_2^{-2j} T_u z_2^{2j}) f(z) = \hat{T}_u f(z) = (\beta z + \bar{\alpha})^{2j} f\left(\frac{\alpha z - \bar{\beta}}{\beta z + \bar{\alpha}}\right). \tag{8.37}$$

We shall now determine the scalar product with respect to which operator T_u is unitary:

$$(\hat{T}_u f, \hat{T}_u g) = (f, g). \tag{8.38}$$

Let us write the right-hand side of (8.38) as

$$(f, g) = \int \overline{f(z')} g(z') \rho(z') d^2 z'$$

and choose

$$w(z) = z' = \frac{\alpha z - \bar{\beta}}{\beta z + \bar{\alpha}}. \tag{8.40}$$

Using Theorems (7.24) and (7.25) of Chapter 7 we obtain,

$$d^2 z' = \left|\frac{dw}{dz}\right|^2 d^2 z = \left|\frac{1}{(\beta z + \bar{\alpha})^2}\right|^2 d^2 z = |\beta z + \bar{\alpha}|^{-4} d^2 z.$$

Hence writing the l.h.s. as $\int \overline{f(z')} g(z') |\beta z + \bar{\alpha}|^{4j} \rho(z) d^2 z$, we have

$$|\beta z + \bar{\alpha}|^{4j} \rho(z) = (\beta z + \bar{\alpha})^{-4} \rho(z')$$

$$\rho(z) = |\beta z + \bar{\alpha}|^{-4j-4} \rho(z'). \tag{8.41}$$

It now immediately follows.

$$\frac{1 + |z|^2}{1 + |z'|^2} = |\beta z + \bar{\alpha}|^2$$

$$\rho(z) = \left(\frac{1 + |z|^2}{1 + |z'|^2}\right)^{-2j-2} \rho(z')$$

$$\rho(z)(1 + |z|^2)^{2j+2} = (1 + |z'|^2)^{2j+2} \rho(z'). \tag{8.42}$$

For this to be true, both sides of (8.42) must be a constant,

$$\rho(z) = C(1 + |z|^2)^{-2j-2}.$$

For later convenience we choose

$$C = \frac{2j + 1}{\pi}$$

so that the scalar product with respect to which T_u is unitary is given by,

$$(f, g) = \int \overline{f(z)} g(z) d\lambda(z) \tag{8.43a}$$

$$d\lambda(z) = \frac{2j+1}{\pi} (1 + |z|^2)^{-2j-2} d^2 z. \tag{8.43b}$$

A complete orthonormal set in this Hilbert space $(B_j(C))$ is given by,

$$u_n(z) = \sqrt{\frac{(2j)!}{(2j-n)!n!}} z^n, \quad n = 0, 1, 2, \ldots, 2j. \tag{8.44}$$

Thus the principal vector in this Hilbert space is given by (see Chapter 5)

$$e_z(\xi) = \sum \overline{u_n(z)} u_n(\xi) = (1 + \bar{z}\xi)^{2j} \tag{8.45}$$

so that

$$f(z) = (e_z, f) = \int (1 + z\bar{\xi})^{2j} f(\xi) d\lambda(\xi). \tag{8.46}$$

8.4   The Clebsch–Gordan Problem of SU(2)

Let us consider the ordered pairs (z_1, z_2), $z_1 \neq z_2$. With each SU(2) matrix

$$\begin{pmatrix} \alpha & \beta \\ -\bar{\beta} & \bar{\alpha} \end{pmatrix}, |\alpha|^2 + |\beta|^2 = 1$$

we associate the transformation

$$(z_1, z_2) \to \left(\frac{\alpha z_1 - \bar{\beta}}{\beta z_1 + \bar{\alpha}}, \frac{\alpha z_2 - \bar{\beta}}{\beta z_2 + \bar{\alpha}} \right) \tag{8.47}$$

on this space and this space is homogeneous with respect to these transformations.

We wish to associate with this space a unitary representation of the SU(2) group. This representation will be constructed in the space of functions $f(z_1, z_2)$ such that

$$\|f\|^2 = \iint |f(z_1, z_2)|^2 d\lambda(z_1) d\lambda(z_2) \tag{8.48}$$

where $d\lambda(z)$ is given by Eq. (8.43b).

The Kronecker product of two irreducible unitary representations $T_u^{j_1}$, $T_u^{j_2}$ given by

$$T_u^{j_1} f(z_1) = (\beta z_1 + \bar{\alpha})^{2j_1} f\left(\frac{\alpha z_1 - \bar{\beta}}{\beta z_1 + \bar{\alpha}}\right) \tag{8.49}$$

$$T_u^{j_2} f(z_2) = (\beta z_2 + \bar{\alpha})^{2j_2} f\left(\frac{\alpha z_2 - \bar{\beta}}{\beta z_2 + \bar{\alpha}}\right) \tag{8.50}$$

will be defined by

$$T_u f(z_1, z_2) = (\beta z_1 + \bar{\alpha})^{2j_1} (\beta z_2 + \bar{\alpha})^{2j_2} f\left(\frac{\alpha z_1 - \bar{\beta}}{\beta z_1 + \bar{\alpha}}, \frac{\alpha z_2 - \bar{\beta}}{\beta z_2 + \bar{\alpha}}\right). \tag{8.51}$$

To decompose the representation of Eq. (8.51) into irreducible representations we shall first choose the canonical SO(2) basis.

The infinitesimal operators of the group for the representations given by Eqs. (8.49)–(8.51) are respectively given by

$$J_+^{(1)} = J_1^{(1)} + i J_2^{(1)} = -z_1^2 \frac{\partial}{\partial z_1} + 2j_1 z_1 \tag{8.52}$$

$$J_-^{(1)} = J_1^{(1)} - i J_2^{(1)} = \frac{\partial}{\partial z_1} \tag{8.53}$$

$$J_3^{(1)} = z_1 \frac{\partial}{\partial z_1} - j_1 \tag{8.54}$$

$$J_+^{(2)} = J_1^{(2)} + i J_2^{(2)} = -z_2^2 \frac{\partial}{\partial z_2} + 2j_2 z_2 \tag{8.55}$$

$$J_3^{(2)} = z_2 \frac{\partial}{\partial z_2} - j_2$$

$$J_+ = -\left(z_1^2 \frac{\partial}{\partial z_1} + z_2^2 \frac{\partial}{\partial z_2}\right) + 2(j_1 z_1 + j_2 z_2) \tag{8.56}$$

$$J_- = \left(\frac{\partial}{\partial z_1} + \frac{\partial}{\partial z_2}\right) \tag{8.57}$$

$$J_3 = \left(z_1 \frac{\partial}{\partial z_1} + z_2 \frac{\partial}{\partial z_2} - j_1 - j_2\right). \tag{8.58}$$

The SO(2) basis for the representations (8.49) and (8.50) is given by

$$f_{m_1}^{j_1} = N_{j_1 m_1} z_1^{j_1 + m_1} \tag{8.59}$$

$$f_{m_2}^{j_2} = N_{j_2 m_2} z_2^{j_2 + m_2} \tag{8.60}$$

while the SO(2) basis for the representation of Eq. (8.51) is given by the simultaneous eigenstates of

$$J_3 = z_1 \frac{\partial}{\partial z_1} + z_2 \frac{\partial}{\partial z_2} - j_1 - j_2 \tag{8.61}$$

$$Q = J_+ J_- + J_3^2 - J_3. \tag{8.62}$$

The SO(2) basis is therefore a homogeneous function of degree $m + j_1 + j_2$ in z_1, z_2.

$$\phi(z_1, z_2) = z_1^{m+j_2+j_1} f(z), \quad z = \frac{z_2}{z_1} \tag{8.63}$$

where $f(z)$ on simplification, satisfies the second order differential equation

$$z(1-z)^2 \frac{d^2 f}{dz^2} + (1-z)[(1-m+j_0) + z(3j_2 + j_1 + m - 1)]\frac{df}{dz}$$
$$+ [(j_1 + j_2)(j_0 - m + 1) - j(j+1) + mj_0$$
$$+ 2zj_2(j_1 + j_2 + m)]f(z) = 0. \tag{8.64}$$

If we now substitute

$$f(z) = (1-z)^{j_1+j_2-j} F(z) \tag{8.65}$$

the function $F(z)$ satisfies the differential equation of the hypergeometric function,

$$z(1-z)\frac{d^2 F}{dz^2} + [(j_0 - m + 1) - z(-2j + j_0 - m + 1)]\frac{dF}{dz}$$
$$+ (j+m)(j_0 - j)F = 0. \tag{8.66}$$

The SO(2) basis for the representation (8.51) is, therefore, given by

$$\phi(z_1, z_2) = N_{j_1 j_2 j m} z_1^{j_1+j_2+m} (1-z)^{j_1+j_2-j}$$
$$\times F(-j - m, j_0 - j; j_0 - m + 1; z). \tag{8.67}$$

Without losing any generality we can choose $j_i > j_2$ so that $j_0 = j_1 - j_2 > 0$. The Clebsch–Gordan coefficient for a particular final j is given by

$$(1-z)^{j_1+j_2-j} F(-j - m, j_0 - j; j_0 - m + 1; z)$$

$$= \sum \frac{\Gamma(1 + j_1 + j_2 - j)}{\Gamma(1 + j_1 + j_2 - j - n)} {}_3F_2 \left[\begin{matrix} -n, -j - m, j_0 - j \\ 1 + j_0 - m, 1 + j_1 + j_2 - j - n \end{matrix} z^n \right] \tag{8.68}$$

so that

$$C\begin{pmatrix} j_1 & j_2 & j \\ m_1 & m_2 & m \end{pmatrix} = \frac{N_{j_1 j_2 jm}}{N_{j_1 m_1 j_2 m_2}} \Gamma \begin{bmatrix} 1 + j_1 + j_2 - j \\ 1 + j_2 + m_2, 1 + j_1 - j - m_2 \end{bmatrix}$$

$$\times \, {}_3F_2 \begin{bmatrix} -j - m, j_0 - j, -j_2 - m_2 \\ j + j_0 - m, 1 + j_1 - j - m_2 \end{bmatrix}. \qquad (8.69)$$

To get the spectrum of j-values we derive the Clebsch–Gordan series. We start from the formula,

$$z^n = \sum (-1)^r \frac{(a)_r (b)_r}{(c + r - 1)_r r!} {}_3F_2 \begin{bmatrix} -n, c + r - 1, -r \\ a, b \end{bmatrix}$$

$$\times \, (1 - z)^r F(a + r, b + r; c + 2r; 1 - z). \qquad (8.70)$$

Setting,

$$n = m_2 + j_2; \quad r = j_1 + j_2 - j$$

$$a = -j_1 - j_2 - m, \quad b = -2j_2$$

$$c = -2(j_1 + j_2)$$

and using the transformation,

$$F(-j - m, j_0 - j; -2j; 1 - z)$$

$$= \frac{\Gamma(m - j_0)\Gamma(-2j)}{\Gamma(-j + m)\Gamma(-j_0 - j)} F(-j - m, j_0 - j; j_0 - m + 1; z) \qquad (8.71)$$

we obtain after some calculations,

$$z^{j_2 + m_2} = \sum_{j=j_0}^{j_1 + j_2} (-1)^{j_1 + j_2 - j} \Gamma$$

$$\times \begin{bmatrix} -j - m, j_0 - j, -j_1 - j_2 - j - 1, m - j_0 - 2j \\ -j + m, -j_0 - j, j_1 + j_2 - j + 1, \\ -j_1 - j_2 - m - 2j - 1, -2j_2 \end{bmatrix}$$

$$\times \, {}_3F_2 \begin{bmatrix} -j_2 - m_2, -j_1 - j_2 - j - 1, j - j_1 - j_2 \\ -j_1 - j_2 - m, -2j_2 \end{bmatrix} (1 - z)^{j_1 + j_2 - j}$$

$$\times \, F(-j - m, j_0 - j; j_0 - m + 1; z). \qquad (8.72)$$

We shall now use the identity

$$_3F_2\begin{bmatrix} a,b,c \\ d,e \end{bmatrix} = (-1)^c\Gamma\begin{bmatrix} s,1-b,d,e \\ s+c,1+c-b,d-c,e-c \end{bmatrix} {}_3F_2\begin{bmatrix} c,d-b,e-b \\ s+c,1+c-b \end{bmatrix}$$

(8.73)

$$s = d+e-a-b-c$$

which is valid when c is a negative integer. We finally obtain after simplification,

$$z^{j_2+m_2} = \sum_{j=j_0}^{j_1+j_2} \Gamma\begin{bmatrix} m-j_0, j_0-j, -j_1-j_2-j-1, m_1+j_1+1, -2j \\ j+m+1, -j_0-j, -j+m, j_1-j-m_2+1, -2j-1 \end{bmatrix}$$

$$(-)^{j_2+m_2} {}_3F_2\begin{bmatrix} -j_2-m_2, -j-m, j_0-j \\ j_0-m+1, j_1-j-m_2+1 \end{bmatrix}(1-z)^{j_1+j_2-j}$$

$$\times F(-j-m, j_0-j; j_0-m+1; z).$$

(8.74)

From the above equation we get the C-G series

$$D_{j_1} \times D_{j_2} = \sum_{j_0}^{j_1+j_2} D_j$$

as well as the normalizer

$$\left| \frac{N_{jm}}{N_{j_1 m_1} N_{j_2 m_2}} \right| = \left\{ \Gamma\begin{bmatrix} m-j_0, j_0-j, -j_1-j_2-j-1, \\ j_2+m_2+1, -2j, m-j_0 \\ -j_2+m_2, -j_0-j, -j+m, j+m+1, \\ -j_1+m_1, 1+j_1+j_2-j \end{bmatrix} \right\}^{\frac{1}{2}}.$$

(8.75)

We, therefore, obtain ignoring a phase

$$C\begin{pmatrix} j_1 & j_2 & j \\ m_1 & m_2 & m \end{pmatrix} = \Gamma\begin{bmatrix} 1+j_1+j_2-j \\ m_2+j_2+1, 1-j+j_1-m_2 \end{bmatrix}$$

$$\times \left\{ \Gamma\begin{bmatrix} m-j_0, m-j_0, j_0-j, -j_1-j_2-j-1, \\ m_2+j_2+1, -2j \\ m+j+1, -j_2+m_2, -j_1+m_1, -j_0-j, \\ -j+m, 1+j_1+j_2-j, -2j-1 \end{bmatrix} \right\}^{\frac{1}{2}}$$

$$\times {}_3F_2\begin{bmatrix} -j-m, j_0-j, -j_2-m_2 \\ j_0-m+1, 1+j_1-j-m_2 \end{bmatrix}.$$

(8.76)

The analysis based on the hypergeometric function presented above is due to Majumdar.[4] The advantage of this method is that it can be generalized to cover the infinite dimensional representations of the $SU(1,1)$ group.

8.5   The Clebsch–Gordan Coefficients in the Coherent State Basis[5]

The $SU(2)$ coherent states are defined by

$$f_\rho^j = (1 + |\rho|^2)^{-j} e^{\rho J_+} f_{-j}^j \tag{8.77}$$

where ρ is a complex number specifying the state and f_m^j are the canonical states introduced above.

The coherent states constitute an overcomplete set with a measure where

$$(f, g) = \int (f, f_\rho^j)(f_\rho^j, g) d\mu(\rho) \tag{8.78a}$$

where

$$d\mu(\rho) = \frac{2j + 1}{\pi} \frac{d^2\rho}{(1 + |\rho|^2)^2}. \tag{8.78b}$$

The $SU(2)$ coherent state can now be easily constructed using Eqs. (8.77) and (8.37). Thus

$$f_\sigma^j = (1 + |\sigma|^2)^{-j} (1 + \sigma z)^{2j}. \tag{8.79}$$

Hence the coherent state basis in the product representation is given by

$$f_{\sigma_1 \sigma_2}^{f_1 f_2} = (1 + |\sigma_1|^2)^{-j_1} (1 + |\sigma_2|^2)^{-j_2} (1 + \sigma_1 z_1)^{2j_1} (1 + \sigma_2 z_2)^{2j_2}. \tag{8.80}$$

The normalized coupled state in this basis is given by,

$$f_{j\sigma}^{j_1 j_2} = (1 + |\sigma|^2)^{-j} e^{\sigma J_+} f_{j,-j} \tag{8.81}$$

where f_{jm} is the orthonormal coupled states as given by Eq. (8.67).

Since by Eq. (8.67)

$$f_{j,-j} = a_j (z_1 - z_2)^{j_1 + j_2 - j} \tag{8.82}$$

where a_j is the normalizer,

$$a_j = \left[\frac{(2j_1)!(2j_2)!(2j + 1)!}{(j_0 + j)!(-j_0 + j)!(j_1 + j_2 - j)!(j_1 + j_2 + j + 1)!} \right]^{\frac{1}{2}}. \tag{8.83}$$

The normalizer a_j is obtained by directly integrating

$$\iint |z_1 - z_2|^{2j_1+2j_2-2j} d\lambda(z_1) d\lambda(z_2)$$

where

$$d\lambda(z_1) = \frac{(2j_1+1)}{\pi}(1+|z_1|^2)^{-2j_1-2} d^2 z_1, \text{ etc.}$$

Thus using Eq. (8.51) we have

$$f_{j\sigma}^{j_1 j} = a_j(1+|\sigma|^2)^{-j}(1+\sigma z_1)^{j_0+j}(1+\sigma z_2)^{-j_0+j}(z_1-z_2)^{j_1+j_2-j}$$

$$j_0 = j_1 - j_2. \tag{8.84}$$

The Clebsch–Gordan coefficient is the overlap,

$$C\begin{pmatrix} j_1 & j_2 & j \\ \sigma_1 & \sigma_2 & \sigma \end{pmatrix} = N_j \int (1+\overline{\sigma_1}\,\overline{z_1})^{2j_1}(1+\overline{\sigma_2}\,\overline{z_2})^{2j_2}(1+\sigma z_1)^{j_0+j}$$

$$\times (1+\sigma z_2)^{-j_0+j}(z_1-z_2)^{j_1+j_2-j}$$

where

$$N_j = a_j(1+|\sigma_1|^2)^{-j_1}(1+|\sigma_2|^2)^{-j_2}(1+|\sigma|^2)^{-j}.$$

The first two factors in the integral is the reproducing kernel (with z replaced by $\overline{\sigma_1}$ and z' replaced by $\overline{\sigma_2}$). Thus

$$C\begin{pmatrix} j_1 & j_2 & j \\ \sigma_1 & \sigma_2 & \sigma \end{pmatrix} = N_j(1+\sigma\overline{\sigma_1})^{j_0+j}(1+\sigma\overline{\sigma_2})^{-j_0+j}(\overline{\sigma_1}-\overline{\sigma_2})^{j_1+j_2-j}.$$

$$\tag{8.85}$$

The CG series expressed in terms of the representation matrices has the traditional appearance:

$$t_{\rho_1\sigma_1}^{j_1}(u)t_{\rho_2\sigma_2}^{j_2}(u)$$

$$= \sum_{j=|j_0|}^{j_1+j_2} \int \overline{C\begin{pmatrix} j_1 & j_2 & j \\ \sigma_1 & \sigma_2 & \sigma \end{pmatrix}} C\begin{pmatrix} j_1 & j_2 & j \\ \rho_1 & \rho_2 & \rho \end{pmatrix} t_{\rho\sigma}^j(u)d\mu(\rho)d\mu(\sigma).$$

$$\tag{8.86}$$

The integral of the product of three representation functions over the group is the product of two Clebsch–Gordan coefficients

$$\int t_{\rho_1\sigma_1}^{j_1}(u)t_{\rho_2\sigma_2}^{j_2}(u)\overline{t_{\rho\sigma}^j(u)}d\mu(u) = \overline{C\begin{pmatrix} j_1 & j_2 & j \\ \sigma_1 & \sigma_2 & \sigma \end{pmatrix}} C\begin{pmatrix} j_1 & j_2 & j \\ \rho_1 & \rho_2 & \rho \end{pmatrix}.$$

$$\tag{8.87}$$

We can also introduce the irreducible tensor operator in this basis

$$T_\rho^j = \sum_{m=-j}^{j} \left(f_m^j, f_\rho^j \right) T_m^j \tag{8.88}$$

where T_m^j stands for irreducible tensor operator in the canonical basis. T_ρ^j has the expected transformation property

$$T_u T_\rho^j T_u^\dagger = \int t_{\sigma\rho}^j(u) T_\sigma^j d\mu(\sigma) \tag{8.89}$$

and satisfies the Wigner–Eckart theorem,

$$\left(f_\beta^{j'}, T_\rho^J f_\alpha^j \right) = C \begin{pmatrix} j & J & j' \\ \alpha & \rho & \beta \end{pmatrix} \langle j' \| T^J \| j \rangle. \tag{8.90}$$

8.6 A Short Introduction to SU(3) Group and the Octet Model

8.6.1 *Tensors of* SU(2) *and* SU(3)

We first recapitulate some basic facts about the SU(2) group. Since in SU(2), tr$u = \alpha + \bar\alpha$ is real, the complex conjugate representation is equivalent to the unconjugated and there is, therefore, no need for tensors of SU(2) to incorporate indices that transform according to the complex conjugate representation. The tensors of SU(2) can, therefore, be written

$$\xi^{p_1 p_2 \cdots p_n} p_1, p_2 \cdots p_n = 1, 2. \tag{8.91}$$

There is only one invariant tensor of SU(2), namely,

$$\epsilon^{p_r p_s}; \quad p_r, p_s = 1, 2.$$

the two-dimensional Levi-civita symbol.

Thus an irreducible tensor of SU(2) must satisfy

$$\epsilon^{p_r p_s} \xi^{p_1 p_2 \cdots p_r \cdots p_s \cdots p_n} = 0 \tag{8.92}$$

and is, therefore, symmetric with respect to all the indices.

The number of independent components of the symmetric tensor of rank n, i.e. the dimension of the IR according to which it transforms is given by Bose–Einstein statistics[6]

$$\dim(\text{IR}) = \frac{(n+d-1)!}{d!n!} \tag{8.93}$$

where d is the dimension of the space ($d = 2$ for SU(2), $d = 3$ for SU(3) etc). Thus

$$\dim(\mathrm{IR})_{SU(2)} = (n + 1); \quad n = 2j. \tag{8.94}$$

This property is not shared by tensors of SU(3) for which the unconjugated and conjugated representations are inequivalent.

The group SU(3) consists of all unitary unimodular transformations in the three-dimensional (complex) vector space C_3 over the complex numbers,

$$\xi^i, \quad i = 1, 2, 3. \tag{8.95}$$

Let us introduce its complex conjugate

$$\xi_i = \bar{\xi}^i. \tag{8.96}$$

Under a transformation of the group the vectors ξ^i, ξ_i transform as

$$\xi'^i = \sum_j u^{ij}\xi^j \quad \xi'_i = \sum \bar{u}^{ij}\xi^j = \sum (u^{-1})^{ji}\xi_j. \tag{8.97}$$

The most general tensor of SU(3), therefore, consists of a set of upper indices and a set of lower indices

$$\mathrm{A}^{\alpha\beta\dots}_{ij\dots} \quad \begin{array}{l} \alpha, \beta \dots = 1, 2, 3 \\ i, j \dots = 1, 2, 3. \end{array} \tag{8.98}$$

Very special tensors are

$$\epsilon_{\alpha\beta r}, \epsilon^{ijk}, \delta^i_\alpha, \tag{8.99}$$

which are the invariant tensors of SU(3). Let us consider now

$$\mathrm{A}^{\alpha\beta\dots\delta}_{ij\dots l}, \tag{8.100}$$

with p upper and q lower indices. The previous procedure (on SU(2)) then immediately requires that for irreducibility the tensor (8.100) must be

(a) symmetric in p-upper indices
(b) symmetric in q-lower indices
(c) traceless, i.e. $\delta^i_\alpha \, \mathrm{A}^{\alpha\beta\dots\delta}_{ij\dots l} = 0$

The dimension of IR can again be found from B.E. statistics, Eq. (8.93). Thus the number of components w.r.t. upper and lower indices,

$$\mathrm{N}_1 = \frac{1}{2}(p + 1)(p + 2)$$

$$\mathrm{N}_2 = \frac{1}{2}(q + 1)(q + 2).$$

Number of conditions given by tracelessness

$$N_3 = \frac{1}{2}p(p+1)q(q+1).$$

Thus

$$\dim(\text{IR})_{\text{SU}(3)} = N_1 N_2 - N_3 = \frac{1}{2}(p+1)(q+1)(p+q+2). \qquad (8.101)$$

The following representations will be useful

$(1,0) = \underline{3}; (0,1) = \underline{3}^*$ (quark model)

$(1,1) = \underline{8}$ (pseudoscalar meson and baryon octet)

$(3,0) = \underline{10}$ (baryon decuplet).

8.6.2   *Generators of* SU(3) *in the fundamental representation — the Gell-Mann basis*

For a unitary representation

$$T_g = U = e^{i\text{H}} \qquad (8.102a)$$

$$\det U = e^{i\text{TrH}} \qquad (8.102b)$$

so that the unimodularity condition yields,

$$\text{TrH} = 0. \qquad (8.103)$$

Thus the generators of SU(3) are traceless matrices. Around the identity we may write,

$$u_{ij} = \delta_{ij} + \epsilon_{ij}. \qquad (8.104)$$

Using the unitarity condition we obtain

$$\epsilon_{ij} = -\bar{\epsilon}_{ji}$$

where we have retained terms linear in ϵ_{ij}.

Setting $j = i$,

$$\epsilon_{ii} = -\overline{\epsilon_{ii}} \qquad (8.105)$$

and the diagonal terms are pure imaginary.

Further

$$\text{Tr}\epsilon = 0. \qquad (8.106)$$

Hence for infinitesimal transformation,

$$u = \begin{pmatrix} 1 + \epsilon_{11} & \epsilon_{12} & \epsilon_{13} \\ -\overline{\epsilon_{12}} & 1 + \epsilon_{22} & \epsilon_{23} \\ -\overline{\epsilon_{13}} & -\overline{\epsilon_{23}} & 1 - \epsilon_{11} - \epsilon_{22} \end{pmatrix}. \qquad (8.107)$$

Thus writing $\epsilon_{12} = a + ib$, etc.

$$u = I + \begin{pmatrix} \epsilon_{11} & 0 & 0 \\ 0 & \epsilon_{22} & 0 \\ 0 & 0 & -\epsilon_{11} - \epsilon_{22} \end{pmatrix} + ia \begin{pmatrix} 0 & -i & 0 \\ i & 0 & 0 \\ 0 & 0 & 0 \end{pmatrix} + ib \begin{pmatrix} 0 & 1 & 0 \\ 1 & 0 & 0 \\ 0 & 0 & 0 \end{pmatrix} + \cdots .$$

Identifying the first three generators with the three components of isospin (SU(2)) and writing

$$F_k = \frac{\lambda_k}{2}; \quad F_1 = I_1, \quad F_2 = I_2, \quad F_3 = I_3.$$

$$\lambda_1 = \begin{pmatrix} 0 & 1 & 0 \\ 1 & 0 & 0 \\ 0 & 0 & 0 \end{pmatrix}; \quad \lambda_2 = \begin{pmatrix} 0 & -i & 0 \\ i & 0 & 0 \\ 0 & 0 & 0 \end{pmatrix}$$

$$\lambda_4 = \begin{pmatrix} 0 & 0 & 1 \\ 0 & 0 & 0 \\ 1 & 0 & 0 \end{pmatrix}; \quad \lambda_5 = \begin{pmatrix} 0 & 0 & -i \\ 0 & 0 & 0 \\ i & 0 & 0 \end{pmatrix}$$

$$\lambda_6 = \begin{pmatrix} 0 & 0 & 0 \\ 0 & 0 & 1 \\ 0 & 1 & 0 \end{pmatrix}; \quad \lambda_7 = \begin{pmatrix} 0 & 0 & 0 \\ 0 & 0 & -i \\ 0 & i & 0 \end{pmatrix}. \tag{8.108}$$

To obtain the diagonal generators λ_3 and λ_8 we write

$$\epsilon_{11} = i(y + x); \quad \epsilon_{22} = i(y - x).$$

Thus

$$\begin{pmatrix} \epsilon_{11} & 0 & 0 \\ 0 & \epsilon_{22} & 0 \\ 0 & 0 & -\epsilon_{11} - \epsilon_{22} \end{pmatrix} = ix \begin{pmatrix} 1 & 0 & 0 \\ 0 & -1 & 0 \\ 0 & 0 & 0 \end{pmatrix} + iy \begin{pmatrix} 1 & 0 & 0 \\ 0 & 1 & 0 \\ 0 & 0 & -2 \end{pmatrix}$$

so that

$$\lambda_3 = \begin{pmatrix} 1 & 0 & 0 \\ 0 & -1 & 0 \\ 0 & 0 & 0 \end{pmatrix} \tag{8.109a}$$

$$\lambda_8 = \frac{1}{\sqrt{3}} \begin{pmatrix} 1 & 0 & 0 \\ 0 & 1 & 0 \\ 0 & 0 & -2 \end{pmatrix}. \tag{8.109b}$$

The factor $\frac{1}{\sqrt{3}}$ is appended to render the hypercharge integral for a class of representations which includes octet and decuplet.

The Lie algebra of SU(3) in the Gell-Mann basis is written as

$$[F_\kappa, F_\mu] = i f_{\kappa\mu\nu} F_\nu$$

where summation over the repeated index is assumed. In addition, there exist in the fundamental three-dimensional representation, anti commutation relation of the form

$$\{F_\kappa, F_\mu\} = d_{\kappa\mu\nu} F_\nu + \frac{1}{3}\delta_{\kappa\mu} I \tag{8.110}$$

for $F_\kappa = \frac{\lambda_\kappa}{2}$. Tables of $d_{\kappa\mu\nu}$ are available in Gell-Mann and Neeman's Eightfold way. The above yields

$$\mathrm{Tr}(F_\kappa F_\mu) = \frac{1}{2}\delta_{\kappa\mu}. \tag{8.111}$$

We now prove that the structure constants $f_{k\mu\nu}$ are fully antisymmetric in all indices. Antisymmetry in the first two indices is quite trivial. Now we shall prove the important equality

$$\mathrm{Tr}(F_\kappa[F_\mu, F_\nu]) = \frac{i}{2} f_{\kappa\mu\nu}. \tag{8.112}$$

The l.h.s.

$$= \mathrm{Tr}\left(\frac{F_\kappa F_\mu}{A}\frac{F_\nu}{B} - \frac{F_\kappa}{A}\frac{F_\nu F_\mu}{B}\right)$$

$$= \mathrm{Tr}\left(F_\nu F_\kappa F_\mu - F_\nu F_\mu F_\kappa\right)$$

$$= \mathrm{Tr}(F_\nu[F_\kappa, F_\mu]) = i f_{\kappa\mu\lambda}\mathrm{Tr}(F_\nu F_\lambda)$$

$$= i f_{\kappa\mu\lambda}\frac{\delta_{\nu\lambda}}{2} \quad \text{(using Eq. (8.111))}.$$

Thus

$$\mathrm{Tr}(F_\kappa[F_\mu, F_\nu]) = \frac{i}{2} f_{\kappa\mu\nu}.$$

It now immediately follows that $f_{\kappa\mu\nu}$ is antisymmetric in μ, ν so that it is fully antisymmetric in all the indices and therefore there are only a limited number of them to be evaluated.

8.6.3   *Quantum numbers of states (state labels in an* IR*)*

From Galois' theorem[7] it follows that every symmetric tensor of p indices can be associated with a homogeneous polynomial of degree p. Every IR labeled by (p, q) can, therefore, be associated with a homogeneous polynomial of degree p in (ξ^1, ξ^2, ξ^3) and of degree q in (ξ_1, ξ_2, ξ_3).

Generators of SU(3) can be obtained as in SU(2) from the group action.

$$T_u f(\xi^1, \xi^2, \xi^3; \xi_1, \xi_2, \xi_3) = f(\xi'^1, \xi'^2, \xi'^3; \xi_1', \xi_2', \xi_3')$$

where writing

$$(\xi^1, \xi^2, \xi^3) \equiv \xi$$

$$\xi' = \xi u, \quad \bar{\xi}' = \bar{\xi}\bar{u}$$

$$\bar{\xi} = (\xi_1, \xi_2, \xi_3) \text{ etc.}$$

For getting $F_1 = I_1$ we identify

$$u = \left[I + \frac{i\epsilon}{2}\lambda_1\right].$$

Thus

$$I_1 = \frac{1}{2}\left(\xi^1\frac{\partial}{\partial\xi^2} + \xi^2\frac{\partial}{\partial\xi^1}\right) - \frac{1}{2}\left(\xi_1\frac{\partial}{\partial\xi_2} + \xi_2\frac{\partial}{\partial\xi_1}\right)$$

$$I_2 = -\frac{i}{2}\left(\xi^1\frac{\partial}{\partial\xi^2} - \xi^2\frac{\partial}{\partial\xi^1}\right) - \frac{i}{2}\left(\xi_1\frac{\partial}{\partial\xi_2} - \xi_2\frac{\partial}{\partial\xi_1}\right)$$

$$I_3 = \frac{1}{2}\left(\xi^1\frac{\partial}{\partial\xi^1} - \xi^2\frac{\partial}{\partial\xi^2}\right) - \frac{1}{2}\left(\xi_1\frac{\partial}{\partial\xi_1} - \xi_2\frac{\partial}{\partial\xi_2}\right) \qquad (8.113)$$

$$Y = \frac{2}{\sqrt{3}}F_8 = \frac{1}{3}\left(\xi^1\frac{\partial}{\partial\xi^1} + \xi^2\frac{\partial}{\partial\xi^2}\right) - \frac{1}{3}\left(\xi_1\frac{\partial}{\partial\xi_1} + \xi_2\frac{\partial}{\partial\xi_2}\right)$$

$$- \frac{2}{3}\left(\xi^3\frac{\partial}{\partial\xi^3} - \xi_3\frac{\partial}{\partial\xi_3}\right). \qquad (8.114)$$

We now introduce

$$\frac{\xi^1}{\xi^3} = z_1, \quad \frac{\xi^2}{\xi^3} = z_2, \xi^3 = u$$

$$\frac{\xi_1}{\xi_3} = \overline{z_1}, \frac{\xi_2}{\xi_3} = \overline{z_2}, \xi_3 = \bar{u}. \qquad (8.115)$$

Using the condition of homogeneity we have

$$f(\xi^1, \xi^2, \xi^3; \xi_1, \xi_2, \xi_3) = u^p\,\bar{u}^q g(z_1, z_2; \overline{z_1}, \overline{z_2}). \qquad (8.116)$$

Applying chain rule we obtain the generators in the space of functions $g(z) \equiv g(z_1, z_2; \bar{z}_1, \bar{z}_2)$.

Thus

$$I_1 = \frac{1}{2}\left(z_1\frac{\partial}{\partial z_2} + z_2\frac{\partial}{\partial z_1}\right) - \frac{1}{2}\left(\bar{z}_1\frac{\partial}{\partial \bar{z}_2} + \bar{z}_2\frac{\partial}{\partial \bar{z}_1}\right)$$

$$I_2 = -\frac{i}{2}\left(z_1\frac{\partial}{\partial z_2} - z_2\frac{\partial}{\partial z_1}\right) - \frac{i}{2}\left(\bar{z}_1\frac{\partial}{\partial \bar{z}_2} - \bar{z}_2\frac{\partial}{\partial \bar{z}_1}\right) \quad \text{etc.}$$

$$Y = \left(z_1\frac{\partial}{\partial z_1} + z_2\frac{\partial}{\partial z_2}\right) - \left(\bar{z}_1\frac{\partial}{\partial \bar{z}_1} + \bar{z}_2\frac{\partial}{\partial \bar{z}_2}\right) - \frac{2}{3}(p-q). \quad (8.117)$$

The form of I_1, I_2, I_3 suggests they resemble addition of two angular momenta (isospin) say i_1 and i_2 so that $g(z)$ is a homogeneous polynomial of degree $2i_1$ in z_1, z_2 and $2i_2$ in $\bar{z}_1, \bar{z}_2$. Thus

$$\left(z_1\frac{\partial}{\partial z_1} + z_2\frac{\partial}{\partial z_2}\right)g(z) = 2i_1 g(z)$$

$$\left(\bar{z}_1\frac{\partial}{\partial \bar{z}_1} + \bar{z}_2\frac{\partial}{\partial \bar{z}_2}\right)g(z) = 2i_2 g(z)$$

$$(8.118)$$

(see Eqs. (8.34), (8.35) of SU(2)).

Thus

$$Y = 2(i_1 - i_2) - \frac{2}{3}(p-q). \quad (8.119)$$

Evidently $2i_1 \leq p$, $2i_2 \leq q$. We, therefore, finally obtain Majumdar's formula for SU(3)^3,

$$|i_1 - i_2| \leq I \leq i_1 + i_2 \quad -I \leq I_3 \leq I \quad (8.120)$$

$$Y = 2(i_1 - i_2) - \frac{2}{3}(p-q) \quad p \geq 2i_1, q \geq 2i_2;$$

$$i_1 = 0, \frac{1}{2}\cdots\frac{p}{2}, \quad i_2 = 0, \frac{1}{2}\cdots\frac{q}{2}.$$

From this formula the state labels I, I_3, Y can be calculated for any IR(p,q). For example for the octet representation $p = 1, q = 1$,

$$i_1 = \frac{1}{2}, 0; \quad i_2 = \frac{1}{2}, 0; \quad \text{taking } i_1 = \frac{1}{2} = i_2$$

$$i_1 = \frac{1}{2}, \quad i_2 = 0, \quad I = 1, 0$$

$$I = \frac{1}{2}, \quad I_3 = \pm\frac{1}{2}, \quad I_3 = \pm 1, 0$$

$$Y = 1, \quad i_1 = 0, \quad i_2 = \frac{1}{2}$$

$$Y = -1, \quad I = \frac{1}{2}, \quad I_3 = \pm\frac{1}{2}$$

$$i_1 = i_2 = 0, \quad Y = 0, \quad I = I_3 = 0.$$

The weight-space plot is given below

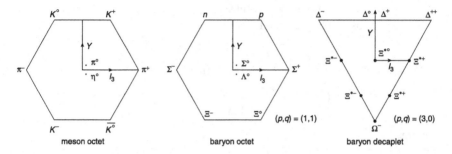

meson octet   baryon octet   baryon decaplet

8.6.4 *Application to mass formula*

We start from the Wigner–Eckart theorem,

$$\left\langle \begin{matrix} \mu \\ \lambda \end{matrix} \middle| T^{\mu_2}_{\lambda_2} \middle| \begin{matrix} \mu_1 \\ \lambda_1 \end{matrix} \right\rangle = \sum_r C_r \begin{pmatrix} \mu_1 & \mu_2 & \mu \\ \lambda_1 & \lambda_2 & \lambda \end{pmatrix} \langle \mu \| T^{\mu_2} \| \mu_1 \rangle_r .$$

If we write

$$H = H_0 + H_I$$

where H_0 is invariant under SU(3) but H_I is not. In the first order

$$M \approx \lim_{p \to 0} \left\langle \begin{matrix} p \\ M \end{matrix} ; \alpha \middle| H_0 \middle| \begin{matrix} p \\ M \end{matrix} ; \alpha \right\rangle + \lim_{p \to 0} \left\langle \begin{matrix} p \\ M \end{matrix} ; \alpha \middle| H_I \middle| \begin{matrix} p \\ M \end{matrix} ; \alpha \right\rangle$$

where α are SU(3) quantum numbers.

According to Gell-Mann's conjecture H_I transforms as,

$$H_I \simeq T^8_{000}.$$

For baryon $\alpha = 8$.

Thus

$$M \simeq M_0 + \left\langle \begin{matrix} 0 \\ M \end{matrix} ; \, 8 \middle| T^8_{000} \middle| \begin{matrix} 0 \\ M \end{matrix} ; 8 \right\rangle$$

Using the CG series

$$8 \times 8 = 1 + 8_1 + 8_2 + 10 + 10^* + 27.$$

The Wigner–Eckart theorem then immediately yields

$$M = M_0 + \sum_{r=1}^{2} a_r C_r \begin{pmatrix} 8 & 8 & 8 \\ II_3 Y & 000 & II_3 Y \end{pmatrix}$$

where a_r are reduced matrix elements.

Using Clebsch–Gordan coefficients

$$C_1 \begin{pmatrix} 8 & 8 & 8 \\ II_3 Y & 000 & II_3 Y \end{pmatrix} = \frac{Y}{2}$$

$$C_2 \begin{pmatrix} 8 & 8 & 8 \\ II_3 Y & 000 & II_3 Y \end{pmatrix} = \left[I(I+1) - \frac{Y^2}{4} \right]$$

we obtain

$$M = M_0 + M_1 Y + M_2 \left[I(I+1) - \frac{Y^2}{4} \right] \quad \left(\text{where } \frac{a_1}{2} = M_1 \right).$$

Eliminating M_0, M_1, M_2 we obtain the celebrated Gell-Mann Okubo mass formula,

$$2(M_N + M_\equiv) = (3M_\wedge + M_\Sigma).$$

For pseudoscalar mesons the term proportional to Y is zero. This is a consequence of the charge conjugation invariance and the fact that meson and antimeson octet are identical.

References

[1] E. P. Wigner, *Group Theory and Its Applications to Quantum Mechanics of Atomic Spectra* (Academic, New York, 1964).
[2] M. Gell-Mann and Y. Neéman, *The Eightfold Way* (New York, Benjamin, 1964).
[3] S. D. Majumdar, *J. Phys.* **A1** (1968) 203.
[4] S. D. Majumdar, *Prog. Theor. Phys.* **19**(4) (1958) 452.
[5] D. Basu and T. K. Kar, *Ann. Phys. (NY)* **207** (1991) 71.
[6] P. Majumdar, *Lectures on SERC Winter School* (Benares), (2007).
[7] P. Majumdar, *Lectures on SERC Winter School* (Benares), (2007).

CHAPTER 9

The Three-Dimensional Lorentz Group

In this chapter we consider the representation theory of the group of real unimodular matrices or equivalently that of pseudounitary unimodular matrices which are the spinor representations of the three-dimensional Lorentz Group.[1] We deal with the construction of irreducible unitary representations as well as some basic problems of harmonic analysis on this group.

The group of real matrices exhibit a more complicated structure than the corresponding group of complex matrices insofar as the former permits irreducible representations acting on analytic function spaces.

It has been shown by Gel'fand and co-workers[2] that every continuous operator that commutes with the operators of a representation is a multiple of unit operator, a property which is called operator irreducibility. However they have pointed out that the operator irreducibility does not always ensure true irreducibility for a representation may be operator irreducible although it has invariant subspaces. For instance, for a class of unitary representations (discrete series) the representation is operator irreducible although the Hilbert space of the representations breaks up into two invariant subspaces consisting of functions analytic in the upper (positive discrete) and lower (negative discrete) half-planes respectively.[3]

One of the basic problems considered in this chapter as well as in the next is the analogue of Peter–Weyl theorem for the infinite dimensional representations. This theorem for finite dimensional unitary representations of compact groups as proved in chapter 7 may be written as

$$F(e) = \sum_p N_p \mathrm{Tr}\,(C_p)$$

where the sum extends over all inequivalent irreducible representations and N_p is the dimension of the corresponding irreducible representation. Therefore, the multiplicity with which each representation appears is equal to its dimensionality.

For infinite dimensional representations however certain class of representations are excluded and N_p is replaced by a suitable measure (Plancherel measure) which differs from group to group. This problem has been solved for this group by employing the invariant concept of character of Gel'fand and Naimark in terms of the integral kernel of the group ring which defines character as a linear functional on the group manifold.

9.1   The Lorentz Group in Two Space and One Time Dimensions and Its Spinor Group

The three-dimensional Lorentz Group is the group of transformations that leaves invariant the indefinite form

$$x_3^2 - x_1^2 - x_2^2. \tag{9.1}$$

We now construct the spinor representation of this group. We introduce the 2×2 pseudounitary unimodular group of matrices

$$u = \begin{pmatrix} \alpha & \beta \\ \bar{\beta} & \bar{\alpha} \end{pmatrix}, \quad |\alpha|^2 - |\beta|^2 = 1 \tag{9.2}$$

satisfying

$$u^\dagger \sigma_3 u = \sigma_3 = u \sigma_3 \, u^\dagger \tag{9.3}$$

where

$$\sigma_3 = \begin{pmatrix} 1 & 0 \\ 0 & -1 \end{pmatrix}. \tag{9.4}$$

Let us now introduce the Hermitian matrix of the form

$$X = \begin{pmatrix} x_3 & x_2 + ix_1 \\ x_2 - ix_1 & x_3 \end{pmatrix} \tag{9.5}$$

satisfying

$$\mathrm{tr}(\sigma_3 X) = 0. \tag{9.6}$$

If we now define

$$X' = uXu^\dagger \tag{9.7}$$

it follows that

$$X'^\dagger = uX^\dagger u^\dagger = uXu^\dagger = X' \tag{9.8}$$

and

$$\text{tr}(\sigma_3 X') = \text{tr}(\sigma_3 u \mathbf{X} u^\dagger) = \text{tr}(u^\dagger \sigma_3 u \mathbf{X}) \tag{9.9}$$

$$= \text{tr}(\sigma_3 \mathbf{X}) = 0. \tag{9.10}$$

Hence X′ has the same form as X and we have

$$X' = \begin{pmatrix} x'_3 & x'_2 + ix'_1 \\ x'_2 - ix'_1 & x'_3 \end{pmatrix}. \tag{9.11}$$

Since, det $X' = \det(u X u^\dagger) = \det X$, i.e.

$$x'^2_3 - x'^2_2 - x'^2_1 = x^2_3 - x^2_2 - x^2_1. \tag{9.12}$$

The transformation

$$X \to X' = u X u^\dagger \tag{9.13}$$

is a Lorentz transformation.

We now note that the transformation Eq. (9.13) can be written as

$$\begin{pmatrix} x'_2 + ix'_1 \\ x'_2 - ix'_1 \\ x'_3 \end{pmatrix} = \begin{pmatrix} \alpha^2 & \beta^2 & 2\alpha\beta \\ \bar\beta^2 & \bar\alpha^2 & 2\bar\alpha\bar\beta \\ \alpha\bar\beta & \bar\alpha\beta & |\alpha|^2 + |\beta|^2 \end{pmatrix} \begin{pmatrix} x_2 + ix_1 \\ x_2 - ix_1 \\ x_3 \end{pmatrix}. \tag{9.14}$$

Since

$$\begin{pmatrix} x_2 + ix_1 \\ x_2 - ix_1 \\ x_3 \end{pmatrix} = \begin{pmatrix} i & 1 & 0 \\ -i & 1 & 0 \\ 0 & 0 & 1 \end{pmatrix} \begin{pmatrix} x_1 \\ x_2 \\ x_3 \end{pmatrix} \tag{9.15}$$

we have

$$x' = T^{-1} U T x \tag{9.16}$$

where

$$x = \begin{pmatrix} x_1 \\ x_2 \\ x_3 \end{pmatrix}, \quad T = \begin{pmatrix} i & 1 & 0 \\ -i & 1 & 0 \\ 0 & 0 & 1 \end{pmatrix}, \quad U = \begin{pmatrix} \alpha^2 & \beta^2 & 2\alpha\beta \\ \bar\beta^2 & \bar\alpha^2 & 2\bar\alpha\bar\beta \\ \alpha\bar\beta & \bar\alpha\beta & |\alpha|^2 + |\beta|^2 \end{pmatrix} \tag{9.17}$$

$$T^{-1} = \begin{pmatrix} \dfrac{-i}{2} & \dfrac{i}{2} & 0 \\ \dfrac{1}{2} & \dfrac{1}{2} & 0 \\ 0 & 0 & 1 \end{pmatrix}. \tag{9.18}$$

Thus the three-dimensional Lorentz transformation L is given by

$$L = T^{-1}UT. \tag{9.19}$$

Substituting T^{-1}, U, T, from Eqs. (9.17) and (9.18) we obtain

$$L = \begin{pmatrix} -\frac{1}{2}\left(\beta^2 + \bar{\beta}^2 - \alpha^2 - \bar{\alpha}^2\right) & \frac{i}{2}\left(\bar{\alpha}^2 + \bar{\beta}^2 - \alpha^2 - \beta^2\right) & i\left(\bar{\alpha}\bar{\beta} - \alpha\beta\right) \\ -\frac{i}{2}\left(\beta^2 - \bar{\beta}^2 - \alpha^2 + \bar{\alpha}^2\right) & \frac{1}{2}\left(\bar{\alpha}^2 + \bar{\beta}^2 + \alpha^2 + \beta^2\right) & \left(\alpha\beta + \bar{\alpha}\bar{\beta}\right) \\ -i\left(\bar{\alpha}\beta - \alpha\bar{\beta}\right) & \left(\bar{\alpha}\beta + \alpha\bar{\beta}\right) & |\alpha|^2 + |\beta|^2 \end{pmatrix}. \tag{9.20}$$

Hence for

$$u = \begin{pmatrix} e^{\frac{i\theta}{2}} & 0 \\ 0 & e^{\frac{-i\theta}{2}} \end{pmatrix}, \tag{9.21}$$

$$L = \begin{pmatrix} \cos\theta & \sin\theta & 0 \\ -\sin\theta & \cos\theta & 0 \\ 0 & 0 & 1 \end{pmatrix} \tag{9.22}$$

which is a pure space rotation.

Similarly for

$$u = \begin{pmatrix} \cos h\frac{\tau}{2} & -\sin h\frac{\tau}{2} \\ -\sin h\frac{\tau}{2} & \cos h\frac{\tau}{2} \end{pmatrix} \tag{9.23}$$

$$L = \begin{pmatrix} 1 & 0 & 0 \\ 0 & \cos h\tau & -\sin h\tau \\ 0 & -\sin h\tau & \cos h\tau \end{pmatrix} \rightarrow J_1 \tag{9.24}$$

which is a Lorentz boost. Similarly for

$$u = \begin{pmatrix} \cos h\frac{\tau}{2} & i\sin h\frac{\tau}{2} \\ -i\sin h\frac{\tau}{2} & \cos h\frac{\tau}{2} \end{pmatrix} \tag{9.25}$$

$$L = \begin{pmatrix} \cos h\tau & 0 & \sin h\tau \\ 0 & 1 & 0 \\ \sin h\tau & 0 & \cos h\tau \end{pmatrix} \rightarrow J_2 \tag{9.26}$$

which is also another Lorentz boost.

In the fundamental spinor representation the generators are given by

$$J_3 = \frac{\sigma_3}{2}, \quad J_2 \rightarrow \frac{i\sigma_2}{2}, \quad J_1 = \frac{i\sigma_1}{2}. \tag{9.27}$$

The Lie algebra of the $SU(1,1)$ group is given by the commutation relations

$$[J_1, J_2] = -iJ_3, \quad [J_2, J_3] = iJ_1 \tag{9.28}$$

$$[J_3, J_1] = iJ_2. \tag{9.29}$$

We shall now note that the $SU(1,1)$ group is isomorphic to the $SL(2,R)$ group. The group $SL(2,R)$ is the group of real unimodular matrices

$$g = \begin{pmatrix} a & b \\ c & d \end{pmatrix}, \quad ad - bc = 1. \tag{9.30}$$

A particular choice of the isomorphism kernel is

$$\eta = \frac{1}{\sqrt{2}} \begin{pmatrix} 1 & i \\ i & 1 \end{pmatrix} \tag{9.31}$$

so that

$$u = \eta^{-1} g \eta. \tag{9.32}$$

Thus

$$\alpha = \frac{1}{2}[(a+d) + i(b-c)] \tag{9.33}$$

$$\beta = \frac{1}{2}[(b+c) + i(a-d)]. \tag{9.34}$$

We now parametrize u as

$$u = \begin{pmatrix} e^{\frac{i\phi}{2}} & 0 \\ 0 & e^{-\frac{i\phi}{2}} \end{pmatrix} \begin{pmatrix} \cos h\frac{\tau}{2} & -\sin h\frac{\tau}{2} \\ -\sin h\frac{\tau}{2} & \cos h\frac{\tau}{2} \end{pmatrix} \begin{pmatrix} e^{\frac{i\theta}{2}} & 0 \\ 0 & e^{-\frac{i\theta}{2}} \end{pmatrix} \tag{9.35}$$

$$= \begin{pmatrix} \cos h\frac{\tau}{2} e^{\frac{i(\phi+\theta)}{2}} & -\sin h\frac{\tau}{2} e^{\frac{i(\phi-\theta)}{2}} \\ -\sin h\frac{\tau}{2} e^{\frac{-i(\phi-\theta)}{2}} & +\cos h\frac{\tau}{2} e^{\frac{-i(\phi+\theta)}{2}} \end{pmatrix}. \tag{9.36}$$

If we define the left invariant differential

$$d\omega = u^{-1} du, \quad d\omega = \begin{pmatrix} d\omega_{11} & d\omega_{12} \\ d\omega_{21} & d\omega_{22} \end{pmatrix} \tag{9.37}$$

it follows

$$dw_{11} = i\left(\frac{d\theta}{2} + \cos h\tau \frac{d\phi}{2}\right) \tag{9.38}$$

$$dw_{12} = -e^{-i\theta}\left(\frac{d\tau}{2} + i\sin h\tau \frac{d\phi}{2}\right). \tag{9.39}$$

Equating the real and imaginary parts

$$dw_{12}^r = -\sin h\tau \sin\theta \frac{d\phi}{2} - \cos\theta \frac{d\tau}{2} \tag{9.40}$$

$$dw_{12}^i = -\sin h\tau \cos\theta \frac{d\phi}{2} + \sin\theta \frac{d\tau}{2}. \tag{9.41}$$

Thus the invariant measure is given by

$$d\mu(u) = c\, dw_{11}^i dw_{12}^r dw_{12}^i = c\frac{\partial\left(\omega_{11}^i, \omega_{12}^r, \omega_{12}^i\right)}{\partial(\tau, \phi, \theta)} d\tau d\theta d\phi \tag{9.42}$$

i.e.

$$d\mu(u) = c\begin{vmatrix} 0 & -\frac{1}{2}\cos\theta & \frac{1}{2}\sin\theta \\ \frac{1}{2}\cos\tau & -\frac{1}{2}\sin\theta\sin h\tau & -\frac{1}{2}\cos\theta\sin h\tau \\ \frac{1}{2} & 0 & 0 \end{vmatrix} d\tau d\theta d\phi \tag{9.43}$$

$$= \frac{c}{8}\sin h\tau d\tau d\theta d\phi = \sin h\tau d\tau d\theta d\phi \text{ setting } c = 8. \tag{9.44}$$

9.1.1 The operators $\mathbf{J_1, J_2, J_3}$ in an irreducible unitary representations[3]

We shall now use the commutation relations, Eqs. (9.28) and (9.29) to determine the general form of the operators $J_i(i = 1, 2, 3)$ in the case of irreducible unitary representations of SU(1, 1). Of course since the group SU(1, 1) is non-compact we can no longer demand the finite dimensionality of the irreducible unitary representations.

As in the case of SU(2) we construct

$$J_\pm = J_1 \pm iJ_2 \tag{9.45}$$

so that

$$[J_3, J_+] = J_+, [J_3, J_-] = -J_- \tag{9.46}$$

$$[J_+, J_-] = -2J_3. \tag{9.47}$$

The invariant operator is given by

$$Q = J_1^2 + J_2^2 - J_3^2 \qquad (9.48)$$

so that

$$[Q, J_1] = [Q, J_2] = [Q, J_3] = 0. \qquad (9.49)$$

We now choose the compact SO(2) basis, i.e. the basis in which the generator J_3 of the SO(2) subgroup is diagonal

$$J_3 f_m^q = m f_m^q \qquad (9.50)$$

$$Q f_m^q = q f_m^q. \qquad (9.51)$$

Following Bargmann we shall assume

$$\left(f_m^q, e^{iJ_3\theta} f_m^q\right) = e^{im\theta}, \quad 0 \le \theta \le 4\pi \qquad (9.52)$$

to be single valued so that m is integral or half-integral

$$m = \varepsilon \pm n \quad \varepsilon = 0, \frac{1}{2} \quad n = 0, 1, 2 \dots. \qquad (9.53)$$

Thus we have the representations C_q^0, $C_q^{\frac{1}{2}}$.

For the representations of the universal covering group of $SU(1,1)$ (Pukanszky) however

$$0 \le \varepsilon < 1.$$

We shall however consider only single and double valued representations, $C_q^0, C_q^{\frac{1}{2}}$. Henceforth we shall suppress the superscript q in f_m^q

$$Q f_m = q f_m \qquad (9.54)$$

$$J_3 f_m = m f_m. \qquad (9.55)$$

From the commutation relation

$$[J_3, J_+] f_m = J_+ f_m \qquad (9.56)$$

we have

$$J_3 J_+ f_m = J_+ J_3 f_m + J_+ f_m = (m+1) J_+ f_m. \qquad (9.57)$$

Thus $J_+ f_m$ is an eigenvector of J_3 with the eigenvalue $m+1$

$$J_+ f_m = \beta_{m+1} f_{m+1}. \qquad (9.58)$$

Similarly

$$J_- f_m = \alpha_m f_{m-1}.$$ (9.59)

Now

$$Q = J_1^2 + J_2^2 - J_3^2 = \frac{1}{2}(J_+ J_- + J_- J_+) - J_3^2$$ (9.60)

so that we have

$$\frac{1}{2}(\alpha_m \beta_m + \alpha_{m+1} \beta_{m+1}) f_m = (q + m^2) f_m.$$ (9.61)

Similarly from the commutation relation we have

$$[J_+, J_-] = -2J_3$$ (9.62)

we have

$$\frac{1}{2}(\alpha_m \beta_m - \alpha_{m+1} \beta_{m+1}) f_m = -m f_m.$$ (9.63)

Hence we have the following equations for the determination of α_m, β_m

$$\frac{1}{2}(\alpha_m \beta_m + \alpha_{m+1} \beta_{m+1}) = q + m^2$$ (9.64)

$$\frac{1}{2}(\alpha_m \beta_m - \alpha_{m+1} \beta_{m+1}) = -m.$$ (9.65)

We shall now demand the unitarity,

$$(J_+ f_{m-1}, f_m) = (f_{m-1}, J_- f_m)$$ (9.66)

$$(f_m, f_{m'}) = \delta_{mm'} \quad \text{(orthonormality of } f_m)$$ (9.67)

which yields

$$\alpha_m = \bar{\beta}_m, \text{ i.e. } \beta_m = \overline{\alpha_m}.$$ (9.68)

The Eq. (9.68) in conjunction with Eqs. (9.64) and (9.65) therefore yield

$$\frac{1}{2}(|\alpha_m|^2 + |\alpha_{m+1}|^2) = q + m^2$$ (9.69)

$$\frac{1}{2}(|\alpha_m|^2 - |\alpha_{m+1}|^2) = -m$$ (9.70)

$$|\alpha_m|^2 = q + m^2 - m$$ (9.71)

$$|\alpha_{m+1}|^2 = q + m^2 + m.$$ (9.72)

We first take $C_q^{\frac{1}{2}}$. For this UIR (Unitary irreducible representation)

$$|\alpha_{\frac{1}{2}}|^2 = q + \frac{1}{4} - \frac{1}{2} = \left(q - \frac{1}{4}\right) > 0 \tag{9.73}$$

$$q > \frac{1}{4}, \quad \therefore \quad q = \frac{1}{4} + s^2 \tag{9.74}$$

$$-\infty < s < \infty$$

$$m = \frac{1}{2} \pm n; \quad m = \pm\frac{1}{2}, \pm\frac{3}{2} \cdots \pm\infty.$$

This is called the principal series of representation of the half integral class $C_s^{\frac{1}{2}}$.

We next take C_q^0: Setting $m = 0$

$$|\alpha_0|^2 = q > 0.$$

This condition is sufficient to make all $|\alpha_m|^2$ positive.

We shall consider in C_q^0 two representations

$$0 < q < \frac{1}{4}, \tag{9.75}$$

$$q > \frac{1}{4}. \tag{9.76}$$

For the former

$$q = \frac{1}{4} - s^2, \quad -\frac{1}{2} < s < \frac{1}{2}. \tag{9.77}$$

For the latter

$$q = \frac{1}{4} + s^2. \tag{9.78}$$

The former will be called the exceptional series of representations C_E^0 and the latter will be called principal series of representations of the integral class C_s^0.

In both the cases $m = \pm 1, \pm 2 \cdots \pm \infty$. All the representations $C_s^{\frac{1}{2}}$, C_s^0, C_E^0 are infinite dimensional.

9.1.2 *Discrete representations* $\mathbf{D}_k^{\pm}$

We have from Eqs. (9.58), (9.59) and (9.68)

$$J_+ f_m = \beta_{m+1} f_{m+1} = \bar{\alpha}_{m+1} f_{m+1} \tag{9.79}$$

$$J_- f_m = \alpha_m f_{m-1}. \tag{9.80}$$

(a) Positive discrete series D_k^+

Let us suppose m has a lower bound, i.e. a lowest value, say m_0, $\alpha_{m_0} = 0$. Then Eq. (9.70) yields,

$$-\frac{1}{2}|\alpha_{m_0+1}|^2 = -m_0 \tag{9.81}$$

$$m_0 = \frac{1}{2}|\alpha_{m_0+1}|^2 > 0. \tag{9.82}$$

Hence m_0 is a positive integer or half integer, i.e. we write.

$$m_0 = k, \quad k = \frac{1}{2}, 1, \frac{3}{2} \ldots. $$

Now

$$|\alpha_{m_0}|^2 = 0 = q + m_0^2 - m_0 = q + k^2 - k \tag{9.83}$$

$$q = k(1-k), \quad k = \frac{1}{2}, 1 \ldots. \tag{9.84}$$

The allowed value of m is

$$m = m_0, m_0 + 1 \ldots$$

$$= k, k+1 \ldots. \tag{9.85}$$

This representation is called positive discrete series D_k^+

$$m = k + n, \quad n = 0, 1, 2 \ldots. \tag{9.86}$$

Negative discrete series D_k^-:

We now assume m has an upper bound, i.e.

$$m_{\max} = \mu \tag{9.87}$$

then,

$$\alpha_{\mu+1} = 0. \tag{9.88}$$

Now

$$\frac{1}{2}(|\alpha_\mu|^2 - |\alpha_{\mu+1}|^2) = -\mu \tag{9.89}$$

$$\frac{1}{2}|\alpha_\mu|^2 = -\mu. \tag{9.90}$$

Thus $\mu < 0$, setting

$$\mu = -k, \quad k = \frac{1}{2}, 1, \frac{3}{2} \ldots \tag{9.91}$$

We have

$$q + \mu^2 + \mu = \frac{1}{2}|\alpha_{\mu+1}|^2 = 0$$

$$q = -\mu(1+\mu) = k(1-k), \quad k = \frac{1}{2}, 1, \frac{3}{2} \dots \qquad (9.92)$$

The allowed value m is

$$m = \mu, \mu - 1, \mu - 2, \dots -\infty$$

$$= -k, -k-1, -k-2, \dots -\infty. \qquad (9.93)$$

Thus

$$m = -k - n, \quad = 0, 1, 2 \dots \qquad (9.94)$$

This is the negative series D_k^-.

We shall now prove the following theorem:

Theorem. *There are no finite dimensional unitary irreducible representations of* $SU(1,1)$.

If there is such a representation, then m has an upper bound μ and a lower bound m_0. Then the allowed values of m are

$$m = \mu, \mu - 1, \mu - 2, \dots, m_0 + 1, m_0 \qquad (9.95)$$

which must be a finite set and

$$\alpha_{\mu+1} = 0, \quad \alpha_{m_0} = 0. \qquad (9.96)$$

From the condition of unitarity

$$\frac{1}{2}|\alpha_\mu|^2 = -\mu > 0. \qquad (9.97)$$

thus $\mu < 0, \mu = -k, k = \frac{1}{2}, 1, \dots$. Also since

$$\frac{1}{2}|\alpha_{m_0+1}|^2 = m_0 > 0. \qquad (9.98)$$

We have

$$\mu - m_0 = -(k + m_0) < 0. \qquad (9.99)$$

But the finite set Eq. (9.95) requires

$$\mu - N = m_0, \quad N > 0 \qquad (9.100)$$

$$\mu - m_0 = N > 0 \qquad (9.101)$$

which the condition Eq. (9.99) contradicts. Hence there is no finite dimensional unitary representations.

Summary. We therefore obtain the following families of irreducible unitary representations.

1. Principal series C_s^0:

$$q = \frac{1}{4} + s^2, \quad -\infty < s < \infty.$$

If we write $q = k(1-k), \ k = \frac{1}{2} + is, \ m = 0, \pm 1, \pm 2 \ldots$

2. Exceptional series C_E^0:

$$q = \frac{1}{4} - s^2, \quad -\frac{1}{2} < s < \frac{1}{2}$$

$$k = \frac{1}{2} + s, m = 0, \pm 1, \pm 2 \ldots$$

3. Principal series $C_s^{\frac{1}{2}}$:

$$q = \frac{1}{4} + s^2, \quad -\infty < s < \infty$$

$$k = \frac{1}{2} + is, \ m = \pm\frac{1}{2}, \pm\frac{3}{2}, \ldots$$

4. Positive discrete series $D_k^+ : k = \frac{1}{2}, 1, \frac{3}{2} \ldots$

$$q = k(1-k), \quad m = k, k+1, \ldots \infty.$$

5. Negative discrete series $D_k^- : k = \frac{1}{2}, 1, \frac{3}{2} \ldots$

$$q = k(1-k), m = -k, -k-1, \ldots -\infty$$

$$J_+ f_m^k = \sqrt{(m+k)(m-k+1)} f_{m+1}^k \tag{9.102}$$

$$J_- f_m^k = \sqrt{(m-k)(m+k-1)} f_{m-1}^k \tag{9.103}$$

where

$$k = \frac{1}{2}, 1, \frac{3}{2} \ldots \text{ for } D_k^{\pm}$$

$$k = \frac{1}{2} + is \text{ for } C_s^0, C_s^{\frac{1}{2}} \tag{9.104}$$

$$k = \frac{1}{2} + s \text{ for } C_E^0. \tag{9.105}$$

9.1.3 Explicit form of the representations in the canonical carrier space

The carrier space for all families of representations of the $SU(1,1)$ group may be taken to be the functions $\phi(\xi_1, \xi_2)$ where ξ_1, ξ_2 are spinors transforming according to the fundamental representation of $SU(1,1)$

$$(\xi_1', \xi_2') = (\xi_1, \xi_2)u \tag{9.106}$$

where

$$u = \begin{pmatrix} \alpha & \beta \\ \bar{\beta} & \bar{\alpha} \end{pmatrix}. \tag{9.107}$$

Since the fundamental representation of $SU(1,1)$ and its complex conjugate are equivalent the functions $\phi(\xi_1, \xi_2)$ are required to satisfy

$$\frac{\partial \phi}{\partial \bar{\xi}_1} = \frac{\partial \phi}{\partial \bar{\xi}_2} = 0. \tag{9.108}$$

So that $\phi(\xi_1, \xi_2)$ is an analytic function of ξ_1 and ξ_2. The generators of $SU(1,1)$ in this realization are

$$J_1 = \frac{i}{2}\left(\xi_1 \frac{\partial}{\partial \xi_2} + \xi_2 \frac{\partial}{\partial \xi_1}\right) \tag{9.109}$$

$$J_2 = \frac{1}{2}\left(\xi_1 \frac{\partial}{\partial \xi_2} - \xi_2 \frac{\partial}{\partial \xi_1}\right) \tag{9.110}$$

$$J_3 = \frac{1}{2}\left(\xi_1 \frac{\partial}{\partial \xi_1} - \xi_2 \frac{\partial}{\partial \xi_2}\right) \tag{9.111}$$

where J_3 is the space rotation and J_1, J_2 are pure Lorentz boosts.

Explicit calculation yields

$$Q = J_1^2 + J_2^2 - J_3^2 = K(1 - K) \tag{9.112}$$

where

$$K = -\frac{1}{2}\left(\xi_1 \frac{\partial}{\partial \xi_1} + \xi_2 \frac{\partial}{\partial \xi_2}\right). \tag{9.113}$$

Since K commutes with J_1, J_2, J_3, in an irreducible representation, $\phi(\xi_1, \xi_2)$ is a homogeneous function of degree - 2 k

$$\phi(\xi_1, \xi_2) = \xi_2^{-2k} f(z) \tag{9.114}$$

$$z = \frac{\xi_1}{\xi_2} \tag{9.115}$$

or equivalently

$$\phi(\xi_1, \xi_2) = \xi_1^{-2k} f(z), \quad z = \frac{\xi_2}{\xi_1}. \tag{9.116}$$

It can now be easily shown (say, taking the first definition, Eq. (9.114))

$$\frac{1 - |z|^2}{1 - |z'|^2} = |\beta z + \bar{\alpha}|^2 > 0 \tag{9.117}$$

where

$$z' = \frac{\alpha z + \bar{\beta}}{\beta z + \bar{\alpha}}. \tag{9.118}$$

Therefore, under the action of $SU(1,1)$ the complex z-plane is foliated into three orbits

(a) $|z| < 1 (\mathrm{K}_1)$
(b) $|z| > 1 (\mathrm{K}_2)$
(c) $|z| = 1 (\mathrm{K}_3)$

and each orbit is suitable for the realization of representation of a particular family (Principal, Exceptional or Discrete).

The unitary representations can be realized in the Hilbert space of function regular in K_i, $i = 1, 2, 3$ depending upon the nature of the representation $\mathrm{C}_s^0, \mathrm{C}_s^{\frac{1}{2}}, \mathrm{C}_E^0, \mathrm{D}_k^+, \mathrm{D}_k^-$ under consideration.

As in $SU(2)$ we construct

$$\mathrm{T}_u f(\xi_1, \xi_2) = f(\xi_1', \xi_2') \tag{9.119}$$

where (ξ_1', ξ_2') is given by Eq. (9.106).

Using the homogeneous nature of the function, we have

$$\mathrm{T}_u \xi_2^{-2k} f(z) = \xi_2' f(z') = \xi^{-2k} (\beta z + \bar{\alpha})^{-2k} f\left(\frac{\alpha z + \bar{\beta}}{\beta z + \bar{\alpha}}\right). \tag{9.120}$$

Setting

$$\hat{\mathrm{T}}_u = \xi_2^{2k} \mathrm{T}_u \xi_2^{-2k}. \tag{9.121}$$

We have

$$\hat{\mathrm{T}}_u f(z) = (\beta z + \bar{\alpha})^{-2k} f\left(\frac{\alpha z + \bar{\beta}}{\beta z + \bar{\alpha}}\right). \tag{9.122}$$

9.1.4 *Principal series of representations* $\mathbf{C}_s^{0,\frac{1}{2}}$ 1

We shall now prove the following theorem.

Theorem. *For the principal series of representations* $(k = \frac{1}{2} + is, m = \varepsilon \pm n, n = 0, 1, 2 \ldots)$ *an invariant Hermitian functional* (f, g) *satisfying* $(T_u f, T_u g) = (f, g)$ *does not exist unless* $f(z)$ *is defined on* K_3 (*i.e.* $|z| = 1$).

Proof. If possible let us suppose $K = K_1$ in Eq. (9.122). The invariant density $\omega(z)$, if it exists, then satisfies the functional equation

$$\tilde{\omega}(z') = \left(\frac{\beta z + \bar{\alpha}}{\bar{\beta}\bar{z} + \alpha}\right)^{1-2k} \tilde{\omega}(z) \tag{9.123}$$

where z' is given by Eq. (9.118) and

$$\tilde{\omega}(z) = \left(1 - |z|^2\right)\omega(z). \tag{9.124}$$

Setting $z = 0$ we have

$$\tilde{\omega}(\bar{\beta}\bar{t}) = \left(\frac{t}{\bar{t}}\right)^{1-2k} \tilde{\omega}(0), \quad t = \frac{1}{\alpha}. \tag{9.125}$$

Setting $\beta = 0$ we immediately obtain

$$\tilde{\omega}(0) = \omega(0) = 0. \tag{9.126}$$

So that

$$\tilde{\omega}(\bar{\beta}\bar{t}) = 0 = \omega(z). \tag{9.127}$$

Therefore there is no invariant Hermitian bilinear functional in $K_1(|z| < 1)$ for the principal series of representations. A similar result holds good for functions regular in K_2.

We now consider the space H of functions $f(z)$ regular in K_3, $|z| = 1$. The Euclidean measure $d\tau$ degenerates into the line element $d\phi$ in K_3 and $\omega(z)$ is determined by

$$\omega(z') = \left(\frac{\beta z + \bar{\alpha}}{\bar{\beta}\bar{z} + \alpha}\right)^{1-2k} \omega(z) = \left(\frac{z}{z'}\right)^{1-2k} \omega(z) \tag{9.128}$$

whence

$$\omega(z) = Cz^{2k-1}, \quad k = \frac{1}{2} + is. \tag{9.129}$$

Setting $C = \frac{1}{2\pi}$, we have

$$(f, g) = \frac{1}{2\pi} \int_0^{2\pi} \bar{f}(z)g(z)z^{2k-1}d\phi. \tag{9.130}$$

9.1.5 *Exceptional series*
$$C_E^o, k = \tfrac{1}{2} + s, -\tfrac{1}{2} < s < \tfrac{1}{2}, m = 0, \pm 1, \pm 2 \dots \pm \infty^1$$

For the exceptional series we have the following theorem.

Theorem. *For the exceptional series of representation an invariant Hermitian bilinear form (f, g) with a non-local metric exists in $K_3(|z| = 1)$.*

Since the condition of positive definiteness in Eq. (9.130) cannot be satisfied unless $(2k - 1)$ is pure imaginary, for the exceptional series we seek a Hermitian bilinear functional of the form

$$(f, g) = \int \int \overline{f(z_1)} \, g(z_2) \omega(z_1, z_2) d\phi_1 d\phi_2 \qquad (9.131)$$

which is a natural generalization of the local bilinear form considered above. The kernel $\omega(z_1, z_2)$ is now determined by

$$\frac{\omega(z_1', z_2')}{z_1' z_2'} = \left(\bar{\beta}\bar{z}_1 + \alpha\right)^{2-2k} \left(\beta z_2 + \bar{\alpha}\right)^{2-2k} \left(\frac{\beta z_1 + \bar{\alpha}}{\bar{\beta}\bar{z}_1 + \alpha}\right)^2 \frac{\omega(z_1, z_2)}{z_1 z_2}. \qquad (9.132)$$

We now define

$$\tilde{\omega}(z_1, z_2) = \frac{\omega(z_1, z_2)}{z_1 z_2} \qquad (9.133)$$

and set $\beta = 0, \alpha = e^{\frac{i\phi}{2}}$. Then

$$\tilde{\omega}(z_1 e^{i\phi}, z_2 e^{i\phi}) = e^{-2i\phi}\tilde{\omega}(z_1, z_2). \qquad (9.134)$$

Since $|z_1| = |z_2| = 1$, setting $z_1 = e^{-i\phi}$ we have

$$\tilde{\omega}(z_1, z_2) = \frac{1}{z_1^2}\tilde{\omega}\left(1, \frac{z_2}{z_1}\right) = \frac{1}{z_1^2}\tilde{\omega}_1\left(\frac{z_2}{z_1}\right). \qquad (9.135)$$

Finally setting $z_1 = -z_2 = 1$ and combining Eqs. (9.132) and (9.135) we obtain

$$\tilde{\omega}_1\left(\frac{\bar{\alpha}\bar{\beta} - \alpha\beta - 1}{\bar{\alpha}\bar{\beta} - \alpha\beta + 1}\right) = \left(\bar{\alpha}\bar{\beta} - \alpha\beta + 1\right)^{2-2k} \tilde{\omega}_1(-1). \qquad (9.136)$$

We now set

$$\alpha = \frac{e^{\frac{-i\pi}{4}}}{\sqrt{2}}(a + 1)^{\frac{1}{2}}\sqrt{z_1} \qquad (9.137)$$

$$\beta = \frac{e^{\frac{-i\pi}{4}}}{\sqrt{2}}(a - 1)^{\frac{1}{2}}\sqrt{\bar{z}_2} \qquad (9.138)$$

where

$$a = \left[1 - \frac{4z_1 z_2}{(z_2 - z_1)^2}\right]^{\frac{1}{2}} > 1 \qquad (9.139)$$

$$|z_1| = |z_2| = 1.$$

Using Eqs. (9.137), (9.138) and (9.139) we obtain

$$\bar{\alpha}\bar{\beta} - \alpha\beta + 1 = \frac{2z_1}{z_1 - z_2} \qquad (9.140)$$

$$\bar{\alpha}\bar{\beta} - \alpha\beta - 1 = \frac{2z_2}{z_1 - z_2}. \qquad (9.141)$$

Substituting Eqs. (9.140), (9.141) in Eq. (9.136) we obtain

$$\tilde{\omega}_1\left(\frac{z_2}{z_1}\right) = C\left(1 - \frac{z_2}{z_1}\right)^{2k-2}. \qquad (9.142)$$

Following Bargmann we choose C as

$$C = \frac{1}{4\pi} \frac{2^{2-2k} e^{i\pi(k-1)}}{\beta\left(\frac{1}{2}, k - \frac{1}{2}\right)}. \qquad (9.143)$$

9.1.6 The positive discrete series $\mathbf{D}_k^+ \, m = k + n; k = \frac{1}{2}, 1 \ldots$ [1]

Setting

$$d\lambda(z) = \omega(z)d^2 z. \qquad (9.144)$$

We obtain the condition of invariance of (f_1, f_2) as

$$\omega(z')\left|\frac{dz'}{dz}\right|^2 = |\beta z + \bar{\alpha}|^{-4k}\,\omega(z). \qquad (9.145)$$

Now

$$\frac{dz'}{dz} = \frac{1}{(\beta z + \bar{\alpha})^2}. \qquad (9.146)$$

Thus

$$\omega(z') = |\beta z + \bar{\alpha}|^{-4k+4}\,\omega(z). \qquad (9.147)$$

Using Eq. (9.117) we have

$$\omega(z') = \left(\frac{1 - |z|^2}{1 - |z'|^2} \right)^{-2k+2} \omega(z) \tag{9.148}$$

$$\frac{\omega(z')}{(1 - |z'|^2)^{2k-2}} = \frac{\omega(z)}{(1 - |z|^2)^{2k-2}} \tag{9.149}$$

which yields

$$\omega(z) = C(1 - |z|^2)^{+2k-2}. \tag{9.150}$$

Following Bargmann we choose the constant C as

$$C = \frac{2k - 1}{\pi}. \tag{9.151}$$

Thus the representations D_k^+ are unitary with respect to the scalar product

$$(f, g) = \int_{|z|<1} \overline{f(z)} g(z) d\lambda(z) \tag{9.152}$$

where

$$d\lambda(z) = \frac{(2k - 1)}{\pi} \left(1 - |z|^2 \right)^{2k-2} d^2 z. \tag{9.153}$$

For the universal covering group of SU(1, 1) k may assume any value $k > 0$. For $0 < k < \frac{1}{2}$ the integral in Eq. (9.152) is to be understood in the sense of its regularization (analytic continuation) as explained in Chapter 3. Thus

$$(f, g) = \frac{2k - 1}{2\pi(1 - e^{4\pi i k})} \int_{\sum} (1 - t)^{2k-2} \int_0^{2\pi} \overline{f(z)} g(z) d\theta \quad z = \sqrt{t}\, e^{i\theta} \tag{9.154}$$

where $\sum$ is a contour that starts from the origin along the positive real axis, encircles the point $+1$ once counterclockwise and returns to the origin along the positive real axis.

9.1.7 *Connection with Hardy spaces*

The carrier space of the discrete series D_k^+ will be shown to be isomorphic to the Hardy spaces. We start from the Hilbert space H_k of functions analytic within the unit disc:

$$(f,g) = \int\limits_{|z|<1} \overline{f(z)}g(z)d\lambda(z) \tag{9.155}$$

$$d\lambda(z) = \frac{2k-1}{\pi}(1-|z|^2)^{2k-2}d^2z. \tag{9.156}$$

A complete orthonormal set in H_k is given by

$$u_n(z) = N_n z^n \tag{9.157}$$

$$N_n = \left[\frac{\Gamma(2k+n)}{n!\Gamma(2k)}\right]^{\frac{1}{2}} \tag{9.158}$$

so that

$$f(z) = \sum_{n=0}^{\infty} a_n u_n(z). \tag{9.159}$$

Thus

$$\sum_{n=0}^{\infty} |a_n|^2 = (f,f) < \infty. \tag{9.160}$$

Hence the series in Eq. (9.159) is absolutely convergent and we have

$$\left|\frac{a_{n+1}}{a_n}\right| \le 1. \tag{9.161}$$

Let us now define

$$f(\theta) = \lim_{|z|\to 1} \sum_{n=0}^{\infty} a_n N_n e^{in\theta} = \sum_{n=0}^{\infty} b_n e^{in\theta} \quad \text{so that} \quad \left|\frac{b_{n+1}}{b_n}\right| \le 1. \tag{9.162}$$

We can therefore write

$$f(z) = \frac{1}{2\pi i} \oint\limits_{|z_1|=1} \frac{f(z_1)}{z-z_1}dz_1. \tag{9.163}$$

Thus

$$(f,g) = \frac{1}{(2\pi)^2} \oint\limits_{|z_1|=1} \oint\limits_{|z_2|=1} \overline{f(z_1)}g(z_2)d\bar{z}_1 dz_2 \int (\bar{z}-\bar{z}_1)^{-1}(z-z_2)^{-1}d\lambda(z). \tag{9.164}$$

Since $|z| < 1$ we easily obtain

$$\int \frac{d\lambda(z)}{(\bar{z}_1 - \bar{z})(z_2 - z)} = \frac{1}{\bar{z}_1 z_2} \int d\lambda(z) \frac{1}{\left[1 - re^{-i(\theta - \phi_1)}\right]\left[1 - re^{+i(\theta - \phi_2)}\right]}.$$

$$(9.165)$$

Simplifying we immediately obtain

$$\int \frac{d\lambda(z)}{(\bar{z}_1 - \bar{z})(z_2 - z)} = \frac{1}{e^{i(\phi_2 - \phi_1)}} F(1, 1; 2k; e^{i(\phi_1 - \phi_2)}).$$

$$(9.166)$$

Thus

$$(f.g) = \int \int \overline{f(z_1)} g(z_2) F\left(1, 1; 2k; \frac{z_1}{z_2}\right) d\phi_2 d\phi_1.$$

$$(9.167)$$

Writing

$$z_1 = e^{i\phi_1}, \quad f(e^{i\phi_1}) = f(\phi_1), \text{etc.}$$

$$(9.168)$$

we have

$$f(\phi_1) = \sum_{n=0}^{\infty} a_n e^{in\phi_1}.$$

$$(9.169)$$

This should be compared and contrasted with the Fourier expansion of a general periodic function in $L^2(0; 2\pi)$

$$\phi(\theta) = \sum_{n=-\infty}^{\infty} a_n e^{in\theta}$$

$$(9.170)$$

where n takes both positive and negative values. The function $\phi(\theta)$ unlike $f(\phi_1)$ is not a boundary function in the Hardy space.

It can be easily ascertained that the scalar product can be defined even for $0 < k < \frac{1}{2}$ using the regularized form given by Eq. (9.154). Thus the Hilbert space H_k of functions analytic within the unit disc is isomorphic to the Hardy space with a non-local metric.

9.1.8 *The Clebsch–Gordan Problem of* $SU(1, 1)^4$

Let us consider the ordered pairs $(z_1, z_2), z_1 \neq z_2$, where (z_1, z_2) are defined on the unit disc $|z| < 1$ or on the boundary $|z| = 1$. With each $SU(1, 1)$

matrix

$$u = \begin{pmatrix} \alpha & \beta \\ \beta & \alpha \end{pmatrix}, \quad |\alpha|^2 - |\beta|^2 = 1. \tag{9.171}$$

We associate the transformation

$$(z_1, z_2) \rightarrow \left(\frac{\alpha z_1 + \bar{\beta}}{\beta z_1 + \bar{\alpha}}, \frac{\alpha z_2 + \bar{\beta}}{\beta z_2 + \bar{\alpha}} \right) \tag{9.172}$$

on this space and this space is homogeneous with respect to these transformations.

We wish to associate with this space a unitary representation of the $SU(1,1)$ group. This representation will be constructed in the space of functions $f(z_1, z_2)$ such that

$$\|f\|^2 = \iint |f(z_1, z_2)|^2 d\lambda(z_1) d\lambda(z_2) \tag{9.173}$$

where $d\lambda(z)$ is defined by Eq. (9.130) for $C_s^{(0,\frac{1}{2})}$ by Eq. (9.131) for C_E^0 and by Eq. (9.156) for D_k^+. For D_k^- we shall slightly alter the description and we shall treat this representation at the end.

The Kronecker product of two irreducible unitary representations $T_u^{k_1}, T_u^{k_2}$ given by

$$T_u^{k_1} f(z_1) = (\beta z_1 + \bar{\alpha})^{-2k_1} f\left(\frac{\alpha z_1 + \bar{\beta}}{\beta z_1 + \bar{\alpha}} \right) \tag{9.174}$$

$$T_u^{k_2} f(z_2) = (\beta z_2 + \bar{\alpha})^{-2k_2} f\left(\frac{\alpha z_2 + \bar{\beta}}{\beta z_2 + \bar{\alpha}} \right) \tag{9.175}$$

will be defined by

$$T_u f(z_1, z_2) = (\beta z_1 + \bar{\alpha})^{-2k_1} (\beta z_2 + \bar{\alpha})^{-2k_2}$$

$$f\left(\frac{\alpha z_1 + \bar{\beta}}{\beta z_1 + \bar{\alpha}}, \frac{\alpha z_2 + \bar{\beta}}{\beta z_2 + \bar{\alpha}} \right). \tag{9.176}$$

To decompose the representation of Eq. (9.176) into irreducible representations we shall first choose the canonical $SO(2)$ basis.

The infinitesimal operators or the generators of the group for the representations given by Eqs. (9.174), (9.175) and (9.176) are respectively given by

$$J_+^{(1)} = J_1^{(1)} + iJ_2^{(1)} = -iz_1^2 \frac{\partial}{\partial z_1} - 2ik_1 z_1 \tag{9.177}$$

$$J_-^{(1)} = J_1^{(1)} - iJ_2^{(1)} = i\frac{\partial}{\partial z_1} \tag{9.178}$$

$$J_3^{(1)} = z_1 \frac{\partial}{\partial z_1} + k_1 \tag{9.179}$$

$$J_+^{(2)} = J_1^{(2)} + iJ_2^{(2)} = -iz_2^2 \frac{\partial}{\partial z_2} - 2ik_2 z_2 \tag{9.180}$$

$$J_-^{(2)} = i\frac{\partial}{\partial z_2} \tag{9.181}$$

$$J_3^{(2)} = z_2 \frac{\partial}{\partial z_2} + k_2 \tag{9.182}$$

$$J_+ = -i\left(z_1^2 \frac{\partial}{\partial z_1} + z_2^2 \frac{\partial}{\partial z_2}\right) - 2i\left(k_1 z_1 + k_2 z_2\right) \tag{9.183}$$

$$J_- = i\left(\frac{\partial}{\partial z_1} + \frac{\partial}{\partial z_2}\right) \tag{9.184}$$

$$J_3 = \left(z_1 \frac{\partial}{\partial z_1} + z_2 \frac{\partial}{\partial z_2} + k_1 + k_2\right). \tag{9.185}$$

The SO(2) bases for the representations Eqs. (9.174) and (9.175) are given by

$$f_{m_1}^{k_1} = N_{k_1 m_1} z_1^{m_1 - k_1} \tag{9.186}$$

$$f_{m_2}^{k_2} = N_{k_2 m_2} z_2^{m_2 - k_2} \tag{9.187}$$

while the SO(2) basis for the representation Eq. (9.176) is given by the simultaneous eigenstates of

$$J_3 = z_1 \frac{\partial}{\partial z_1} + z_2 \frac{\partial}{\partial z_2} + k_1 + k_2 \tag{9.188}$$

$$Q = J_+ J_- - J_3^2 + J_3. \tag{9.189}$$

The canonical compact basis is therefore a homogeneous function of degree $m - k_1 - k_2$ in z_1, z_2:

$$\phi(z_1, z_2) = z_1^{m - k_1 - k_2} f(z) \tag{9.190}$$

where $f(z)$, on simplification, satisfies the second order differential equation

$$z(1-z)^2\frac{d^2f}{dz^2} + (1-z)[(1-m-k_0) - z(3k_2 + k_1 - m + 1)]$$

$$\frac{df}{dz} + [k(1-k) + (m+k_0-1)(k_1+k_2) - mk_0 - 2zk_2(m-k_1-k_2)]f(z) = 0.$$
(9.191)

If we now substitute

$$f(z) = (1-z)^{k-k_1-k_2}F(z).$$
(9.192)

The function $F(z)$ satisfies the differential equation of the hypergeometric function

$$z(1-z)\frac{d^2F}{dz^2} + [(1-m-k_0) - z(2k-m-k_0+1)]\frac{dF}{dz} - (k-m)(k-k_0)F = 0.$$
(9.193)

Thus the SO(2) basis for the representation Eq. (9.176) is given by

$$\phi(z_1, z_2) = z_1^{m-k_1-k_2}(1-z)^{k-k_1-k_2}F(k-m, k-k_0; 1-m-k_0; z).$$
(9.194)

Other solutions of hypergeometric equation may also occur depending upon the representation Eqs. (9.174) and (9.175).

9.1.9 $D_{k_1}^+ \times D_{k_2}^+$

We shall first consider representations Eqs. (9.174) and (9.175) both belonging to the positive discrete series. Without loosing any generality we can choose $k_1 > k_2$ so that $k_0 = k_1 - k_2 > 0$. The Clebsch–Gorden coefficient (CGC) for a particular final k is given by

$$(1-z)^{k-k_1-k_2}F(k-m, k-k_0; 1-m-k_0; z) = \sum \frac{\Gamma(k_1+k_2-k+n)}{\Gamma(k_1+k_2-k)n!}$$

$$_3F_2\begin{bmatrix} k-m & k-k_0, & -n \\ 1-m-k_0, & 1+k-k_1-m \end{bmatrix}z^n$$
(9.195)

so that

$$C\begin{pmatrix} k_1 & k_2 & k \\ m_1 & m_2 & m \end{pmatrix} = \frac{N_{km}}{N_{k_1m_1}N_{k_2m_2}}\Gamma\begin{bmatrix} 1+k-k_1-k_2 \\ m_2-k_2+1, & 1+k-k_1-m_2 \end{bmatrix}$$

$$\times {}_3F_2\begin{bmatrix} k-m, & k-k_0, & k_2-m_2 \\ 1-m-k_0, & 1+k-k_1-m_2 \end{bmatrix}.$$
(9.196)

To get the spectrum of k values we derive the Clebsch–Gordan series. We start from the formula

$$z^n = \sum (-1)^r \frac{(a)_r (b)_r}{(c+r-1)_r r!} {}_3F_2 \begin{bmatrix} -n, & c+r-1 & -r \\ a, & b & \end{bmatrix} (1-z)^r$$

$$F(a+r, b+r; c+2r; 1-z). \tag{9.197}$$

Setting

$$n = m_2 - k_2; \quad r = k - k_1 - k_2$$
$$a = k_1 + k_2 - m,$$
$$b = 2k_2, \qquad\qquad c = 2k_1 + 2k_2.$$

We have

$$z^{m_2-k_2} = \sum_{k=k_1+k_2}^{\infty} (-1)^{k-k_1-k_2} \frac{\Gamma(k-m)\Gamma(k-k_0)\Gamma(k_1+k_2+k-1)}{\Gamma(2k-1)\Gamma(k-k_1-k_2+1)\Gamma(2k_2)}$$

$$\times \frac{1}{\Gamma(k_1+k_2-m_2)}$$

$$\times {}_3F_2 \begin{bmatrix} k_2 - m_2 & k_1+k_2+k-1 & k_1+k_2-k \\ k_1+k_2-m, & 2k_2 & \end{bmatrix}$$

$$\times (1-z)^{k-k_1-k_2} F(k-m, k-k_0; 2k; 1-z). \tag{9.198}$$

Using the transformation

$$F(k-m, k-k_0; 2k; 1-z)$$

$$= \Gamma \begin{bmatrix} k_0+m, & 2k \\ k+m, & k+k_0 \end{bmatrix} F(k-m, k-k_0; 1-m-k_0; z). \tag{9.199}$$

We obtain

$$z^{m_2-k_2} = \sum_{k=k_1+k_2}^{\infty} (-1)^{k-k_1-k_2}$$

$$\Gamma \begin{bmatrix} k-m, & k-k_0, & k_1+k_2+k-1, \\ 2k-1, & k-k_1-k_2+1, & k+m, \end{bmatrix}$$

$$\begin{matrix} k_0+m, & 2k, \\ k+k_0 & k_1+k_2-m & 2k_2 \end{matrix}$$

$$\times {}_3F_2 \begin{bmatrix} k_2-m_2, & k_1+k_2+k-1, & k_1+k_2-k \\ k_1+k_2-m, & 2k_2 & \end{bmatrix}$$

$$\times (1-z)^{k-k_1-k_2} F(k-m, k-k_0; 1-m-k_0; z). \tag{9.200}$$

We shall now use the identity

$$
{}_3F_2 \begin{bmatrix} a, & b, & c \\ d, & e \end{bmatrix} = (-1)^{+c} \Gamma \begin{bmatrix} s, & 1-b, & d & e \\ s+c, & 1+c-b, & d-c, & e-c \end{bmatrix}
$$

$$
{}_3F_2 \begin{bmatrix} c, & d-b, & e-b \\ s+c, & 1+c-b \end{bmatrix}, \quad s = d+e-a-b-c \qquad (9.201)
$$

which is valid when c is a negative integer. Hence

$$
{}_3F_2 \begin{bmatrix} k_1 + k_2 + k - 1, \; k_1 + k_2 - k, \; k_2 - m_2 \\ k_1 + k_2 - m, \qquad 2k_2 \end{bmatrix}
$$

$$
= (-1)^{-m_2 + k_2}
$$

$$
\times \Gamma \begin{bmatrix} 1 - k_1 - m_1, & 1 + k - k_1 - k_2 & k_1 + k_2 - m, & 2k_2 \\ 1 - k_0 - m, & 1 + k - k_1 - m_2, & k_1 - m_1, & k_2 + m_2 \end{bmatrix}
$$

$$
\times {}_3F_2 \begin{bmatrix} k - m, & k - k_0, & k_2 - m_2 \\ 1 - k_0 - m, & 1 + k - k_1 - m_2 \end{bmatrix}. \qquad (9.202)
$$

We finally obtain after simplification

$$
z^{m_2 - k_2} = \sum_{k=k_1+k_2}^{\infty} (-1)^{k_2 - m_2} \Gamma \begin{bmatrix} k_0 + m, & k - k_0, & k_1 + k_2 + k - 1, \\ m - k + 1, & 2k - 1, & (1) \end{bmatrix}
$$

$$
\begin{matrix} k_0 + m, & 2k & 1 + m_1 - k_1, & 1 \\ 1 + k - k_1 - m_2 & k_2 + m_2. & k_1 + m_1, & k + m, & k_0 + k \end{matrix}
$$

$$
\times {}_3F_2 \begin{bmatrix} k_2 - m_2, & k - m, & k - k_0 \\ 1 - k_0 - m, & 1 + k - k_1 - m_2 \end{bmatrix}
$$

$$
\times (1 - z)^{k - k_1 - k_2} F(k - m, k - k_0; 1 - m - k_0; z). \qquad (9.203)
$$

From this we get the CG series

$$
D_{k_1}^+ \times D_{k_2}^+ = \sum_{k=k_1+k_2}^{\infty} D_k^+ \qquad (9.204)
$$

as well as the normalizer

$$
\left| \frac{N_{km}}{N_{k_1 m_1} N_{k_2 m_2}} \right| = \left\{ \Gamma \begin{bmatrix} k_0 + m, & k_0 + m, & k - k_0, & k_1 + k_2 + k - 1, \\ m - k + 1, & 2k - 1, & k_2 + m_2, & k_1 + m_1 \end{bmatrix} \right.
$$

$$
\left. \begin{matrix} 2k, & m_2 - k_2 + 1 \\ k + m, & k + k_0, & 1 + k - k_1 - k_2 \end{matrix} \right] \right\}^{\frac{1}{2}}.
$$

$$
\qquad (9.205)
$$

We therefore obtain, ignoring a phase

$$C \begin{pmatrix} k_1 & k_2, & k \\ m_1 & m_2, & m \end{pmatrix}$$

$$= \Gamma \begin{bmatrix} 1 + k - k_1 - k_2 \\ m_2 - k_2 + 1, & 1 + k - k_1 - m_2 \end{bmatrix}$$

$$\times \Gamma \begin{bmatrix} (k_0 + m), & (k_0 + m), & k - k_0, \\ m - k + 1, & 2k - 1, & k_2 + m_2, \end{bmatrix}$$

$$\begin{matrix} k_1 + k_2 + k - 1, & 2k, & m_2 - k_2 + 1 \\ k_1 + m_1, & k + m, & k + k_0, & 1 + k - k_1 - k_2 \end{matrix} \Bigg]^{\frac{1}{2}}$$

$$\times {}_3F_2 \begin{bmatrix} k - m, & k - k_0, & k_2 - m_2 \\ 1 - m - k_0, & 1 + k - k_1 - m_2 \end{bmatrix}. \qquad (9.206)$$

9.1.10 $C_s^{0,\frac{1}{2}} \times D_{k_2}^+$

We shall now consider representations Eqs. (9.174) and (9.175) to belong to the principal series

$$C_{s_1}^{(0,\frac{1}{2})} \left(\text{i.e. } k_1 = \frac{1}{2} + is_1, |z_1| = 1 \right), \quad m_1 = \varepsilon \pm n_1, \varepsilon = 0, \frac{1}{2}$$

and positive discrete series respectively. Thus $|z_2| < 1$ and $m_2 = k_2 + n, k_2 = \frac{1}{2}, 1, \frac{3}{2} \ldots n = 0, 1, 2, \ldots$. Here $k_0 = k_1 - k_2 = $ complex and the variable $z = \frac{z_2}{z_1}$ in the hypergeometric function satisfies $|z| < 1$.

We take the solution regular at the origin,

$$\phi(z_1, z_2) = z_1^{m-k_1-k_2} (1-z)^{k-k_1-k_2} F(k - m, k - k_0; 1 - m - k_0; z). \qquad (9.207)$$

The non-normalized Clebsch–Gordan coefficient for a particular final k is given by the Taylor expansion of Eq. (9.207)

$$(1-z)^{k-k_1-k_2} F(k - m, k - k_0; 1 - m - k_0; z)$$

$$= \sum \frac{\Gamma(k_1 + k_2 - k + n)}{\Gamma(k_1 + k_2 - k)n!}$$

$$\times {}_3F_2 \begin{bmatrix} -n, & k - m, & k - k_0 \\ 1 - m - k_0, & 1 + k - k_1 - m_2 \end{bmatrix} z^n. \qquad (9.208)$$

Hence we obtain the CGC as

$$C\begin{pmatrix} k_1 & k_2 & k \\ m_1 & m_2 & m \end{pmatrix} = \frac{N_{km}(-)^{m_2-k_2}}{N_{k_1m_1}N_{k_2m_2}}\Gamma\begin{bmatrix} 1+k-k_1-k_2 \\ 1+k-k_1-m_2, & m_2-k_2+1 \end{bmatrix}$$

$$\times {}_3F_2\begin{bmatrix} k_2-m_2, & k-m, & k-k_0 \\ 1-k_0-m, & 1+k-k_1-m_2 \end{bmatrix}.$$

$$(9.209)$$

To determine the CG series we write formula Eq. (9.197) as

$$z^{m_2-k_2} = \sum(-)^r\Gamma\frac{(k_1+k_2-m+r)\Gamma(2k_2+r)\Gamma(2k_1+2k_2+r-1)}{\Gamma(k_1+k_2-m)\Gamma(2k_1+2k_2+2r-1)r!\,\Gamma(2k_2)}$$

$$\times {}_3F_2\begin{bmatrix} k_2-m_2, & 2k_1+2k_2+r-1, & -r \\ k_1+k_2-m, & 2k_2 \end{bmatrix}(1-z)^r$$

$$\times F(k_1+k_2-m+r, 2k_2+r, 2k_1+2k_2+2r; 1-z). \qquad (9.210)$$

We now introduce the meromorphic function

$$\chi(k) = \frac{\Gamma(k-m)\Gamma(k-k_0)\Gamma(k_1+k_2-k)\Gamma(k_1+k_2+k-1)}{\Gamma(2k-1)\Gamma(k_1+k_2-m)\Gamma(2k_2)}$$

$$\times {}_3F_2\begin{bmatrix} k_2-m_2, & k_1+k_2+k-1, & k_1+k_2-k \\ k_1+k_2-m, & 2k_2 \end{bmatrix}$$

$$\times (1-z)^{k-k_1-k_2}F(k-m, k-k_0 2k; 1-z). \qquad (9.211)$$

The only singularities of $\chi(k)$ on the right half of the complex k plane are simple poles at

$$k_1+k_2-k = -r$$

i.e. $k = r+k_1+k_2, \quad r = 0,1,2,\ldots\infty$ \qquad (9.212)

and for $m > 0$ simple poles at

$$k-m = -r \qquad (9.213)$$

$k = m-r, r = 0,1,2\ldots m-1$, for integer m and $m-\dfrac{3}{2}$ for half integer m.

The sum of the residues of $\chi(k)$ at the simple poles at $k = k_1+k_2+r$, $r = 0,1,2\ldots$ equals the negative of the infinite series (9.210). Thus remembering the direction of the contour,

$$\frac{1}{2\pi i} \int_{\frac{1}{2}-i\infty}^{\frac{1}{2}+i\infty} \chi(k)dk + \frac{1}{2\pi i} \int_s \chi(k)dk = z^{m_2-k_2} - \sum_{r=0}^m \operatorname{Re} s \left[\chi(k)\right]_{k=m-r}$$
(9.214)

where S is a large semicircle on the right. Using Stirling's formula and the behavior of the hypergeometric function for large $|k|$, with $\operatorname{Re} k > 0$ the integral over S can be seen to vanish. Hence

$$\frac{1}{2\pi i} \int_{\frac{1}{2}-i\infty}^{\frac{1}{2}+i\infty} \chi(k)dk = z^{m_2-k_2} - \sum_{k=1, \text{ or } \frac{3}{2}}^m D_k^+$$

where

$$D_k^+ = \frac{(-)^{m-k}\Gamma(k-k_0)\Gamma(k_1+k_2-k)\Gamma(k_1+k_2+k-1)}{\Gamma(m-k+1)\Gamma(2k-1)\Gamma(k_1+k_2-m)\Gamma(2k_2)}$$

$$\times {}_3F_2\left[\begin{array}{ccc} k_2-m_2, & k_1+k_2+k-1, & k_1+k_2-k \\ k_1+k_2-m, & 2k_2 \end{array}\right]$$

$$\times (1-z)^{k-k_1-k_2} F(k-m, k-k_0; 2k; 1-z).$$
(9.215)

Thus

$$z^{m_2-k_2} = \sum_{k=1/\frac{3}{2}}^m \frac{(-)^{(m-k)}}{(m-k)!}\Gamma^+(k) + \frac{1}{2\pi i}\int_{\frac{1}{2}-i\infty}^{\frac{1}{2}+i\infty}\chi(k)dk$$
(9.216)

where

$$\Gamma^+(k) = \frac{\Gamma(k-k_0)\Gamma(k_1+k_2-k)\Gamma(k_1+k_2+k-1)}{\Gamma(2k-1)\Gamma(k_1+k_2-m)\Gamma(2k_2)}$$

$$\times {}_3F_2\left[\begin{array}{ccc} k_2-m_2, & k_1+k_2+k-1, & k_1+k_2-k \\ k_1+k_2-m, & 2k_2 \end{array}\right]$$

$$\times (1-z)^{k-k_1-k_2} F(k-m, k-k_0; 2k; 1-z)$$
(9.217)

using the identities

$$F(k-m, k-k_0; 2k; 1-z) = \Gamma\left[\begin{array}{cc} k_0+m, & 2k \\ k+m, & k_0+k \end{array}\right]$$

$$\times F(k-m, k-k_0; 1-m-k_0; z)$$
(9.218)

valid for $k - m = -r, r = 0, 1, 2 \ldots m$ and

$$
{}_3F_2 \left[\begin{array}{ccc} k_2 - m_2, & k_1 + k_2 + k - 1, & k_1 + k_2 - k \\ k_1 + k_2 - m, & 2k_2 & \end{array} \right]
$$

$$
= \Gamma \left[\begin{array}{cc} k_0 + k, & 2k_2 \\ k_2 + m_2, & k_1 + k - m_2 \end{array} \right]
$$

$$
\times {}_3F_2 \left[\begin{array}{ccc} k - m, & k_1 + k_2 + k - 1, & k_2 - m_2 \\ k_1 + k_2 - m, & k_1 + k - m_2 & \end{array} \right] \qquad (9.219)
$$

valid for $k_2 - m_2 = -n, n = 0, 1, 2 \ldots$, we easily obtain

$$
\Gamma^+(k) = (-)^{m_2 - k_2} \Gamma \left[\begin{array}{ccc} 2k, & k - k_0, k_1 + k_2 + k - 1, & \\ 2k - 1, & k + m, & k_1 + k_2 + k - 1, \end{array} \right.
$$

$$
\left. \begin{array}{ccc} k_1 + k_2 - k, & 1 - k_0 - k, & k_0 + k \\ 1 - k_0 - m, & k_2 + m_2, & k_1 + k - m_2 \end{array} \right]
$$

$$
\times {}_3F_2 \left[\begin{array}{cc} k - m, & k_1 + k_2 + k - 1, k_2 - m_2 \\ k_1 + k_2 - m, & k_1 + k - m_2 \end{array} \right]
$$

$$
\times (1 - z)^{k - k_1 - k_2} F(k - m, k - k_0; 1 - m - k_0; z). \qquad (9.220)
$$

Folding the integral about the real axis we obtain

$$
\int_{\frac{1}{2} - i\infty}^{\frac{1}{2} + i\infty} \chi(k) dx = \int_{\frac{1}{2}}^{\frac{1}{2} + i\infty} [\chi(k) + \chi(1 - k)] \, dk. \qquad (9.221)
$$

Using the identity (behavior around $z = 1$)

$$
F(k - m, k - k_0; 1 - m - k_0; z)
$$

$$
= \frac{\Gamma(1 - m - k_0)\Gamma(1 - 2k)F(k - m, k - k_0; 2k; 1 - z)}{\Gamma(1 - k - m)\Gamma(1 - k - k_0)}
$$

$$
+ \frac{\Gamma(1 - k - k_0)\Gamma(2k - 1)(1 - z)^{1 - 2k}}{\Gamma(k - m)\Gamma(k - k_0)}.
$$

We obtain after simplification

$$
\int_{\frac{1}{2} - i\infty}^{\frac{1}{2} + i\infty} \chi(k) dk = \int_{\frac{1}{2}}^{\frac{1}{2} + i\infty} \Gamma \left[\begin{array}{cccc} k - m, & k - k_0, & 1 - k - m, & 1 - k - k_0, \\ 2k - 1, & 1 - 2k, & k_1 + k_2 - m, & 2k_2, \end{array} \right.
$$

$$
\left. \begin{array}{cc} k_1 + k_2 - k, & k_1 + k_2 + k - 1 \\ 1 - k_0 - m & \end{array} \right]
$$

$$
\times {}_3F_2 \left[\begin{array}{ccc} k_2 - m_2, & k_1 + k_2 + k - 1, & k_1 + k_2 - k \\ k_1 + k_2 - m, & 2k_2 & \end{array} \right]
$$

$$
\times (1 - z)^{k - k_1 - k_2} F(k - m, k - k_0, 1 - k_0 - m; z).
$$

$$
\qquad (9.222)
$$

Combining these results we obtain

$$z^{m_2-k_2} = \sum_{k=1(\text{or}\frac{3}{2})}^{m} (-)^{m-k} \frac{\Gamma^+(k)}{(m-k)!}$$

$$+ \frac{1}{2\pi i} \int_{\frac{1}{2}}^{\frac{1}{2}+i\infty} dk \Gamma \begin{bmatrix} k-m, & k-k_0, & 1-k-m, \\ 2k-1, & 1-2k, & k_1+k_2-m, \end{bmatrix}$$

$$\begin{matrix} 1-k-k_0, & k_1+k_2-k, & k_1+k_2+k-1 \\ 2k_2, & 1-k_0-m \end{matrix} \Bigg]$$

$$\times {}_3F_2 \begin{bmatrix} k_2-m_2 & k_1+k_2+k-1, & k_1+k_2-k \\ k_1+k_2-m & & 2k_2 \end{bmatrix}$$

$$\times (1-z)^{k-k_1-k_2} F(k-m, k-k_0; 1-m-k_0; z). \qquad (9.223)$$

To get the normalized CGC's in the SO(2) basis we use the identity (valid for $k_2 - m_2 = -n, n = 0, 1, 2, \ldots$).

$${}_3F_2 \begin{bmatrix} k_1+k_2+k-1, & k_1+k_2-k, & k_2-m_2 \\ k_1+k_2-m, & 2k_2 \end{bmatrix}$$

$$= (-)^{m_2-k_2}\Gamma \begin{bmatrix} 1-k+k_0, & 2k_2 \\ k_2+m_2, & 1+k_1-k-m_2 \end{bmatrix}$$

$$\times {}_3F_2 \begin{bmatrix} 1-k-m, & k_1+k_2-k, & k_2-m_2 \\ k_1+k_2-m, & 1+k_1-k-m_2 \end{bmatrix} \qquad (9.224)$$

so that

$$z^{m_2-k_2} = \sum \frac{(-)^{m-k}}{(m-k)!}\Gamma^+(k)$$

$$+ \frac{(-)^{m_2-k_2}}{2\pi i} \int_{\frac{1}{2}}^{\frac{1}{2}+i\infty} dk \Gamma \begin{bmatrix} k-m, & 1-k-m, & k-k_0, \\ 2k-1, & 1-2k, & k_1+k_2-m, \end{bmatrix}$$

$$\begin{matrix} k_1+k_2-k, & k_1+k_2+k-1, & 1-k_0-k, & 1-k+k_0 \\ 1-k_0-m, & k_2+m_2, & 1+k_1-k-m_2, \end{matrix} \Bigg]$$

$$\times {}_3F_2 \begin{bmatrix} 1-k-m, & k_1+k_2-k, & k_2-m_2 \\ k_1+k_2-m, & 1+k_1-k-m_2 \end{bmatrix}$$

$$\times (1-z)^{k-k_1-k_2} F(k-m, k-k_0; 1-k_0-m; z). \qquad (9.225)$$

From this we immediately obtain the CG series

$$C_{s_1}^{(0)} \times D_{k_2}^+ = \sum_{k=1(\frac{3}{2})}^{\infty} D_k^+ + \int_0^{\infty} C_s^{(\epsilon)} ds \quad \begin{matrix} \epsilon = 0 \text{ for } k_2 = \text{integral} \\ \epsilon = \frac{1}{2} \text{ for } k_2 = \text{half integral} \end{matrix}$$

$$(9.226)$$

$$C_{s_1}^{(\frac{1}{2})} \times D_{k_2}^+ = \sum_{k=1(\frac{3}{2})}^{\infty} D_k^+ + \int_0^{\infty} C_s^{(\varepsilon)} ds \begin{array}{l} \varepsilon = 0 \text{ for } k_2 = \text{half integral} \\ \varepsilon = \frac{1}{2} \text{ for } k_2 = \text{integral.} \end{array}$$

(9.227)

The normalization constant is easily obtained by comparing the direct and inverse expansions. Thus for $k \epsilon D_k^+$

$$C\begin{pmatrix} k_1 & k_2 & k \\ m_1 & m_2 & m \end{pmatrix} = \frac{N_{km}(-)^{m_2-k_2}}{N_{k_1m_1}N_{k_2m_2}} \Gamma\begin{bmatrix} 1+k-k_1-k_2 \\ 1+k-k_1-m_2, & m_2-k_2+1 \end{bmatrix}$$

$$\times (-)^{m_2-k_2} {}_3F_2\begin{bmatrix} k-m, & k-k_0, & k_2-m_2 \\ 1-k_0-m, & 1+k-k_1-m_2 \end{bmatrix}$$

$$= \left\{ \Gamma\begin{bmatrix} 2k, & k_1+k_2+k-1, & k-k_0, \\ 2k-1, & k+m, & m-k+1, & k_1+k_2-m, \end{bmatrix}\right.$$

$$\left. \begin{array}{cc} k_1+k_2-k, & 1-k_0-k \\ 1-k_0-m, & 1-k_0-m, & k_2+m_2, & m_2-k_2+1 \end{array} \right]\right\}^{\frac{1}{2}}$$

$$\times \Gamma\begin{bmatrix} 1+k-k_1-k_2 \\ 1+k-k_1-m_2 \end{bmatrix}$$

$$\times {}_3F_2\begin{bmatrix} k-m, & k-k_0, & k_2-m_2 \\ 1-k_0-m, & 1+k-k_1-m_2 \end{bmatrix}.$$

(9.228)

Similarly for $k \epsilon C_s^{\varepsilon}$

$$C\begin{pmatrix} k_1 & k_2 & k \\ m_1 & m_2 & m \end{pmatrix} = \left\{ \Gamma\begin{bmatrix} k-m, & 1-k-m, & k-k_0, \\ 2k-1, & 1-2k, & k_1+k_2-m, \end{bmatrix}\right.$$

$$\left. \begin{array}{ccc} k_1+k_2-k & k_1+k_2+k-1, & 1-k_0-k \\ 1-k_0-m, & k_2+m_2 \end{array} \right]\right\}^{\frac{1}{2}}$$

$$\times \Gamma\begin{bmatrix} 1+k-k_1-k_2 \\ 1+k-k_1-m_2, & m_2-k_2+1 \end{bmatrix}$$

$$\times {}_3F_2\begin{bmatrix} k-m, & k-k_0, & k_2-m_2 \\ 1-k_0-m, & 1+k-k_1-m_2 \end{bmatrix}.$$

(9.229)

9.1.11 $C_{s_1}^{(\varepsilon_1)} \times C_{s_2}^{(\varepsilon_2)}$

We now consider both the representations Eqs. (9.174) and (9.175) to belong to the principal series of SU(1,1). Thus $k_1 = \frac{1}{2} + is_1, m_1 = \varepsilon_1 \pm n_1, k_2 = \frac{1}{2} + is_2, m_2 = \varepsilon_2 \pm n_2, n_1 = 0, 1, 2, \ldots, n_2 = 0, 1, 2, \ldots |z_1| = |z_2| = 1$. Thus $z = z_2/z_1$ should lie on the unit circle.

We first find the solutions of Eq. (9.193) for discrete k. For $k \epsilon D_k^+$ the solution to be taken is

$$e_{km}^{(1)} = (1-z)^{k-k_1-k_2} F(k-m, k-k_0; 1-k_0-m; z). \qquad (9.230)$$

On the other hand if $k \epsilon D_k^-$

$$e_{km}^{(5)} = z^{m+k_0}(1-z)^{k-k_1-k_2} F(k+m, k+k_0; 1+m+k_0; z). \qquad (9.231)$$

If $k \epsilon C_s^\epsilon$ then any two independent solutions of Eq. (9.193) correspond to possible Clebsch–Gordan series and one is led to the problem of choosing from the infinite possibilities a pair of solutions which are orthogonal on the unit circle. The Clebsch–Gordan coefficient generated by a pair of orthonormalized basis functions are then automatically orthonormal.

When k is complex ($k = \frac{1}{2} + is$) there are, in general, two linearly independent solutions of Eq. (9.193) which must be chosen so that they satisfy the requirement of orthonormality. The pair of solutions,

$$e_{km}^{(1)} = (1-z)^{k-k_1-k_2} F(k-m, k-k_0; 1-k_0-m; z) \qquad (9.232)$$

$$e_{km}^{(2)} = (-z)^{k_0-k}(1-z)^{k-k_1-k_2} F\left(k+m, k-k_0; 1+m-k_0; \frac{1}{z}\right) \qquad (9.233)$$

as shown later satisfies

$$\left(e_{km}^{(r)}, e_{k'm}^{(s)}\right) = N_{km}^{(r)} \delta_{rs} \delta(\operatorname{Im} k - \operatorname{Im} k'). \qquad (9.234)$$

The Hilbert space therefore decomposes into two mutually orthogonal subspaces leading to a pair of orthogonal Clebsch–Gordan coefficients.

The solution $e_{km}^{(2)}$ can be expressed as a linear combination of the first and second solution of the hypergeometric equation.

$$
e_{km}^{(2)} = \Gamma \begin{bmatrix} m+k_0, & m+1-k_0 \\ m+1-k, & m+k \end{bmatrix} e_{km}^{(1)}
$$
$$
- e^{-i\pi(k_0+m)}\Gamma \begin{bmatrix} 1-m-k_0, m+k_0, & m+1-k_0 \\ m+1+k_0, & k-k_0, & 1-k-k_0 \end{bmatrix} e_{km}^{(5)}
$$
$$\qquad (9.235)$$

where $e_{km}^{(5)}$ stands for the second solution of the hypergeometric equation and is given by Eq. (9.231).

However when $k \epsilon D_k^\pm$, the values of m are bounded above or below and Eq. (9.193) admits only one independent solution which is given by Eq. (9.230) for D_k^+ and Eq. (9.231) for D_k^-. There is therefore no duplicity for representations belonging to $D_k^\pm$.

Since k_1, k_2 and m are fixed, the monomials z^{m_2} can be regarded as representing the bases in the product representation. Further m_2 takes all integer or half-integer values on the real line. The non-normalized Clebsch–Gordan coefficients are, therefore, the coefficients of Fourier expansion of

$$z^{k_2} e_{km}^{(1)}, \quad z^{k_2} e_{km}^{(2)} \tag{9.236}$$

$$z^{k_2} e_{km}^{(r)} = \sum_{m_2 = -\infty}^{\infty} a_{m_2}^{(r)} z^{m_2} \tag{9.237}$$

$$a_{m_2}^{(1)} = -\frac{1}{2\pi i} \int_s dz\, z^{k_2 - m_2 - 1} (1 - z)^{k - k_1 - k_2} F(k - m, k - k_0; 1 - m - k; z). \tag{9.238}$$

The singularities of the integrand are the branch points at $z = 0$ and $z = 1$. With the standard choice of the cut for the hypergeometric function (part 1) integrand in Eq. (9.238) is single valued and analytic in the entire z plane assumed cut along the positive real axis from zero to infinity. Since Re $k = \frac{1}{2}$ for $k \epsilon C_s^\varepsilon$ we have

$$a_{m_2}^{(1)} = \frac{1}{2\pi i} \int_c dz\, z^{k_2 - m_2 - 1} (1 - z)^{k - k_1 - k_2} F(k - m, k - k_0; 1 - m - k_0; z) \tag{9.239}$$

where C stands for the part of the contour formed by the small circle of radius ε around the origin and the part of the branch cut from ε to 1-0.

Comparing Eq. (9.238) with the standard single loop contour integral representation of the generalized hypergeometric function.

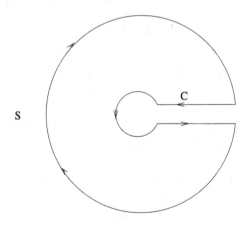

Fig. 9.1

We obtain

$$a_{m_2}^{(1)} = e^{-i\pi(m_2-k_2)}\Gamma\begin{bmatrix} k - k_1 - k_2 + 1 \\ m_2 - k_2 + 1, \quad k + 1 - m_2 - k_1 \end{bmatrix}$$

$$\times\,_3F_2\begin{bmatrix} k_2 - m_2, & k - k_0, & k - m \\ 1 - k_0 - m, & k + 1 - k_1 - m_2 \end{bmatrix}. \tag{9.240}$$

Similarly

$$a_{m_1,m_2}^{(2)} = e^{-2\pi i m_2}a_{-m_1,-m_2}^{(1)}. \tag{9.241}$$

The $_3F_2^{(1)}$ functions appearing in Eqs. (9.240) and (9.241) are absolutely convergent for all values of m_1, m_2, m.

We shall also require the function $z^{k_2}e_{km}^{(5)}$ as given by and its Fourier coefficient.

$$a_{m_2}^{(5)} = a_{m_2}^{(1)\mathcal{I}} \tag{9.242}$$

where the superscript $\mathcal{I}$ corresponds to the simultaneous interchange

$$k_1 \leftrightarrow k_2, \quad m_1 \leftrightarrow -m_2.$$

9.2   The Clebsch–Gordan Series

We start from the formula 9.210 with

$$a = k_1 + k_2 - |m|, \quad b = 2k_2, \quad c = 2(k_1 + k_2).$$

We now consider for $m > 0$ the analytic function

$$\chi(k) = \frac{\Gamma(k - m)\Gamma(k - k_0)\Gamma(k_1 + k_2 - k)(k_1 + k_2 + k - 1)}{\Gamma(2k - 1)\Gamma(k_1 + k_2 - m)\Gamma(2k_2)}$$

$$\times\,_3F_2\begin{bmatrix} k_2 - m_2, & k_1 + k_2 - k, & k_1 + k_2 + k - 1 \\ k_1 + k_2 - m & 2k_2 \end{bmatrix}$$

$$\times (1 - z)^{k-k_1-k_2}F(k - m, k - k_0; 2k; 1 - z). \tag{9.243}$$

Let us now enclose the poles at $k = k_1 + k_2 + r, r = 0, 1, 2, \ldots$ by a contour C consisting of the infinite semicircle s on the right half of the k-plane and the line $\text{Re}\,k = \frac{1}{2}$. Since C contains additional poles at $k = m - n, (n = 0, 1, 2, \ldots)$ for, $m > 0\,(m_1 < 0)$ and since the integral over the semicircular part vanishes, we have, as before

$$z^{m_2-k_2} = \frac{1}{2\pi i} \int_{\frac{1}{2}-i\infty}^{\frac{1}{2}+i\infty} \chi(k)dk + \sum_{k=1(\frac{3}{2})}^{m} D_k^+ \qquad (9.244)$$

where D_k^+ stands for the residue of $\chi(k)$ at

$$k = m - n; \quad n = 0, 1 \ldots m - 1 \text{ or } \left(m - \frac{3}{2}\right).$$

When k_1, k_2 both belong to $C_s^{(\varepsilon)}$ the double multiplicity phenomenon discussed in the preceding section makes it necessary to transform the right-hand side of Eq. (9.244) into a form involving the basis functions $e_{km}^{(1)} e_{km}^{(2)}$ and the complex conjugates of the Fourier coefficients $a^{(1)}, a^{(2)}$.

First we use the Thomae Whipple identity.

$$F_p(0,4,5,) = \Gamma \begin{bmatrix} \alpha_{045} & \alpha_{024} & \alpha_{034} \\ \alpha_{235} & \beta_{10} & 1 - \beta_{10} \end{bmatrix} F_n(4;3,5)$$

$$+ \Gamma \begin{bmatrix} \alpha_{045} & \alpha_{024} & \alpha_{034} \\ \alpha_{145} & \alpha_{124} & \alpha_{134} \end{bmatrix} F_p(1;0,3) \qquad (9.245)$$

which connects the three $_3F_2(1)$ series

$$_3F_2^{(0)} = {_3F_2} \begin{bmatrix} k_2 - m_2, & k_1 + k_2 - k, & k_1 + k_2 + k - 1 \\ k_1 + k_2 - m, & 2k_2 \end{bmatrix} \qquad (9.246)$$

$$_3F_2^{(1)} = {_3F_2} \begin{bmatrix} 1 - k_2 - m_2, & 1 - k + k_0, & 1 - k - m \\ 1 + k_0 - m, & 1 + k_1 - k - m_2 \end{bmatrix} \qquad (9.247)$$

$$_3F_2^{(2)} = {_3F_2} \begin{bmatrix} 1 - k_2 + m_2, & 1 - k + k_0, & 1 - k + m \\ 1 + k_0 + m, & 1 - k + k_1 + m_2 \end{bmatrix}. \qquad (9.248)$$

The last two occur in the complex conjugates of the Clebsch–Gordan coefficients $a_{m_2}^{(1)}, a_{m_2}^{(2)}$. In the present case we obtain

$$_3F_2^{(0)} = \Gamma \begin{bmatrix} k_1 + m_1, & 1 - k_1 - m_1, & 1 - k + k_0, \\ 1 - k - m, & k_2 - m_2, & k - m, \end{bmatrix}$$

$$\begin{matrix} k + k_0, & 2k_2 & k_1 + k_2 - m \\ k_1 + k_2 - k, & 1 + m + k_0, & k_1 + k + m_2 \end{matrix} \Bigg] \, _3F_2^{(2)}$$

$$- \Gamma \begin{bmatrix} k_1 + m_1, & 1 - k_1 - m_1, & 1 - k + k_0, & k + k_0, \\ k_2 + m_2, & 1 + m + k_0, & -k_0 - m, & k_1 + k_2 + k - 1, \end{bmatrix}$$

$$\begin{matrix} 2k_2 & k_1 + k_2 - m \\ 1 + k_0 - m, & 1 - k + k_1 - m_2 \end{matrix} \Bigg] \, _3F_2^{(1)}. \qquad (9.249)$$

The $_3F_2(1)$ functions appearing above are invariant under Legendre transformation $k \to 1 - k$, for instance

$$\Gamma \begin{bmatrix} k_1 + k_2 - k \\ 1 - k + k_1 - m_2 \end{bmatrix} {}_3F_2^{(1)}(k) = \Gamma \begin{bmatrix} k_1 + k_2 - k - 1 \\ k_1 + k - m_2 \end{bmatrix} {}_3F_2^{(1)}(1 - k).$$

(9.250)

Substituting Eqs. (9.250) and (9.249) in Eq. (9.244) folding the integral about the real axis and using the relation

$$\Gamma \begin{bmatrix} 1 - m - k_0, & 1 - 2k \\ 1 - m - k, & 1 - k_0 - k \end{bmatrix} F(k - m, k - k_0; 2k; 1 - z)$$

$$+ (1 - z)^{1 - 2k} \Gamma \begin{bmatrix} 2k - 1 \ 1 - m - k_0 \\ k - m, \quad k - k_0 \end{bmatrix}$$

$$\times F(1 - k - m, 1 - k - k_0; 2 - 2k; 1 - z)$$

$$= F(k - m, k - k_0; 1 - k_0 - m; z).$$

(9.251)

We have

$$z^{m_2 - k_2} = \sum_{k=1(\frac{3}{2})}^{m} D_k^+ - \frac{1}{2\pi i} \int_{\frac{1}{2}}^{\frac{1}{2} + i\infty} dk$$

$$\Gamma \begin{bmatrix} k_1 + m_1, & 1 - k_1 - m_1, & k_1 + k_2 - k, \\ k_2 + m_2, & k_0 + m + 1, & -k_0 - m, \end{bmatrix}$$

$$k_0 + k, \qquad k_0 + 1 - k,$$

$$k_0 - m + 1, \quad k_1 + 1 - k - m_2$$

$$\begin{matrix} k - k_0, & 1 - k - k_0, & k - m, & 1 - k - m \\ 2k - 1, & 1 - 2k & 1 - k_0 - m, \end{matrix} \Bigg] {}_3F_2^{(1)} e_{km}^{(1)}$$

$$+ \frac{1}{2\pi i} \int dk \ \Gamma \begin{bmatrix} k_1 + m_1, & 1 - k_1 - m, & k_0 + k, \\ k_2 - m_2, & k_0 + m + 1, & k + 1 - k_0, \end{bmatrix}$$

$$k_0 + 1 - k, \quad k - k_0, \quad 1 - k - k_0, \quad k_1 + k_2 + k - 1,$$

$$1 - k_0 - m, \quad k_0 + m, \quad k_1 + k + m_2, \qquad 2k - 1,$$

$$\begin{matrix} k + m, & 1 - k + m \\ 1 - 2k \end{matrix} \Bigg] {}_3F_2^{(2)} e_{km}^{(2)} + Y$$

(9.252)

where

$$Y = \frac{e^{-i\pi(k_0+m)}}{2\pi i} \int_{\frac{1}{2}}^{\frac{1}{2}+i\infty} dk \; \Gamma \begin{bmatrix} k_1 + m_1, & 1 - m_1 - k_1, & k_0 + k, \\ k_2 - m_2, & k_0 + m + 1, & m + 1 + k_0 \end{bmatrix}$$

$$\begin{matrix} k_0 + 1 - k, & k_1 + k_2 + k - 1, & k + m & 1 - k + m \\ k_1 + k + m_2, & 2k - 1 & 1 - 2k \end{matrix} \bigg] \; _3F_2^{(2)} e_{km}^{(5)}$$

$$= \frac{e^{-i\pi(k_0+m)}}{2\pi i} \int_{\frac{1}{2}-i\infty}^{\frac{1}{2}+i\infty} dk \; \Gamma \begin{bmatrix} k_1 + m_1, & 1 - k_1 - m_1, & k_1 + k_2 + k - 1 \\ k_2 - m_2, & k_0 + m + 1, & k_1 + k + m_2, \end{bmatrix}$$

$$\begin{matrix} k + m & k_0 + k \\ 2k - 1 \end{matrix} \bigg] \; _3F_2^{(2)} (1-z)^{k-k_1-k_2} F(k - m, k - k_0; 2k; 1 - z).$$

$$(9.253)$$

Replacing the hypergeometric function appearing in Eq. (9.253) by

$$z^{m+k_0} F(k + m, k + k_0; 2k; 1 - z) \qquad (9.254)$$

and the $_3F_2(1)$ function by

$$\Gamma \begin{bmatrix} k_0 + m + 1, & k_1 + k + m_2, & k_1 + k_2 - k, \\ 2k_1, & k_1 + k_2 + m, & m_2 - k_2 + 1, \end{bmatrix}$$

$$_3F_2 \begin{bmatrix} k_1 + m_1, & k_1 + k_2 + k - 1, & k_1 + k_2 - k, \\ 2k_1, & k_1 + k_2 + m, \end{bmatrix}. \qquad (9.255)$$

We now have

$$Y = \frac{e^{-i\pi(k_0+m)}}{2\pi i} \Gamma \begin{bmatrix} k_1 + m_1, & 1 - k_1 - m_1 \\ k_2 - m_2, & m_2 - k_2 + 1 \end{bmatrix}$$

$$\times \int_{\frac{1}{2}-i\infty}^{\frac{1}{2}+i\infty} dk \; \Gamma \begin{bmatrix} k + m, & k + k_0, \\ 2k_1, & k_1 + k_2 + m, \end{bmatrix}$$

$$\begin{matrix} k_1 + k_2 + k - 1, & k_1 + k_2 - k \\ 2k - 1 \end{matrix} \bigg]$$

$$\times \, _3F_2 \begin{bmatrix} k_1 + m_1, & k_1 + k_2 + k - 1, & k_1 + k_2 - k \\ 2k_1, & k_1 + k_2 + m \end{bmatrix} z^{k_0+m}$$

$$\times (1-z)^{k-k_1-k_2} F(k + m, k + k_0; 2k; 1 - z). \qquad (9.256)$$

The integrand in this expression has simple poles at

$$k = k_1 + k_2 + r, \quad r = 0, 1, 2, \ldots$$

but at no other points on the right of the line $\operatorname{Re} k = \frac{1}{2}$ of the complex k-plane. Therefore for integral m:

$$Y = e^{-i\pi k_0} \frac{\sin \pi k_2}{\sin \pi k_1} z^{k_0 + m} z^{-k_1 - m_1}$$

$$= e^{-i\pi k_0} \frac{\sin \pi k_2}{\sin \pi k_1} z^{m_2 - k_2}. \tag{9.257}$$

Using this value of Y we obtain

$$z^{m_2 - k_2} = e^{-i\pi k_2} \frac{\sin \pi k_1}{\sin \pi k_0} \sum_k D_k^+$$

$$+ \frac{1}{2\pi i} \int_{\frac{1}{2}}^{\frac{1}{2} + i\infty} dk \, \Gamma \begin{bmatrix} k_0 + k, & k_0 + 1 - k, & k - k_0, \\ 1 - k_0 - m, & k_0 - m + 1, & 2k - 1, \end{bmatrix}$$

$$\begin{matrix} 1 - k - k_0, & k - m, & 1 - k - m \\ 1 - 2k & & \end{matrix} \Bigg] e^{-i\pi(2k_2 - 1)} \overline{a_{m_2}^{(1)}} e_{km}^{(1)}$$

$$+ \frac{1}{2\pi i} \int_{\frac{1}{2}}^{\frac{1}{2} + i\infty} dk \, \Gamma \begin{bmatrix} k_0 + k, & k_0 + 1 - k, & k - k_0, \\ 1 + m + k_0, & 1 + m - k_0, & 2k - 1 \end{bmatrix}$$

$$\begin{matrix} 1 - k - k_0, & k + m, & 1 - k + m \\ 1 - 2k & & \end{matrix} \Bigg] e^{-i\pi(2k_2 - 1)} \overline{a_{m_2}^{(2)}} e_{km}^{(2)}. \tag{9.258}$$

By introducing $\eta(k, m)$ which is 1 for discrete k and $\cot \pi(k - m)$ for continuous k the discrete and the continuous parts can be given the same formal appearance.

$$z^{m_2 - k_2} = \left(\sum_{k=1(\frac{3}{2})}^{m} + i \int_{\frac{1}{2}}^{\frac{1}{2} + i\infty} dk \right) b^{(1)}(k_1, k_2, k; m_1, m_2) e_{km}^{(1)}$$

$$+ i \int_{\frac{1}{2}}^{\frac{1}{2} + i\infty} dk \, b^{(2)}(k_1, k_2, k; m_1, m_2) e_{km}^{(2)} \tag{9.259}$$

where

$$b^{(1)}(k_1, k_2, k; m_1, m_2)$$

$$= \bar{\eta}(k, m) e^{i\pi(m_2 - k_2)} \Gamma \begin{bmatrix} k_0 + k, & k_0 + 1 - k, & k - k_0, \\ m_2 + k_2, & k_0 - m + 1, & 1 - k_0 - m, \end{bmatrix}$$

$$\begin{matrix} 1 - k - k_0, & 2k, & k_1 + k_2 - k & 1 \\ 2k - 1, & k + m, & 1 - k + m & k_1 + 1 - k - m_2 \end{matrix} \Bigg]$$

$$\times {}_3F_2 \begin{bmatrix} 1 - k_2 - m_2, & 1 - k + k_0, & 1 - k - m \\ 1 + k_0 - m, & k_1 + 1 - k - m_2 & \end{bmatrix} \tag{9.260}$$

$$b^{(2)}(k_1, k_2, k; m_1, m_2) = e^{2\pi i m_2} \, b^{(1)}(k_1, k_2, k; -m_1, -m_2) \qquad (9.261)$$

the normalization factors for the CG coefficients are obtained by comparing the direct and inverse expansions. If $A_{km}^{(r)}(r = 1, 2)$ is the normalization factor then

$$A_{km}^{(1)} = \frac{1}{2\pi} e^{-i\pi\left(k_2 - \frac{1}{2}\right)} \left\{ \Gamma \begin{bmatrix} k_0 + k, & 1 - k - k_0, \\ 1 + k_0 - m & 1 - k_0 - m, \end{bmatrix} \right.$$

$$\left. \begin{bmatrix} k - k_0, & 1 - k + k_0, & k - m, & 1 - k - m \\ 2k - 1, & 1 - 2k \end{bmatrix} \right\}^{\frac{1}{2}} \qquad (9.262)$$

$$A_{km}^{(2)} = A_{k,-m}^{(1)}.$$

These are in complete agreement with the normalization factors determined in the next section by a different method. Using Eq. (9.262) we obtain the normalized Clebsch–Gordan coefficients.

$$C_r \begin{pmatrix} k_1, & k_2, & k \\ m_1, & m_2, & m \end{pmatrix} = A_{km}^{(r)} \, a_{m_2}^{(r)} \qquad (9.263)$$

for k lying in the continuous spectrum. Here $a_{m_2}^{(r)}(r = 1, 2)$ is given by Eq. (9.240) and Eq. (9.241).

For the discrete spectrum the Clebsch–Gordan coefficients are given by

$$C_1 \begin{pmatrix} k_1, & k_2, & k \\ m_1, & m_2, & m \end{pmatrix}$$

when k belongs to D_k^+ and by

$$C_2 \begin{pmatrix} k_1, & k_2, & k \\ m_1, & m_2, & m \end{pmatrix}$$

when k belongs to D_k^-.

The formula Eq. (9.258) along with the corresponding formula for $m < 0$ expresses the Clebsch–Gordan series.

$$C_{s_1}^{(\varepsilon_1)} \times C_{s_2}^{(\varepsilon_2)} = \sum_{k=1\left(\frac{3}{2}\right)}^{\infty} (D_k^+ + D_k^-) + 2 \int_0^{\infty} C_s^{(\varepsilon)} \, ds \qquad (9.264)$$

$$\varepsilon = |\varepsilon_1 - \varepsilon_2|.$$

9.2.1   Orthogonality of the eigenfunctions and the Clebsch–Gordan coefficients

For the derivation of the orthogonality condition of the Clebsch–Gordan coefficients we write down the scalar product with respect to which the functions e_{km} are orthogonal.

$$(e_{km}, e_{lm}) = \int\limits_0^{2\pi} \overline{e_{km}(z)}\, e_{lm} z^{2k_2 - 1} d\phi \qquad (9.265)$$

$$z = e^{i\phi}.$$

The orthogonality of the eigenfunctions can now be established in the traditional way by explicit evaluation of this scalar product with the help of Sturm–Liouville theory of the second order differential equation.

We shall first show that if k is in the principal series, the functions $e_{km}^{(1)}, e_{km}^{(2)}$ constitute two independent orthogonal sets. Using the standard formula

$$F(a, b; c; z) = \Gamma \begin{bmatrix} c, & b - a \\ b, & c - a \end{bmatrix} (-z)^{-a} F\left(a, 1 - c + a; 1 - b + a; \frac{1}{z}\right)$$

$$+ \Gamma \begin{bmatrix} c, & a - b \\ a, & c - b \end{bmatrix} (-z)^{-b} F\left(b, 1 - c + b; 1 - a + b; \frac{1}{z}\right)$$

$$(9.266)$$

and following the method of the previous sections the integral over the unit circle can now be expressed as a single loop contour integral

$$\left(e_{km}^{(1)}, e_{lm}^{(1)}\right) = e^{i\pi(m + k_1 + k_2 - 2)} \Gamma \begin{bmatrix} 1 + k_0 - m, & m + k_0 \\ k_0 + k, & k_0 + 1 - k \end{bmatrix}$$

$$\times (-i) \int_1^{0+} \phi_{km}\, \phi_{lm} z^{-k_0 - m} dz$$

$$+ \Gamma \begin{bmatrix} 1 + k_0 - m, & -k_0 - m \\ k - m, & 1 - k - m \end{bmatrix} (-i) e^{i\pi(2k_2 - 2)} \int_1^{0+} \phi_{km}^I \phi_{lm} dz$$

$$(9.267)$$

where the path of integration (as explained in Chapter 3) is a contour that starts from $z = 1$ along the positive real axis encircles the point $z = 0$ once counter clockwise and returns to $z = 1$ along the positive real axis.

In Eq. (9.267)

$$\phi_{km} = (1 - z)^{k-1} F(k - m, k - k_0; 1 - k_0 - m; z) \qquad (9.268)$$

$$\phi_{km}^I = (1 - z)^{k-1} F(k - m, k + k_0; 1 + k_0 + m; z). \qquad (9.269)$$

The integrand in the second term on the r.h.s (9.267), of which is regular at $z = 0$ is continuous across the branch cut and the integral therefore vanishes. The only contribution to the scalar product therefore comes from the first term and we have

$$\left(e_{km}^{(1)}, e_{lm}^{(1)}\right) = e^{i\pi(m+k_1+k_2-2)} \Gamma \begin{bmatrix} 1 + k_0 - m, & m + k_0 \\ k_0 + k, & 1 - k + k_0 \end{bmatrix} G(l, k; m)$$

$$(9.270)$$

where

$$G(l, k; m) = -i \lim_{\epsilon \to 0} \left[\left(e^{-2\pi i k_0} - 1\right)\right] \int_{\epsilon}^{1} z^{-(k_0+m)} \phi_{km}(z) \phi_{lm}(z)$$

$$+ \int_{s} z^{-(k_0+m)} \phi_{km}(z) \phi_{lm}(z). \qquad (9.271)$$

Here "s" stands for the small circle of radius ϵ centered at the origin.

The integral appearing in the r.h.s of the above equation can be evaluated by using the differential equation satisfied by the eigen functions ϕ_{km} and ϕ_{lm} and we have

$$\int_{\epsilon}^{1} z^{-(k_0+m)} \phi_{km}(z)\ \phi_{lm}(z) dz$$

$$= \frac{\lim_{z \to 1}}{(k - l)(k + l - 1)} \left\{ \left[(1 - z)^2 \left(\phi_{lm} \frac{d\phi_{km}}{dz} - \phi_{km} \frac{d\phi_{lm}}{dz}\right)\right] \right.$$

$$\left. - \frac{\epsilon^{1-m-k_0}}{(1 - k_0 - m)} \right\} \int_{s} z^{-(k_0+m)} \phi_{km}(z)\phi_{lm}(z) dz$$

$$= \frac{\left(e^{-2\pi i k_0} - 1\right) \epsilon^{1-k_0-m}}{1 - k_0 - m}. \qquad (9.272)$$

Combining Eqs. (9.271), (9.272) we finally obtain

$$G(l, k; m) = \frac{2e^{-i\pi k_0} \sin \pi k_0}{(k - l)(k + l - 1)} \lim_{z \to 1} \left[(1 - z)^2 \times \left(\phi_{lm} \frac{d\phi_{km}}{dz} - \phi_{km} \frac{d\phi_{lm}}{dz}\right)\right].$$

$$(9.273)$$

Explicit evaluation of the r.h.s (using the behavior of the hypergeometric function around $z = 1$ (Sec. Chap. 4, Part 1) yields.

$$\left(e_{km}^{(1)}, e_{lm}^{(1)}\right) = 4\pi^2 e^{i\pi(2k_2-1)} \Gamma \begin{bmatrix} 1 + k_0 - m, & 1 - k_0 - m, \\ k_0 + k, & k_0 + 1 - k, \end{bmatrix}$$

$$\begin{bmatrix} 2k - 1, & 1 - 2k & 1 \\ k - k_0, & 1 - k - k_0, & k - m & 1 - k - m \end{bmatrix}$$

$$\delta(\text{Im } k - \text{Im } l) \quad \text{for Im } k, \text{Im } l > 0. \tag{9.274}$$

The normalization factor as given by Eq. (9.274) agrees exactly with that obtained in the previous section by comparing the direct and inverse expansions. Similarly.

$$\left(e_{km}^{(2)}, e_{lm}^{(1)}\right) = -ie^{-i\pi 2J_2} \int_1^{0+} \phi_{km}^I(z)\phi_{lm}(z)dz. \tag{9.275}$$

Since the integrand is regular at $z = 0$, the integral vanishes

$$\text{i.e. } \left(e_{km}^{(2)}, e_{lm}^{(1)}\right) = 0 \tag{9.276}$$

the normalized Clebsch–Gordan coefficients $C_r \begin{pmatrix} k_1 & k_2, & k \\ m_1, & m_2, & m \end{pmatrix}$ $r = 1, 2$
are then coefficients of Fourier expansion of the orthonormalized eigen functions

$$f_{km}^{(r)} = A_{km}^{(r)} z^{k_2} e_{km}^{(r)} \tag{9.277}$$

satisfying

$$\left(f_{km}^{(r)}, f_{lm}^{(s)}\right) = \delta_{rs} \delta(\text{Im } k - \text{Im } l). \tag{9.278}$$

The orthogonality condition for the Clebsch–Gordan coefficients now follows immediately from that of the coupled states Eq. (9.277) and Eq. (9.278)

$$\sum_{m_2=-\infty}^{\infty} \overline{C_r \begin{pmatrix} k_1 & k_2, & k \\ m_1 & m_2, & m \end{pmatrix}} C_s \begin{pmatrix} k_1 & k_2, & k \\ m_1 & m_2, & m \end{pmatrix} = \delta_{rs}\delta(\text{Im } k - \text{Im } l). \tag{9.279}$$

When $k \epsilon D_k^{\pm}$, the orthonormality condition can be derived essentially in the same way and we have

$$\sum_{m_2} \overline{C \begin{pmatrix} k_1 & k_2, & k \\ m - m_2, & m_2, & m \end{pmatrix}} C \begin{pmatrix} k_1 & k_2, & k \\ m - m_2, & m_2, & m \end{pmatrix} = \delta_{kk'}. \tag{9.280}$$

9.2.2 $\mathbf{C}^{\epsilon}_{s_1} \times \mathbf{C}^{(0)}_E$

We now consider $k_1 = \frac{1}{2} + is_1$

$$-\infty < s_1 < \infty, \quad k_2 = \frac{1}{2} + s_2, -\frac{1}{2} < s_2 < \frac{1}{2}.$$

We define

$$\sigma = 1 - k_1 - k_2$$
$$\bar{\sigma} = k_1 - k_2 = k_0$$
$$\bar{k}_0 = \sigma.$$

As before we take the solution for $k\epsilon D^+_k$,

$$g^{(1)}_{km} = z^{k_2} e^{(1)}_{km}. \tag{9.281}$$

For $k\epsilon D^-_k$,

$$g^{(5)}_{km} = z^{k_2} e^{(5)}_{km} = z^{k_1+m}(1-z)^{k-k_1-k_2}F(k+m, k-k_0; 1+m+k_0; z). \tag{9.282}$$

Since for $\mathbf{C}^{\epsilon}_s, k$ is complex, there are in general two linearly independent solutions of the hypergeometric equation which must be chosen so that they satisfy the condition of orthonormality. The pair of solutions.

$$g^{(1)}_{km} = z^{k_2} e^{(1)}_{km} = z^{k_2}(1-z)^{k-k_1-k_2}F(k-m, k-k_0; 1-m-k; z) \tag{9.283}$$

$$g^{(2)}_{km} = (-z)^{k_0-k} z^{k_2}(1-z)^{k-k_1-k_2}F\left(k+m, k-k_0; 1+m-k; \frac{1}{z}\right) \tag{9.284}$$

as shown later satisfy

$$\left(g^{(r)}_{km}, g^{(s)}_{lm}\right) = N^{(r)}_{km}\delta_{rs}\delta(\operatorname{Im} k - \operatorname{Im} l). \tag{9.285}$$

The Hilbert space H under the reduction

$$\mathbf{T}^{k_1} \times \mathbf{T}^{k_2} = \sum_{\oplus k} \sigma_k \mathbf{T}_k. \tag{9.286}$$

therefore decomposes as before, into two mutually orthogonal subspaces leading to a pair of orthogonal Clebsch–Gordan coefficients. However for $k\epsilon D^{\pm}_k$ there is as before no duplicity problem.

The non-normalized Clebsch–Gordan coefficients are again coefficients of Fourier expansion of $g_{km}^{(r)}, r = 1, 2$

$$a_{m_2}^{(1)} = \frac{1}{2\pi i} \int_1^{0_+} z^{k_2 - m_2 - 1}(1 - z)^{k - k_1 - k_2} F(k - m, k - k_0; 1 - k_0 - m; z)$$

$$= e^{-i\pi(m_2 - k_2)} \Gamma \begin{bmatrix} k - k_1 - k_2 + 1 \\ m_2 - k_2 + 1, \quad k - k_1 - m_2 + 1 \end{bmatrix}$$

$$\times {}_3F_2 \begin{bmatrix} k_2 - m_2, & k - k_0, & k - m \\ 1 - k_0 - m & k - k_1 - m_2 + 1 \end{bmatrix}. \tag{9.287}$$

Similarly

$$a_{m_1, m_2}^{(2)} = e^{-2\pi i m_2} a_{-m_1, -m_2}^{(1)}. \tag{9.288}$$

9.2.3   *The Clebsch–Gordan Series*

As before we write

$$z^{m_2 - k_2} = \frac{1}{2\pi i} \int_{\frac{1}{2} - i\infty}^{\frac{1}{2} + i\infty} \chi(k) dk + \sum_{k = 1\left(\frac{3}{2}\right)}^{m} D_k^+ \tag{9.289}$$

where for $m > 0, m_1 < 0$

$$\chi(k) = \frac{\Gamma(k - m)\Gamma(k - k_0)\Gamma(k_1 + k_2 - k)\Gamma(k_1 + k_2 + k - 1)}{\Gamma(2k - 1)\Gamma(k_1 + k_2 - m)\Gamma(2k_2)}$$

$${}_3F_2 \begin{bmatrix} k_2 - m_2, & k_1 + k_2 - k, & k_1 + k_2 + k - 1 \\ k_1 + k_2 - m & 2k_2 \end{bmatrix}$$

$$(1 - z)^{k - k_1 - k_2} F(k - m, k - k_0; 2k; 1 - z). \tag{9.290}$$

Folding the integral about the real axis and using the behavior around $z = 1$

$$F(a, b; c; z) = \frac{\Gamma(c)\Gamma(c - a - b)}{\Gamma(c - a)\Gamma(c - b)} F(a, b; a + b - c + 1; 1 - z)$$

$$+ \frac{\Gamma(c)\Gamma(a + b - c)}{\Gamma(a)\Gamma(b)} (1 - z)^{c - a - b}$$

$$\times F(c - a, c - b; c - a - b + 1; 1 - z).$$

We have

$$z^{m_2-k_2} = \int_{\frac{1}{2}}^{\frac{1}{2}+i\infty} dk \; \Gamma \begin{bmatrix} k-m, & k-k_0, & 1-k-m, \\ 2k-1, & 1-2k, & k_1+k_2-m, \end{bmatrix}$$

$$\begin{matrix} 1-k-k_0, & k_1+k_2-k & k_1+k_2+k-1 \\ 2k_2 & 1-k_0-m \end{matrix} \Bigg]$$

$$\times {}_3F_2 \begin{bmatrix} k_2-m_2 & k_1+k_2-k, & k_1+k_2+k-1 \\ k_1+k_2-m, & 2k_2 \end{bmatrix} e_{km}^{(1)}$$

$$+ \sum_k D_k^+. \tag{9.291}$$

We now use the Thomae–Whipple identity

$$F_p(0;4,5) = -\Gamma \begin{bmatrix} \alpha_{045}, & \alpha_{034}, & \alpha_{014} \\ \alpha_{135}, & \beta_{02}, & 1-\beta_{02} \end{bmatrix} F_n(4;0,1)$$

$$+ \Gamma \begin{bmatrix} \alpha_{045}, & \alpha_{034}, & \alpha_{014} \\ \alpha_{245}, & \beta_{234}, & 1-\beta_{124} \end{bmatrix} F_p(2;1,5) \tag{9.292}$$

which yields

$${}_3F_2 \begin{bmatrix} k_2-m_2, & k_1+k_2-k, & k_1+k_2+k-1 \\ k_1+k_2-m, & 2k_2 \end{bmatrix}$$

$$= -\Gamma \begin{bmatrix} k_1-m_1, & k_0+1-k, & 1-k_1-m_1, & 2k_2 \\ k_2+m_2, & k_0+m+1, & -k_0-m, & k_1-k-m_2+1 \end{bmatrix}$$

$$\times {}_3F_2 \begin{bmatrix} k_2-m_2, & k_1+k_2-k & 1-k-m \\ k_1+k_2-m, & k_1-k-m_2+1 \end{bmatrix}$$

$$+ \Gamma \begin{bmatrix} k_1-m_1, & k_0+1-k, & 1-k_1-m_1, \\ k_2-m_2, & 1-k-m, & k-m, \end{bmatrix}$$

$$\begin{matrix} k_1+k_2-m, & 2k_2 \\ k_1+k_2+m & k_1-k+m_2+1 \end{matrix} \Bigg]$$

$$\times {}_3F_2 \begin{bmatrix} k_2+m_2, & k_1+k-k, & 1-k+m \\ k_1+k_2+m, & k_1-k+m_2+1 \end{bmatrix}. \tag{9.293}$$

Substituting the above identity and using

$$e_{km}^{(1)} = \Gamma \begin{bmatrix} 1-k+m, & k+m \\ k_0+m, & 1-k_0+m \end{bmatrix} e_{km}^{(2)}$$

$$+ e^{-i\pi(k_0-m)} \Gamma \begin{bmatrix} 1-k+m, & k+m, & 1+m-k_0, \\ 1+k_0+m, & k-k_0, & 1-k-k_0, \end{bmatrix}$$

$$\begin{matrix} m+k_0 & 1-m-k_0 \\ m+k_0, & 1+m-k_0 \end{matrix} \Bigg] e_{km}^{(5)}. \tag{9.294}$$

We obtain as before

$$z^{m_2-k_2} = \sum_{k=1\left(\frac{3}{2}\right)}^{m} D_k^+ - \frac{1}{2\pi i} \int_{\frac{1}{2}}^{\frac{1}{2}+i\infty} dk$$

$$\times \Gamma \begin{bmatrix} k_1 + m_1, & 1 - k_1 - m, & k_0 + 1 - k, & k - m & 1 - k - m, \\ k_2 - m_2, & 1 + k_0 + m, & -k_0 - m & 2k - 1 & 1 - 2k \end{bmatrix}$$

$$\begin{matrix} k + k_0, & 1 - k - k_0, & k_1 + k_2 - k, & k_1 + k_2 + k - 1 \\ k_1 + k_2 - m & 1 - k_0 - m & k_1 - k - m_2 + 1 \end{matrix} \Bigg]$$

$$\times {}_3F_2 \begin{bmatrix} k_2 - m_2, & k_1 + k_2 - k, & 1 - k - m \\ k_1 + k_2 - m, & k_1 - k - m_2 + 1 \end{bmatrix} e_{km}^{(1)}$$

$$+ \frac{1}{2\pi i} \int_{\frac{1}{2}}^{\frac{1}{2}+i\infty} dk$$

$$\times \Gamma \begin{bmatrix} k_1 - m, & 1 - k_1 - m_1, & k_0 + 1 - k, & k + m, & 1 - k + m \\ k_2 - m_2, & k_1 + k_2 + m, & 2k - 1, & 1 - 2k \end{bmatrix}$$

$$\begin{matrix} k + k_0 & 1 - k - k_0, & k_1 + k_2 + k - 1, & k_1 + k_2 - k \\ k_1 - k + m_2 + 1 \end{matrix} \Bigg]$$

$${}_3F_2 \begin{bmatrix} k_2 + m_2, & k_1 + k_2 - k, & 1 - k + m \\ k_1 + k_2 + m, & k_1 - k_2 + m_2 + 1 \end{bmatrix} e_{km}^{(2)} + Y \tag{9.295}$$

where

$$Y = \frac{e^{-i\pi(k_0+m)}}{2\pi i} \int_{\frac{1}{2}}^{\frac{1}{2}+i\infty} \Gamma \begin{bmatrix} k_1 + m_1, & 1 - k_1 - m_1, & 1 - k - m \\ k_2 - m_2, & k_1 + k_2 + m, & 2k - 1, 1 - 2k, \end{bmatrix}$$

$$\begin{matrix} k + m, & k_1 + k_2 - k & k_1 + k_2 + k - 1 & 1 - k + k_0 \\ 1 + m - k_0, & k_1 - k + m_2 + 1 & 1 + k_0 + m \end{matrix} \Bigg]$$

$$\times {}_3F_2 \begin{bmatrix} k_2 + m_2, & k_1 + k_2 - k, & 1 - k + m \\ k_1 + k_2 + m, & k_1 - k + m_2 + 1 \end{bmatrix}$$

$$\times z^{k_0+m}(1-z)^{k-k_1-k_2} F(k+m, k+k_0; 1+m+k_0; z). \tag{9.296}$$

We now use the identity

$${}_3F_2 \begin{bmatrix} k_2 + m_2, & k_1 + k_2 - k, & 1 - k + m \\ k_1 + k_2 + m, & k_1 - k + m_2 + 1 \end{bmatrix}$$

$$= \Gamma \begin{bmatrix} k_0 + k, & k_1 - k + m_2 + 1 \\ m_2 - k_2 + 1, & 2k \end{bmatrix}$$

$${}_3F_2 \begin{bmatrix} k_1 + m_1, & k_1 + k_2 - k, & k_1 + k_2 + k - 1 \\ k_1 + k_2 + m, & 2k_1 \end{bmatrix} \tag{9.297}$$

which yields after simplification

$$Y = \frac{e^{-i\pi(k_0+m)}}{2\pi i} z^{k_0+m} \int_{\frac{1}{2}-i\infty}^{\frac{1}{2}+i\infty} \Gamma \begin{bmatrix} k_1 + m, & 1 - k_1 - m_1 \\ k_2 - m_2, & m_2 - k_2 + 1, \end{bmatrix} \chi(k) dk$$

(9.298)

where

$$\chi(k) = \frac{\Gamma(k+m)\Gamma(k+k_0)\Gamma(k_1+k_2-k)\Gamma(k_1+k_2+k-1)}{\Gamma(2k-1)\Gamma(k_1+k_2+m)\Gamma(2k_1)}$$

$$\times {}_3F_2 \begin{bmatrix} k_1 + m_1, & k_1 + k_2 - k, & k_1 + k_2 + k - 1 \\ k_1 + k_2 + m, & 2k \end{bmatrix}$$

$$\times (1-z)^{k-k_1-k_2} F(k+m, k+k_0; 2k; 1-z).$$ (9.299)

Simplifying as before as obtain

$$z^{m_2-k_2} = \frac{\sin \pi(k_1+m_1)}{\sin \pi(k_0+m)} e^{-i\pi(k_2-m_2)} \sum_{k=1\left(\frac{3}{2}\right)}^{m} D_k^+ + \frac{1}{2\pi i} \int_{\frac{1}{2}}^{\frac{1}{2}+i\infty} dk$$

$$\Gamma \begin{bmatrix} k_0 + 1 - k, & k - m, & 1 - k - m, & k - k_0 \\ k_2 + m_2, & 2k - 1, & 1 - 2k, & k_1 + k_2 - m, \end{bmatrix}$$

$$\begin{matrix} 1 - k - k_0, & k_1 + k_2 - k & k_1 + k_2 + k - 1 \\ 1 - k_0 - m, & k_1 - k - m_2 + 1 \end{matrix} \Bigg] e^{-i\pi(k_2-m_2)}$$

$${}_3F_2 \begin{bmatrix} k_2 - m_2, & k_1 + k_2 - k, & 1 - k - m \\ k_1 + k_2 - m, & k_1 - k - m_2 + 1 \end{bmatrix} e_{km}^{(1)}$$

$$+ \frac{1}{2\pi i} \int_{\frac{1}{2}}^{\frac{1}{2}+i\infty} dk$$

$$\times \Gamma \begin{bmatrix} k_0 + 1 - k, & k + m, & 1 - k + m & 1 - k - k_0, \\ k_2 - m_2, & k_1 + k_2 + m & 2k - 1, & 1 - 2k, \end{bmatrix}$$

$$\begin{matrix} k - k_0 & k_1 + k_2 - k, & k_1 + k_2 + k - 1 \\ 1 - k_0 + m, & k_1 - k + m_2 + 1 \end{matrix} \Bigg] e^{-i\pi(k_2+m_2)}$$

$${}_3F_2 \begin{bmatrix} k_2 + m_2, & k_1 + k_2 - k, & 1 - k + m \\ k_1 + k_2 + m, & k_1 - k + m_2 + 1 \end{bmatrix} e_{km}^{(2)}.$$

(9.300)

By introducing $\eta(k, m)$ which is 1 for discrete k and $\cot \pi(k-m)$ for continuous k the discrete and the continuous parts can be given the some formal appearance.

$$z^{m_2-k_2} = \left(\sum_{k=1\left(\frac{3}{2}\right)}^{m} +i \int_{\frac{1}{2}}^{\frac{1}{2}+i\infty} dk \right) b^{(1)} \begin{pmatrix} k_1 & k_2, & k \\ m_1, & m_2, & m \end{pmatrix} e_{km}^{(1)}$$

$$+ i \int_{\frac{1}{2}}^{\frac{1}{2}+i\infty} dk \, b^{(2)} \begin{pmatrix} k_1 & k_2, & k \\ m_1, & m_2, & m \end{pmatrix} e_{km}^{(2)} \tag{9.301}$$

$$b^{(1)} \begin{pmatrix} k_1 & k_2, & k \\ m_1, & m_2, & m \end{pmatrix}$$
$$= \bar{\eta}(k, m) e^{i\pi(m_2-k_2)}$$
$$\times \Gamma \begin{bmatrix} k_0 + 1 - k, & k - k_0, & 1 - k - k_0, & k_1 + k_2 - k, \\ 2k - 1, & k_2 + m_2, & k_1 + k_2 - m, & 1 - k_0 - m, \end{bmatrix}$$
$$\begin{matrix} k_1 + k_2 + k - 1, & 2k \\ k_1 - k - m_2 + 1 \end{matrix} \Bigg]$$
$$\times {}_3F_2 \begin{bmatrix} k_2 - m_2, & k_1 + k_2 - k, & 1 - k - m \\ k_1 + k_2 - m, & k_1 - k - m_2 + 1 \end{bmatrix} \tag{9.302}$$

$$b^{(2)} \begin{pmatrix} k_1 & k_2, & k \\ m_1, & m_2, & m \end{pmatrix} = b^{(1)} \begin{pmatrix} k_1 & k_2, & k \\ -m_1, & -m_2, & -m \end{pmatrix}. \tag{9.303}$$

Equation (9.301) yields

$$C_{s_1}^{(\epsilon)} \times C_{E_2}^{(0)} = \sum_{k=1\left(\frac{3}{2}\right)}^{\infty} \left(D_k^+ + D_k^- \right) + 2 \int_0^{\infty} C_s^{(\epsilon)} \, ds. \tag{9.304}$$

The normalization factor obtained by comparing the direct and inverse expansion is given by

$$A_{km}^{(1)} = \frac{e^{-i\pi\left(k_1 - \frac{1}{2}\right)}}{2\pi} \left\{ \Gamma \begin{bmatrix} k - k_0, & k + k_2 - k & k_1 + k_2 + k - 1, \\ 2k - 1, & 1 - 2k, \end{bmatrix} \right.$$
$$\left. \begin{matrix} 1 - k - k_0, & m_2 - k_2 + 1, & k - m, & 1 - k - m \\ k_2 + m_2, & k_1 + k_2 - m, & 1 - k_0 - m \end{matrix} \Bigg] \right\}^{\frac{1}{2}} \tag{9.305}$$

$$A_{km}^{(2)} = A_{k,-m}^{(1)}. \tag{9.306}$$

9.3   The Character Problem and the Plancherel Formula for SU(1,1)

A major tool in group representation theory is the theory of character. For finite dimensional representations character is traditionally defined as the sum of the eigenvalues of the representation matrix. It should be pointed out that the unitary operators of an infinite dimensional Hilbert space do not have character in this sense since the infinite sum consists of numbers of unit modulus; for example for $g = e$ one has $D(e) = I$ and the sum of the diagonal elements of the infinite dimensional unit matrix is ∞. A major step ahead of the traditional concept of character was taken by Gel'fand and Naimark[5] in their definition in terms of the integral kernel of the group ring which defines character as a linear functional on the group manifold.

We denote by X the set of infinitely differentiable functions $x(g)$ on the group, which are equal to zero outside a bounded set. If $g \to T_g$ is a representation of the group G we set

$$T_x = \int d\mu(g)x(g)T_g \tag{9.307}$$

where $d\mu(g)$ is the left and right invariant measure (assumed coincident) on G and the integration extends over the entire group manifold.

The product $T_{x_1}T_{x_2}$ can be written in the form

$$T_{x_1}T_{x_2} = \int d\mu(g)x(g)T_g$$

where

$$x(g) = \int x_1(g_1)x_2(g_1^{-1}g)d\mu(g_1). \tag{9.308}$$

The function $x(g)$ defined by Eq. (9.308) will be called the product of the functions x_1, x_2 and denoted by $x_1x_2(g)$

Let us suppose that $g \to T_g$ is a unitary representation of the group G realized in the Hilbert space H of the functions $f(z)$ with the scalar product

$$(f.g) = \int \overline{f(z)}g(z)d\lambda(z)$$

where $d\lambda(z)$ is the measure in H.

Then the operator T_x is an integral operator with a kernel:

$$T_xf(z) = \int K(z,z_1)f(z_1)d\lambda(z_1).$$

It then follows that $K(z, z_1)$ is positive definite Hilbert–Schmidt kernel, satisfying

$$\int |K(z, z_1)|^2 d\lambda(z) d\lambda(z_1) < \infty.$$

Such a kernel has a trace

$$T_r(T_x) = \int K(z, z) d\lambda(z).$$

Using the definition in Eq. (9.307) one can prove that $T_r(T_x)$ can be written in the form

$$T_r(T_x) = \int x(g)\pi(g) d\mu(g).$$

The function $\pi(g)$ is the character of the representation $g \to T_g$. It should be noted that in this definition the matrix representation of the group does not appear and as will be shown below it makes a complete synthesis of the finite and infinite dimensional irreducible unitary representations.

The main problem in the Gelfand–Naimark theory of character is the construction of the integral kernel $K(z, z_1)$ which requires a judicious choice of the carrier space of the representation. We first explain the procedure by the simple example of SU(2) taking the traditional Bargmann–Segal space $B_j(c)$ (see Chapter 5) as the carrier space of the representation.

The group SU(2) consists of 2×2 unitary unimodular matrices

$$u = \begin{pmatrix} \alpha & \beta \\ -\bar\beta & \bar\alpha \end{pmatrix} \tag{9.309}$$

$$|a|^2 + |\beta|^2 = 1.$$

We know that every unitary unimodular matrix u can be diagonalized by a unitary unimodular matrix v so that

$$u = v\epsilon(\theta_0)v^{-1} \tag{9.310}$$

where $\epsilon(\theta_0)$ is the diagonal form of u:

$$\epsilon(\theta_0) = \begin{pmatrix} e^{i\theta_0/2} & 0 \\ 0 & e^{-i\theta_0/2} \end{pmatrix}. \tag{9.311}$$

Since v is also an SU(2) matrix it can be factorized in terms of Euler angles as

$$v = \epsilon(\eta)\alpha(\tau)\epsilon(\chi) \quad 0 \le \eta \le 4\pi, 0 \le \tau \le \pi, 0 \le \chi \le 2\pi \tag{9.312}$$

where

$$\alpha(\tau) = \begin{pmatrix} \cos\frac{\tau}{2} & \sin\frac{\tau}{2} \\ -\sin\frac{\tau}{2} & \cos\frac{\tau}{2} \end{pmatrix}. \tag{9.313}$$

We therefore obtain the following parameterization of u:

$$u = \epsilon(\eta)\alpha(\tau)\epsilon(\theta_0)\alpha^{-1}(\tau)\epsilon^{-1}(\eta). \tag{9.314}$$

The parameterization (9.314) yields

$$\alpha = \cos\frac{\theta_0}{2} + i\sin\frac{\theta_0}{2}\cos\tau$$

$$\beta = -ie^{i\eta}\sin\frac{\theta_0}{2}\sin\tau. \tag{9.315}$$

The finite element of SU(2) in $B_j(c)$ is given by

$$T_u f(z) = (\beta z + \bar\alpha)^{2j} f\left(\frac{\alpha z - \bar\beta}{\beta z + \alpha}\right). \tag{9.316}$$

This representation is unitary with respect to the scalar product,

$$(f,g) = \int \overline{f(z)}g(z)d\lambda(z) \tag{9.317}$$

where

$$d\lambda(z) = \frac{2j+1}{\pi}\left(1+|z|^2\right)^{-2j-2} d^2z$$

$$d^2z = dxdy, \quad z = x + iy. \tag{9.318}$$

A complete orthonormal set in $B_j(c)$ is given by

$$u_n(z) = \sqrt{\frac{(2j)!}{(2j-n)!n!}} z^n \quad n = 0,1,2,\ldots 2j. \tag{9.319}$$

Hence the principal vector[6] in this space is given by

$$e_z(z_1) = \sum_{n=0}^{2j} \overline{u_n(z)}u_n(z_1) = (1+\bar z z_1)^{2j} \tag{9.320}$$

so that

$$f(z) = \int (1+z\bar z_1)^{2j} f(z_1)d\lambda(z_1). \tag{9.321}$$

We now construct the group ring which consists of the operators

$$\mathbf{T}_x = \int d\mu(u)x(u)\mathbf{T}_u$$

where $d\mu(u)$ is the invariant measure on SU(2) and $x(u)$ is an arbitrary test function on the group, which vanishes outside a bounded set. The action of the group ring is given by

$$\mathbf{T}_x f(z) = \int x(u)\,(\beta z + \bar\alpha)^{2j}\, f\left(\frac{\alpha z - \bar\beta}{\beta z + \bar\alpha}\right) d\mu(u). \tag{9.322}$$

We now use the reproducing kernel $\overline{e_z(z_1)}$ to write

$$f\left(\frac{\alpha z - \bar\beta}{\beta z + \bar\alpha}\right) = \int \left[1 + \frac{(\alpha z - \bar\beta)\bar z_1}{(\beta z + \bar\alpha)}\right]^{2j} f(z_1)d\lambda(z_1). \tag{9.323}$$

Thus

$$\mathbf{T}_x f(z) = \int \mathbf{K}(z, z_1)f(z_1)d\lambda(z_1) \tag{9.324}$$

where the kernel $\mathbf{K}(z, z_1)$ is given by

$$\mathbf{K}(z, z_1) = \int x(u)\,(\beta z + \bar\alpha)^{2j} \left[1 + \frac{(\alpha z - \bar\beta)\,\bar z_1}{\beta z + \bar\alpha}\right]^{2j} d\mu(u). \tag{9.325}$$

Since the kernel $\mathbf{K}(z, z_1)$ is of the Hilbert–Schmidt type we have

$$\mathbf{T}_r(\mathbf{T}_x) = \int \mathbf{K}(z, z)d\lambda(z). \tag{9.326}$$

Using the definition Eq. (9.325) of the kernel we have

$$\mathbf{T}_r(\mathbf{T}_x) = \int x(u)\pi(u)d\mu(u) \tag{9.327}$$

where

$$\pi(u) = \int (\beta z + \bar\alpha)^{2j} \left[1 + \frac{(\alpha z - \bar\beta)\,\bar z}{\beta z + \bar\alpha}\right]^{2j} d\lambda(z). \tag{9.328}$$

Using the parametrization Eq. (9.315) and the substitution

$$z = \tan\frac{\theta}{2}e^{i\phi} \quad 0 \le \theta < \pi \;\; 0 \le \phi \le 2\pi. \tag{9.329}$$

We obtain after some calculations

$$\pi(u) = \frac{(2j+1)}{4\pi} \int_0^\pi d\theta \sin\theta \int_0^{2\pi} d\phi$$

$$\times \left[\cos\frac{\theta_0}{2} - i\sin\frac{\theta_0}{2}(\cos\tau\cos\theta + \sin\tau\cos\eta\sin\theta\cos\phi)\right.$$

$$\left. - \sin\tau\sin\eta\sin\theta\sin\phi)\right]^{2j}. \tag{9.330}$$

If we now introduce the unit vectors $\hat{n}$ and $\hat{r}$ as

$$\hat{n} = (\sin\tau\cos\eta, -\sin\tau\sin\eta, \cos\tau)$$

$$\hat{r} = (\sin\theta\cos\phi, \sin\theta\sin\phi, \cos\theta) \tag{9.331}$$

the above equation Eq. (9.330) can be written as

$$\pi(u) = \frac{2j+1}{4\pi} \int_0^\pi d\theta \sin\theta \int_0^{2\pi} d\phi \left[\cos\frac{\theta_0}{2} - i\sin\frac{\theta_0}{2}\hat{n}\cdot\hat{r}\right]^{2j}. \tag{9.332}$$

We now rotate the coordinate system such that the 3-axis (z-axis) coincides with the fixed vector $\hat{n}$. Thus

$$\pi(u) = \frac{2j+1}{2} \int_0^\pi d\theta \sin\theta \left(\cos\frac{\theta_0}{2} - i\sin\frac{\theta_0}{2}\cos\theta\right)^{2j}. \tag{9.333}$$

The above integral is quite elementary and yields,

$$\pi(u) = \frac{\sin\left(j+\frac{1}{2}\right)\theta_0}{\sin(\theta_0/2)}. \tag{9.334}$$

9.3.1   The group SU(1, 1)

The elements of the group SU(1,1) may be divided into three subsets (a) elliptic (b) hyperbolic and (c) parabolic. We define them as follows. Let

$$\alpha = \alpha_1 + i\alpha_2 \quad \text{and} \quad \beta = \beta_1 + i\beta_2 \quad \text{so that} \tag{9.335}$$

$$\alpha_1^2 + \alpha_2^2 - \beta_1^2 - \beta_2^2 = 1. \tag{9.336}$$

The elliptic elements are those for which

$$\alpha_2^2 - \beta_1^2 - \beta_2^2 > 0. \tag{9.337}$$

Hence if we set

$$\alpha_2' = \sqrt{\alpha_2^2 - \beta_1^2 - \beta_2^2} \quad \alpha_1^2 + \alpha_2'^2 = 1. \tag{9.338}$$

So that $-1 < \alpha_1 < 1$.

On the other hand the hyperbolic elements of $SU(1,1)$ are those for which

$$\alpha_2^2 - \beta_1^2 - \beta_2^2 < 0.$$

Hence we write

$$\alpha_2' = \sqrt{\beta_1^2 + \beta_2^2 - \alpha_2^2}. \tag{9.339}$$

We have

$$\alpha_1^2 - \alpha_2'^2 = 1 \tag{9.340}$$

so that $|\alpha_1| > 1$.

We exclude the parabolic class corresponding to

$$\alpha_2 = \sqrt{\beta_1^2 + \beta_2^2} \tag{9.341}$$

as this is a submanifold of lower dimensions.

If we diagonalize the $SU(1,1)$ matrix the eigenvalues are given by

$$\lambda = \alpha_1 \pm \sqrt{\alpha_1^2 - 1}. \tag{9.342}$$

We shall consider the elliptic case $-1 < \alpha_1 < 1$ first. Thus setting

$$\alpha_1 = \cos(\theta_0/2) \quad 0 < \theta_0 < 2\pi$$

we have $\lambda = e^{\pm i\theta_0/2}$. We shall now show that every elliptic element of $SU(1,1)$ can be diagonalized by a pseudounitry transformation, i.e.

$$v^{-1}uv = \epsilon(\theta_0) = \begin{pmatrix} \delta_1 & 0 \\ 0 & \delta_2 \end{pmatrix} \delta_1 = \bar{\delta}_2 = e^{i\theta_0/2} \tag{9.343}$$

where $v \in SU(1,1)$.

To prove this we first note that the above equation can be written as

$$uv_1 = \delta_1 v_1$$

$$uv_2 = \delta_2 v_2$$

where

$$v_1 = \begin{pmatrix} v_{11} \\ v_{21} \end{pmatrix}, \quad v_2 = \begin{pmatrix} v_{12} \\ v_{22} \end{pmatrix}. \tag{9.344}$$

Thus v_1 and v_2 are the eienvectors of the matrix u belonging to the eigenvalues δ_1 and δ_2 respectively. Hence v_1 and v_2 are linearly independent so that $\det v \neq 0$.

We normalize the matrix v such that

$$\det v = v_{11}v_{22} - v_{12}v_{21} = 1. \tag{9.345}$$

We now show that the eigenvectors v_1 and v_2 are pseudo orthogonal, i.e. orthogonal with respect to the metric,

$$\sigma_3 = \begin{pmatrix} 1 & 0 \\ 0 & -1 \end{pmatrix}. \tag{9.346}$$

In fact from equation Eq. (9.344) we easily obtain

$$\delta_1^2 v_2^\dagger \sigma_3 v_1 = v_2^\dagger u^+ \sigma_3 u v_1 = v_2^\dagger \sigma_3 v_1. \tag{9.347}$$

Since $\delta_1 \neq 1$ we have

$$v_2^\dagger \sigma_3 v_1 = 0. \tag{9.348}$$

If we further normalize

$$v_1^\dagger \sigma_3 v_1 = 1. \tag{9.349}$$

We easily obtain

$$v_{22} = \frac{\overline{v_{11}}}{11}, v_{21} = \overline{v_{12}}. \tag{9.350}$$

Thus for the elliptic elements of $SU(1,1)$ the transformation matrix v is also an $SU(1,1)$ matrix. Since every matrix $v \in SU(1,1)$ can be written as

$$v = \epsilon(\eta)\alpha(\sigma)\epsilon(\theta_0), 0 \leq \eta \leq 4\pi, 0 \leq \sigma < \infty, 0 \leq \theta_0 \leq 2\pi \tag{9.351}$$

$$\alpha(\rho) = \begin{pmatrix} \cosh\dfrac{\rho}{2} & \sinh\dfrac{\rho}{2} \\ \sinh\dfrac{\rho}{2} & \cosh\dfrac{\rho}{2} \end{pmatrix}. \tag{9.352}$$

We immediately obtain

$$u = \epsilon(\eta)\alpha(\rho)\epsilon(\theta_0)\alpha^{-1}(\rho)\epsilon^{-1}(\eta). \tag{9.353}$$

The above parametrization yields

$$\alpha = \cos\frac{\theta_0}{2} + i\sin\frac{\theta_0}{2}\cosh\rho$$

$$\beta = -ie^{i\eta}\sin\frac{\theta_0}{2}\sinh\rho \quad 0 \leq \eta \leq 4\pi. \tag{9.354}$$

The invariant measure for the elliptic elements are given by

$$d\mu(u)_{\text{elliptic}} = \sin^2\frac{\theta_0}{2}\frac{d\theta_0}{2}\sinh\rho\,d\rho\,d\eta. \tag{9.355}$$

We now consider the hyperbolic elements of $SU(1,1)$. Since now $|\alpha_1| > 1$ the diagonal matrix

$$\epsilon(\sigma) = \begin{pmatrix} \text{sgn}\lambda\, e^{\text{sgn}\lambda\sigma/2} & 0 \\ 0 & \text{sgn}\lambda\, e^{-\text{sgn}\lambda\sigma/2} \end{pmatrix} \tag{9.356}$$

belongs to $SL(2, R)$, it can be regarded as the diagonal form of the matrix g (see Eq. (9.330)) with $|\alpha_1| = |a + d|/2 > 1$.

An analysis parallel to the one for the elliptic elements shows that for $|a + d|/2 > 1$ every matrix $g \in SL(2, R)$ can be diagonalized also by a matrix $v \in SL(2, R)$. Thus

$$g = v\epsilon(\sigma)v^{-1}. \tag{9.357}$$

Since every matrix $v \in SL(2, R)$ can be decomposed as

$$v = e(\theta)\alpha(\rho)\epsilon(\alpha) \tag{9.358}$$

where

$$e(\theta) = \begin{pmatrix} \cos\dfrac{\theta}{2} & \sin\dfrac{\theta}{2} \\[2mm] -\sin\dfrac{\theta}{2} & \cos\dfrac{\theta}{2} \end{pmatrix} \tag{9.359}$$

$$\alpha(\rho) = \begin{pmatrix} \cosh\dfrac{\rho}{2} & \sinh\dfrac{\rho}{2} \\[2mm] \sinh\dfrac{\rho}{2} & \cosh\dfrac{\rho}{2} \end{pmatrix}. \tag{9.360}$$

We obtain the following parametrization of the hyperbolic elements of $SL(2, R)$

$$g = e(\theta)\alpha(\rho)\epsilon(\sigma)\alpha^{-1}(\rho)e^{-1}(\theta). \tag{9.361}$$

The use of the isomorphism kernel then yields

$$\alpha = \operatorname{sgn} \lambda \cosh\frac{\sigma}{2} + i \sinh\frac{\sigma}{2} \sinh\rho$$

$$\beta = -ie^{-i\theta} \sinh\frac{\sigma}{2} \cosh\rho. \tag{9.362}$$

The corresponding invariant measure is then given by

$$d\mu(u)_{\text{hyperbolic}} = \sinh^2\frac{\sigma}{2}\frac{d\sigma}{2} \cosh\rho \, d\rho \, d\theta. \tag{9.363}$$

We shall now consider the irreducible unitary representations of the universal covering group of $SL(2, R)$ (denoted by $\widetilde{SL}(2, R)$) from which Bargmann's representations follow as special cases.

(a) Principal series of representations C_s^ε:

$$J_1^2 + J_2^2 - J_3^2 = k(1 - k), \quad k = \frac{1}{2} + is$$

$$m = \varepsilon \pm n, \quad n = 0, 1, 2 \ldots$$

$$0 \le \varepsilon < 1.$$

(b) Positive discrete series D_k^+:

$$k > 0, \ m = k + n, \ n = 0, 1, 2 \dots.$$

(c) Negative discrete series $D_{\widetilde{k}}^-$:

$$k > 0, \ m = -k - n, \ n = 0, 1, 2 \dots.$$

(d) Exceptional series of representations C_q^ε:

$$\varepsilon(1 - \varepsilon) < q < \frac{1}{4}, \ 0 \le \epsilon < 1$$

$$m = \varepsilon \pm n, \ n = 0, 1, 2 \dots.$$

Bargmann's integral/half-integral representation follows as special cases on setting $\varepsilon = 0, \frac{1}{2}$.

We shall omit the exceptional representation as they will not appear in the calculation.

We now assert that the representations C_s^ε and $C_s^{1-\varepsilon}$ belonging to the principal series of $\widetilde{SL}(2, R)$ are unitarily equivalent. The finite element of the group for the representation C_s^ε is given by (in terms of SU (1, 1) parameters):

$$T_u^{(\varepsilon, s)} f(z) = (\bar{\beta} \, \bar{z} + \alpha)^{-k+\varepsilon} (\beta z + \bar{\alpha})^{-k-\varepsilon} f\left(\frac{\alpha z + \bar{\beta}}{\beta z + \bar{\alpha}}\right) \qquad (9.364)$$

$$k = \frac{1}{2} + is, \ 0 \le \epsilon < 1, \ |z| = 1.$$

We shall show that the representation C_s^ε and $C_s^{1-\varepsilon}$ are equivalent by showing

$$\mathrm{Tr}\,(T_x^{(\varepsilon, s)}) = \mathrm{Tr}\,(T_x^{(1-\varepsilon, s)}). \qquad (9.365)$$

As shown in Sec. A the integral kernel of the group ring is given by

$$[T_x^{(\varepsilon, s)}] g(\theta) = \sum_{\eta = 0, 1} \int_0^\pi K_{(\eta, \varepsilon)}(\theta, \theta_1) g(\theta_1) d\theta_1 \qquad (9.366)$$

where

$$K_{(\eta, \varepsilon)}(\theta, \theta_1) = \frac{1}{4} \int \cos 2\pi \eta \epsilon x (\underline{\theta}^{-1} \underline{k}\, \underline{\theta}_1) e^{-i(\theta - \theta_1)\varepsilon} |k_{22}|^{-2k} d\mu_l(k) \qquad (9.367)$$

and following the traditional procedure of classical analysis we have defined the value of the function on the real axis as

$$f(x) = \frac{1}{2}[f(x + i0) + f(x - i0)]. \qquad (9.368)$$

Now according to Gel'fand and co-workers

$$\text{Tr}\,[T_{x,\eta}^{(\varepsilon,s)}] = \int K_{(\eta,\varepsilon)}(\theta,\theta)d\theta. \tag{9.369}$$

Since the crucial exponential factor drops out from $K_{(\eta,\varepsilon)}(\theta,\theta)$ we have

$$\text{Tr}\,[T_{x,\eta}^{(\varepsilon,s)}] = \text{Tr}\,[T_{x,\eta}^{(1-\varepsilon,s)}]. \tag{9.370}$$

We may therefore write,

$$\text{Tr}\,[T_x^{(\varepsilon,s)}] = \theta(2\varepsilon)\theta(1-2\varepsilon)\text{Tr}\,[T_x^{(\varepsilon,x)}] + \theta(2-2\varepsilon)\theta(2\varepsilon-1)\text{Tr}\,[T_x^{(\varepsilon,s)}]$$
$$= \theta(2\varepsilon)\theta(1-2\varepsilon)\text{Tr}\,[T_x^{(\varepsilon,s)}] + \theta(2-2\varepsilon)\theta(2\varepsilon-1)\text{Tr}\,[T_x^{(1-\varepsilon,s)}]. \tag{9.371}$$

The problem consists of two parts: (1) evaluation of the character of the representations $C_s^\varepsilon, D_k^+, D_k^-$; (2) the inversion problem and the subsequent computations of $x(e)$.

A. The principal series of representations C_s^ε [7]

The principal series C_s^ε is realized in the Hilbert space of functions defined on the unit circle $|z| = 1$. The finite element of the group is given by

$$T_u^{(\varepsilon,s)}f(z) = (\beta z + \bar{\alpha})^{-k-\varepsilon}(\bar{\beta}\,\bar{z} + \alpha)^{-k+\varepsilon}f\left(\frac{\alpha z + \bar{\beta}}{\beta z + \bar{\alpha}}\right) \tag{9.372}$$

with $k = \frac{1}{2} + is, 0 \le \varepsilon < 1$. These representations are unitary with respect to the scalar product:

$$(f,g) = \int_0^{2\pi} \bar{f}(z)g(z)d\theta \quad (z = e^{i\theta}). \tag{9.373}$$

It has already been shown that the representations C_s^ε and $C_s^{1-\varepsilon}$ are unitarily equivalent. Thus the representation sets C_s^ε for $0 \le \varepsilon \le \frac{1}{2}$ and $\frac{1}{2} \le \varepsilon < 1$ are equivalent.

We now construct the operator of the group ring

$$T_x^{(\varepsilon,s)} = \int d\mu(u)x(u)T_u^{(\varepsilon,s)} \tag{9.374}$$

where $x(u)$ is an arbitrary test function on the group which vanishes outside a bounded set. If we define

$$x^\dagger(u) = \overline{x(u^{-1})}$$

we have

$$x_1^\dagger x_2(e) = \int \overline{x_1(u^{-1})} x_2(u^{-1}) d\mu(u) = \int \overline{x_1(u)} x_2(u) d\mu(u). \qquad (9.375)$$

The operator of the group ring is now given by

$$\mathrm{T}_x^{(\varepsilon,s)} \cdot f(z) = \int d\mu(u) x(u)(\beta z + \bar\alpha)^{-k-\varepsilon}(\bar\beta \, \bar z + \alpha)^{-k+\varepsilon} \times f\left(\frac{\alpha z + \bar\beta}{\beta z + \bar\alpha}\right).$$
$$(9.376)$$

Setting $z = +ie^{i\theta}$ and performing the left translation.

$$u \longrightarrow \underline\theta^{-1} u$$

$$\underline\theta = \begin{pmatrix} e^{i\theta/2} & 0 \\ 0 & e^{-i\theta/2} \end{pmatrix}$$

$$\mathrm{T}_x f(ie^{i\theta}) = \int d\mu(u) x(\theta^{-1} u) e^{-i\theta\varepsilon}$$

$$(i\beta + \bar\alpha)^{-k-\varepsilon}(-i\bar\beta + \alpha)^{-k+\varepsilon} f\left(\frac{+i\alpha + \bar\beta}{+i\beta + \bar\alpha}\right). \qquad (9.377)$$

We now map the SU$(1,1)$ matrix u onto the SL$(2,\mathrm{R})$ matrix g by using the isomorphism kernel η (see Eq. (9.32)) and perform the Iwasawa decomposition

$$g = k\theta_1 \qquad (9.378)$$

where

$$k = \begin{pmatrix} k_{11} & k_{12} \\ 0 & k_{22} \end{pmatrix}, \quad k_{11}k_{22} = 1 \qquad (9.379)$$

belongs to the subgroup K of real triangular matrices of determinant unity and $\theta_1 \in \Theta$ where Θ is the subgroup of pure rotation matrices

$$\theta_1 = \begin{pmatrix} \cos(\theta_1/2) & +\sin(\theta_1/2) \\ -\sin(\theta_1/2) & \cos(\theta_1/2) \end{pmatrix}.$$

The above decomposition can also be written as

$$u = \underline{k}\,\underline{\theta_1}$$

where $\underline{k} = \eta^{-1} k \eta$, etc.

We now write

$$(i\beta + \bar\alpha) = (k_{22} + i0) e^{-i\theta_1/2}$$

$$(-i\bar\beta + \alpha) = (k_{22} - i0) e^{i\theta_1/2}$$

and set

$$g(\theta) = f(+ie^{i\theta}). \tag{9.380}$$

Since under the decomposition (9.378) the invariant measure decomposes as

$$d\mu(u) = \frac{1}{2}d\mu_l(g) = \frac{1}{2}d\mu_r(g) = \frac{1}{4}d\mu_l(k)d\theta_1$$

we have

$$T_x^{(\varepsilon,s)}g(\theta) = \int_\Theta K(\theta,\theta_1)g(\theta_1)d\theta_1$$

where

$$T_x^{(\varepsilon,s)}g(\theta) = \sum_{\eta=0,1} \int_\Theta K_\eta(\theta,\theta_1)g(\theta_1)d\theta_1 \tag{9.381}$$

with

$$K_\eta(\theta,\theta_1) = \frac{1}{4}\int x(\underline{\theta}^{-1}\underline{k}\,\underline{\theta}_1)e^{-i(\theta-\theta_1)\varepsilon}|k_{22}|^{-2k}\cos 2\pi\varepsilon\eta\, d\mu_l(k).$$

In obtaining the integral kernel $K_\eta(\theta,\theta_1)$ of the group ring we have written

$$(k_{22} \pm i0) = |k_{22}|e^{\pm i\pi\eta}$$

$$\eta = 0 \quad \text{for} \quad k_{22} > 0$$

$$\eta = 1 \quad \text{for} \quad k_{22} < 0$$

and following the traditional procedure of classical analysis we have defined the value of the function on the real axis as

$$f(x) = \frac{1}{2}[f(x+i0) + f(x-i0)]. \tag{9.382}$$

Since the kernel is of the Hilbert–Schmidt type

$$\text{Tr}\,\left(T_{x,\eta}^{(\varepsilon,s)}\right) = \int_0^{2\pi} K_\eta(\theta,\theta)d\theta \tag{9.383}$$

which can be written in the form,

$$\text{Tr}\,\left[T_{x,\eta}^{(\varepsilon,s)}\right] = \frac{1}{4}\int d\theta \int d\mu_l(k)|k_{22}|^{-2k}x(\underline{\theta}^{-1}\underline{k}\,\underline{\theta}) \times \cos 2\pi\varepsilon\eta$$

$$= \text{Tr}\,[T_{(x,\eta)}^{(1-\varepsilon,s)}]. \tag{9.384}$$

Before proceeding any further, we note that $\underline{\theta}^{-1}\underline{k}\,\underline{\theta}$ represents a hyperbolic element of $SU(1,1)$:

$$u = \underline{\theta}^{-1}\underline{k}\,\underline{\theta}.$$

Calculating the trace of both sides we have

$$k_{22} + 1/k_{22} = 2\alpha_1. \tag{9.385}$$

In the previous section we have seen that for the elliptic elements of $SU(1,1)$ $\alpha_1 = \cos(\theta_0/2) < 1$. Equation (9.58), therefore, for the elliptic case yields

$$k_{22}^2 - 2k_{22}\cos(\theta_0/2) + 1 = 0,$$

which has no real solution. Thus the elliptic elements of $SU(1,1)$ do not contribute to the character of the principal series of representations. We, therefore, assert that for this particular class of unirreps the trace is concentrated on the hyperbolic elements.

We shall now show that every hyperbolic element of $SU(1,1)$ [i.e. those with $|\alpha_1| = |(a+d)|/2 > 1$] can be represented as

$$u = \underline{\theta}^{-1}\underline{k}\,\underline{\theta}, \tag{9.386}$$

or equivalently as

$$g = \theta^{-1}k\theta. \tag{9.387}$$

Here $k_{11} = \lambda^{-1}$, $k_{22} = \lambda$ are the eigenvalues of the matrix g taken in any order.

We recall that every $g\epsilon SL(2,\mathbb{R})$ for the hyperbolic case can be diagonalized as

$$v'gv'^{-1} = \delta, \tag{9.388}$$

where

$$\delta = \begin{pmatrix} \delta_1 & 0 \\ 0 & \delta_2 \end{pmatrix}, \delta_1\delta_2 = 1; \delta_1, \delta_2, \text{ real},$$

belongs to the subgroup D of real diagonal matrices of determinant unity and $v'\epsilon SL(2,\mathbb{R})$. If we write the Iwasawa decomposition for v',

$$v' = k'\theta,$$

then

$$g = \theta^{-1}k'^{-1}\delta k'\theta.$$

Now $k'^{-1}\delta k' \epsilon K$ so that writing $k = k'^{-1}\delta k'$ we have decomposition (9.388) in which

$$k_{11} = \delta_1 = \lambda^{-1}, k_{22} = \delta_2 = \lambda.$$

If these eigenvalues are distinct then for a given ordering of them the matrices k, θ are determined uniquely by the matrix g.

It follows that for a given choice of λ the parameters θ and k_{12} are uniquely determined. We note that there are exactly two representations of the matrix g by means of formula (9.387) corresponding to two distinct possibilities:

$$k_{11} = \operatorname{sgn}\lambda|\lambda|^{-1} = \operatorname{sgn}\lambda e^{\sigma/2}, \ k_{22} = \operatorname{sgn}\lambda|\lambda| = \operatorname{sgn}\lambda e^{-\sigma/2}$$

$$k_{11} = \operatorname{sgn}\lambda|\lambda|^{-1} = \operatorname{sgn}\lambda e^{-\sigma/2}, \ k_{22} = \operatorname{sgn}\lambda|\lambda| = \operatorname{sgn}\lambda e^{\sigma/2}.$$

Let us now remove from K the elements with $k_{11} = k_{22} = 1$. This operation cuts the group K into two connected disjoint components. Neither of these components contain two matrices which differ only by permutation of the two diagonal elements. In correspondence with this partition the integral in the r.h.s. of Eq. (9.384) is represented in the form of a sum of two integrals,

$$\operatorname{Tr}(T_{x,\eta}) = \frac{1}{4}\int_\Theta d\theta \int_{K_1} d\mu_l(k)|k_{22}|^{-2k} x(\underline{\theta^{-1}k\theta})\cos 2\pi\varepsilon\eta$$

$$+\frac{1}{4}\int_\Theta d\theta \int_{K_2} d\mu_l(k)|k_{22}|^{-2k} x(\underline{\theta^{-1}k\theta})\cos 2\pi\varepsilon\eta. \qquad (9.389)$$

As θ runs over the subgroup Θ and k runs over the components K_1 or K_2 the matrix $g = \theta^{-1}k\theta$ runs over the hyperbolic elements of the group $SL(2, R)$ or equivalently $u = \underline{\theta^{-1}k\theta}$ runs over the hyperbolic elements of $SU(1, 1)$. We shall now prove that in K_1 or K_2,

$$d\mu_l(k)d\theta = \frac{4|k_{22}|}{|k_{22} - k_{11}|}d\mu(u). \qquad (9.390)$$

To prove this we start from the left invariant differential element,

$$dw = g^{-1}dg,$$

where $g \in SL(2, R)$ and dg is the matrix of the differentials g_{pq}, i.e. following the notation of Eq. (24),

$$dg = \begin{pmatrix} da & db \\ dc & dd \end{pmatrix}.$$

The elements dw are invariant under the left translation $g \to g_0 g$. Hence choosing a basis in the set of all dg we immediately obtain a differential left invariant measure. For instance choosing $dw_{12}, dw_{21}, dw_{22}$ as the independent invariant differentials we arrive at the left invariant measure on $SL(2, \mathbb{R})$,

$$d\mu_l(g) = dw_{12} dw_{21} dw_{22}. \tag{9.391}$$

In a similar fashion we can define the right invariant differentials,

$$dw' = dg g^{-1},$$

which is invariant under the right translation $g \to g\, g_0$.

To prove the formula (9.390) we write the decomposition (9.387) as

$$\theta g = k\theta,$$

so that

$$d\theta g + \theta dg = dk\theta + kd\theta. \tag{9.392}$$

It easily follows that

$$dw = g^{-1} dg = \theta^{-1} d\mu \theta, \tag{9.393}$$

where

$$d\mu = k^{-1} dk + d\theta\, \theta^{-1} - k^{-1} d\theta \theta^{-1} k. \tag{9.394}$$

In accordance with the choice of the independent elements of dw as mentioned above, we choose the independent elements of $d\mu$ as $d\mu_{12}, d\mu_{21}, d\mu_{22}$. Equation (9.393) then leads to

$$dw_{11} + dw_{22} = d\mu_{11} + d\mu_{22}, \tag{9.395}$$

$$dw_{11} dw_{22} - dw_{12} dw_{21} = d\mu_{11} d\mu_{22} - d\mu_{12} d\mu_{21}. \tag{9.396}$$

Further since $\text{Tr}\,(d\mu) = \text{Tr}\,(dw) = 0$ we immediately obtain

$$dw_{22}^2 + dw_{12} dw_{21} = d\mu_{22}^2 + d\mu_{12} d\mu_{21},$$

which can be written in the form

$$d\eta_1^2 + d\eta_2^2 - d\eta_3^2 = d\eta_1'^2 + d\eta_2'^2 - d\eta_3'^2, \tag{9.397}$$

where

$$d\eta_1 = (dw_{12} + dw_{21})/2, \quad d\eta_1' = (d\mu_{12} + d\mu_{21})/2,$$

$$d\eta_2 = dw_{22}, \quad d\eta_2' = d\mu_{22},$$

$$d\eta_3 = (dw_{12} - dw_{21})/2, \quad d\eta_3' = (d\mu_{12} - d\mu_{21})/2.$$

Equation (9.397) implies that the set $d\eta$ and the set $d\eta'$ are connected by a Lorentz transformation. Since the volume element $d\eta_1 d\eta_2 d\eta_3$ is invariant under such a transformation we have,

$$d\eta_1 d\eta_2 d\eta_3 = d\eta'_1 d\eta'_2 d\eta'_3. \tag{9.398}$$

But the l.h.s. of Eq. (9.398) is $dw_{12} dw_{21} dw_{22}/2$ and r.h.s. is $d\mu_{12} d\mu_{21} d\mu_{22}/2$. Hence we easily obtain

$$d\mu_l(g) = d\mu_{12} d\mu_{21} d\mu_{22}.$$

We now write Eq. (9.394) in the form

$$d\mu = du + dv \tag{9.399}$$

where $du = k^{-1} dk$ is the left invariant differential on K and

$$dv = d\theta\theta^{-1} - k^{-1} d\theta\theta^{-1} k. \tag{9.400}$$

In Eq. (9.399), du is a triangular matrix whose independent non-vanishing elements are chosen to be du_{12}, du_{22} so that

$$d\mu_l(k) = du_{12} du_{22}.$$

On the other hand dv is a 2×2 matrix having one independent element which is chosen to be dv_{21}. Since the Jacobian connecting $d\mu_{12} d\mu_{21} d\mu_{22}$ and $du_{12} du_{22} dv_{21}$ is a triangular determinant having 1 along the main diagonal we obtain

$$d\mu_l(g) = d\mu_l(k) dv_{21}. \tag{9.401}$$

It can now be easily verified that each element $k\epsilon K$ with distinct diagonal elements (which is indeed the case for K_1 or K_2) can be represented uniquely in the form

$$k = \zeta^{-1} \delta \zeta, \tag{9.402}$$

where δ belongs to the subgroup of real diagonal matrices with unit determinant and $\zeta \epsilon Z$, where Z is a subgroup of K consisting of real matrices of the form

$$\zeta = \begin{pmatrix} 1 & \zeta_{12} \\ 0 & 1 \end{pmatrix}.$$

Writing Eq. (9.402) in the form $\zeta k = \delta \zeta$ we obtain

$$k_{pp} = \delta_p, \quad \zeta_{12} = k_{12}/(\delta_1 - \delta_2).$$ (9.403)

Using the decomposition (9.402) we can now write Eq. (9.400) in the form

$$dv = \zeta^{-1} dp \zeta,$$

where

$$dp = d\lambda - \delta^{-1} d\lambda \delta,$$

$$d\lambda = \zeta d\sigma \zeta^{-1}, \quad d\sigma = d\theta \theta^{-1}.$$

From the above equations it now easily follows that

$$dv_{21} = \frac{|\delta_2 - \delta_1|}{|\delta_2|} \frac{d\theta}{2}.$$ (9.404)

Now recalling that in $K_1, |k_{22}| = e^{-\sigma/2}$ and in $K_2, |k_{22}| = e^{\sigma/2}$ we have,

$$\mathrm{Tr}\,(T_{x,\eta}) = \int x(u) \pi_\eta^{(\varepsilon,s)}(u) d\mu(u),$$

where the character $\pi_\eta^{(\varepsilon,s)}(u)$ is given by

$$\pi_\eta^{(\varepsilon,s)}(u) = \frac{e^{(2k-1)\sigma/2} + e^{-(2k-1)\sigma/2}}{|e^{\sigma/2} - e^{-\sigma/2}|} \cos 2\pi\varepsilon\eta$$

and

$$T_r[T_x^{(\varepsilon,s)}] = \sum_{\eta=0,1} \int d\mu(u) x(u) \pi_\eta^{(\varepsilon,s)}(u)$$ (9.405)

where

$$\pi_\eta^{(\varepsilon,s)} = \frac{\cos s\sigma}{\sinh \frac{\sigma}{2}} \cos 2\pi\varepsilon\eta.$$ (9.406)

B. The positive discrete series $D_k^{+\,7}$

The finite element of the group for the representation is given by

$$T_u^{k+} f(z) = (\beta z + \bar{\alpha})^{-2k} f\left(\frac{\alpha z + \bar{\beta}}{\beta z + \bar{\alpha}}\right)$$ (9.407)

where $f(z)$ is an analytic function regular within the unit disc. This representation is unitary with respect to the scalar product

$$(f,g) = \int \overline{f(z)} g(z) d\lambda(z)$$ (9.408)

$$d\lambda(z) = \frac{(2k-1)}{\pi}(1 - |z|^2)^{2k-2}.$$ (9.409)

The integral converges in the usual sense for $k > \frac{1}{2}$. For $0 < k < \frac{1}{2}$ the integral is to be understood in the sense of its regularization (analytic continuation).

Thus

$$(f,g) = \frac{(2k-1)}{2\pi(1 - e^{4\pi ik})} \int_{\Sigma} dt (1-t)^{2k-2} \int_0^{2\pi} \overline{f(z)} g(z) d\theta \qquad (9.410)$$

where $\sum$ is a contour (in the t-plane) that starts from the origin along the positive real axis, encireles the point $+1$ once counterclockwise and returns to the origin along the positive real axis.

A complete orthonormal set is given by

$$u_n(z) = \sqrt{\frac{\Gamma(2k+n)}{\Gamma(2k)}} z^n, \ n = 0, 1, 2 \ldots. \qquad (9.411)$$

The principal vector is therefore given by,

$$e_z(z_1) = \sum_{n=0}^{\infty} \overline{u_n(z)} u_n(z_1) \qquad (9.412)$$

$$= (1 - \bar{z} z_1)^{-2k} \qquad (9.413)$$

so that

$$f(z) = (e_z, f) = \int_{|z_1|<1} (1 - z\bar{z}_1)^{-2k} f(z_1) d\lambda(z_1). \qquad (9.414)$$

The action of the group ring

$$T_x^{k+} = \int d\mu(u) x(u) T_u^{k+}$$

where $d\mu(u)$ is the invariant measure on $SU(1,1)$ is given by

$$T_x^{k+} f(z) = \int d\mu(u) x(u) (\beta z + \bar{\alpha})^{-2k} f\left(\frac{\alpha z + \bar{\beta}}{\beta z + \bar{\alpha}}\right). \qquad (9.415)$$

Now we use Eq. (9.414) to write

$$f\left(\frac{\alpha z + \bar{\beta}}{\beta z + \bar{\alpha}}\right) = \int_{|z_1|<1} \left[1 - \frac{(\alpha z + \bar{\beta})}{(\beta z + \bar{\alpha})} \bar{z}_1\right]^{-2k} f(z_1) d\lambda(z_1).$$

This immediately yields

$$T_x^{k+} f(z) = \int_{|z_1|<1} K(z, z_1) f(z_1) d\lambda(z_1) \qquad (9.416)$$

where

$$K(z, z_1) = \int (\beta z + \bar{\alpha})^{-2k} \left[1 - \frac{(\alpha z + \bar{\beta})}{\beta z + \bar{\alpha}} \bar{z}_1 \right]^{-2k} x(u) d\mu(u). \qquad (9.417)$$

Since the kernel is again of the Hilbert-Schmidt type we have

$$\text{Tr}\,(\text{T}_x^{k+}) = \int K(z, z) d\lambda(z) \qquad (9.418)$$

which can be written in the form

$$\text{Tr}\,(\text{T}_x^{k+}) = \int d\mu(u) x(u) \pi^{k+}(u) \qquad (9.419)$$

where

$$\pi^{k+}(u) = \int_{|z|<1} d\lambda(z)[(\beta z + \bar{\alpha}) - (\alpha z + \bar{\beta})\bar{z}]^{-2k}. \qquad (9.420)$$

We now substitute

$$z = \tanh \frac{\tau}{2} e^{i\theta} \qquad (9.421)$$

$$0 \le \tau < \infty; \quad 0 \le \theta \le 2\pi.$$

For the elliptic elements we have then

$$\pi^{k+}(u) = \frac{2k-1}{4\pi} \int_0^\infty d\tau \sinh \tau \int_0^{2\pi} d\phi \left[\cos \frac{\theta_0}{2} - i \sin \frac{\theta_0}{2} \hat{n}, \hat{r} \right]^{-2k} \qquad (9.422)$$

where $\hat{n}$ and $\hat{r}$ are unit time-like vectors

$$\hat{n} = (\cosh \rho, -\sin h\rho \sin \eta, \sinh \rho \cos \eta) \qquad (9.423a)$$

$$\hat{r} = (\cosh \tau, \sinh \tau \sin \phi, \sinh \tau \cos \phi) \qquad (9.423b)$$

and $\hat{n} \cdot \hat{r}$ is the Lorentz invariant form

$$\hat{n} \cdot \hat{r} = \hat{n}_3 \hat{r}_3 - \hat{n}_2 \hat{r}_2 - \hat{n}_1 \hat{r}_1.$$

We now perform a Lorentz transformation such that the time-axis coincides with the time like vector $\hat{n}$. Thus

$$\hat{n} \cdot \hat{r} = \cosh \tau$$

and we have

$$\pi^{k+}(u) = \frac{2k-1}{2} \int_0^\infty d\tau \sinh \tau \left[\cos \frac{\theta_0}{2} - i \sin \frac{\theta_0}{2} \cosh \tau \right]^{-2k}.$$

The integration is quite elementary and we have

$$\pi^{k+}(u) = \frac{e^{\frac{i\theta_0}{2}(2k-1)}}{\left(e^{\frac{-i\theta_0}{2}} - e^{\frac{i\theta_0}{2}}\right)}. \tag{9.424}$$

For the hyperbolic elements we substitute Eq. (9.362) and the transformation (9.421a) so that

$$\pi^{k+}(u) = \frac{2k-1}{4\pi} \int_0^\infty d\tau \sinh\tau \int_0^{2\pi} d\phi \left[\text{sgn}\lambda \cosh\frac{\sigma}{2} - i\sinh\frac{\sigma}{2}\hat{n}\cdot\hat{r}\right]^{-2k} \tag{9.425}$$

where $\hat{n}$ is the space-like vector

$$\hat{n} = (\sinh\rho, \, -\cosh\rho\cos\eta, \, -\cosh\rho\sin\eta)$$

and

$$\hat{r} = (\cosh\tau, \, \sinh\tau\sin\phi, \, \sinh\tau\cos\phi)$$

is as usual time-like. If we perform a Lorentz transformation such that the 1st space axis coincides with the fixed space like vector $\hat{n}$

$$\hat{n}\cdot\hat{r} = \sinh\tau\cos\phi$$

and define following Eq. (9.382),

$$\pi^{k+}(u) = \frac{1}{2}\left[\pi^{k+}(x+i0) + \pi^{k+}(x-i0)\right]$$

where $x = \text{sgn}\lambda\cosh\frac{\sigma}{2}$. Thus

$$\pi^{k+}(u) = \int_0^\infty d\tau \sinh\tau \int_0^{2\pi} d\phi \left[\cosh\frac{\sigma}{2} - i\sinh\frac{\sigma}{2}\sinh\tau\cos\phi\right]^{-2k}$$
$$\times \cos 2\pi\epsilon'\eta. \tag{9.426}$$

The integral can now be evaluated as in Ref. (7) and we have

$$\pi^{k+}(u) = \frac{e^{-\frac{\sigma}{2}(2k-1)}}{e^{\frac{\sigma}{2}} - e^{-\frac{\sigma}{2}}} \cos 2\pi\epsilon'\eta \tag{9.427}$$

where ϵ' is defined by

$$k = n+1-\epsilon' \quad 0 \le \epsilon' \le \frac{1}{2}$$
$$= n+2-\epsilon', \quad \frac{1}{2} \le \epsilon' < 1$$

and $\eta = 0, 1$.

C. *Negative discrete series* D_k^-

The finite element of the group in this case is taken to be

$$T_u^{k-} f(z) = (\alpha + \bar{\beta} z)^{-2k} f\left(\frac{\beta + \bar{\alpha} z}{\alpha + \bar{\beta} z}\right) \tag{9.428}$$

where $f(z)$ is analytic within the unit disc. The scalar product with respect which T_u^{k-} is unitary is given by

$$\begin{aligned}
(f, g) &= \int \overline{f(z)} g(z) d\lambda(z) \\
d\lambda(z) &= \frac{2k-1}{\pi} (1 - |z|^2)^{2k-2}.
\end{aligned} \tag{9.429}$$

The principal vector is as before

$$e_z(z_1) = (1 - \bar{z} z_1)^{-2k}$$

so that

$$f(z) = \int_{|z_1| < 1} (1 - z \bar{z}_1)^{-2k} f(z_1) d\lambda(z_1). \tag{9.430}$$

Proceeding in the same way as before

$$\pi^{k-}(u) = \int_{|z| < 1} [(\alpha + \bar{\beta} z) - (\beta + \bar{\alpha} z)\bar{z}]^{-2k} d\lambda(z). \tag{9.431}$$

Setting $z = \tanh \frac{\tau}{2} e^{i\phi}$ we have as before, for the elliptic elements

$$\pi^{k-}(u) = \frac{2k-1}{2} \int d\tau \sinh \tau \left[\cos \frac{\theta_0}{2} + i \sin \frac{\theta_0}{2} \cosh \tau\right]^{-2k} = \frac{e^{-\frac{i\theta_0}{2}(2k-1)}}{e^{\frac{i\theta_0}{2}} - e^{-\frac{i\theta_0}{2}}}. \tag{9.432}$$

For the hyperbolic elements a parallel calculation gives

$$\pi^{k-}(u) = \frac{e^{-\frac{\sigma}{2}(2k-1)} \cos 2\pi \epsilon' \eta}{[e^{\frac{\sigma}{2}} - e^{-\frac{\sigma}{2}}]}, \quad \eta = 0, 1 \tag{9.433}$$

and ϵ' is defined in the same way as before.

The Problem of Inversion and the Plancherel Formula

Let us start from,

$$\begin{aligned}
\mathrm{Tr}(T_x^{(\epsilon,s)}) = \sum_{\eta=0,1} \Bigg[&\int_{\text{elliptic}} x(u) \pi^{(\epsilon,s)}(u) d\mu(u) \\
&+ \int_{\text{hyperbolic}} x(u) \pi_\eta^{(\epsilon,s)}(u) d\mu(u) \Bigg].
\end{aligned} \tag{9.434}$$

For the principal series (as indicated by the index s) the first term is zero. For the hyperbolic elements

$$d\mu(u) = \sinh^2 \frac{\sigma}{2} \frac{d\sigma}{2} \cosh \rho d\rho d\theta \tag{9.435a}$$

$$\pi_\eta^{(\epsilon,s)}(u) = \frac{\cos s\sigma}{\sinh \frac{\sigma}{2}} \cos 2\pi\epsilon\eta. \tag{9.435b}$$

Thus if we define

$$\phi_\eta(t) = \int x(\theta, \rho; \eta, t) \cosh \rho d\rho d\theta$$

we have, with $\frac{\sigma}{2} = t$, $\eta = 0, 1$

$$\text{Tr}(T_x^{(\epsilon,s)}) = \int_0^\infty [\phi_0(t) + \cos 2\pi\epsilon\phi_1(t)] \sinh t \cos 2st dt. \tag{9.436}$$

Now we shall divide ϵ into two regions, $0 \le \epsilon \le \frac{1}{2}$, and $\frac{1}{2} < \epsilon < 1$. Thus

$$\text{Tr}(T_x^{\epsilon,s}) = \theta(2\epsilon)\theta(1 - 2\epsilon)\text{Tr}(T_x^{(\epsilon,s)}) + \theta(2\epsilon - 1)\theta(2 - 2\epsilon)\text{Tr}(T_x^{(\epsilon,s)}).$$

These two regions yield two sets of representations which are unitarily equivalent.

We now have

$$\text{Tr}(T_x^{k+} + T_x^{k-}) = \int [\phi_0(t) + \cos 2\pi\epsilon'\phi_1(t)] \times \sinh t dt e^{-\nu t}$$
$$- \int_0^\pi d\theta \sin^2 \theta \int \int x(\eta, \rho; \theta) \frac{\sin \nu\theta}{\sin \theta} \times \sinh \rho d\rho d\eta$$

where $\nu = \nu(\epsilon') = 2k - 1$.

We now set

$$\int \int x(\eta, \rho; \theta) \sinh \rho d\rho d\eta = \text{F}(\theta) \tag{9.437}$$

so that

$$\int_0^\pi \text{F}(\theta) \sin \theta \sin \nu\theta d\theta = \int_0^\infty [\phi_0(t) + \cos 2\pi\epsilon'\phi_1(t)] \sinh t e^{-\nu t} dt$$
$$-\text{Tr}(T_x^{k+} + T_x^{k-}). \tag{9.438}$$

Applying the inversion formula for the Fourier cosine transform

$$[\phi_0(t) + \cos 2\pi\epsilon'\phi_1(t)] \sinh t = \frac{4}{\pi} \int_0^\infty ds \cos 2st$$
$$\times [\theta(1 - 2\epsilon')\text{Tr}(T_x^{(\epsilon',s)})\theta(2\epsilon') + \theta(2\epsilon' - 1)\theta(2 - 2\epsilon')\text{Tr}(T_x^{(\epsilon',s)})] \tag{9.439}$$

we obtain

$$\int F(\theta) \sin\theta \sin\nu\theta d\theta$$

$$= \frac{4}{\pi} \int ds \int dt \, \text{Tr}(T_x^{(\epsilon',s)})\theta(1 - 2\epsilon')e^{-\nu t}\cos 2st\theta(2\epsilon')$$

$$+ \frac{4}{\pi} \int ds \int dt \, \text{Tr}(T_x^{(1-\epsilon',s)})\theta(2 - 2\epsilon')\theta(2\epsilon' - 1)e^{-\nu t}\cos 2st$$

$$- \text{Tr}(T_x^{k+} + T_x^{k-}). \tag{9.440}$$

The above formula implies that in the Fourier sine transform of the function $F(\theta)\sin\theta$,

$$F(\theta)\sin\theta = \frac{2}{\pi}\int_0^\infty c(\nu)\sin\nu\theta d\nu$$

$$c(\nu) = \frac{4}{\pi}\int ds \int dt \, \text{Tr}(T_x^{(\epsilon,s)})e^{-\nu t}\theta(1 - 2\epsilon')\cos 2st\theta(2\epsilon')$$

$$+ \frac{4}{\pi}\int ds \int dt \, \text{Tr}(T_x^{(1-\epsilon',s)})e^{-\nu t}\theta(2 - 2\epsilon')\theta(2\epsilon' - 1)\cos 2st$$

$$- \text{Tr}(T_x^{k+} + T_x^{k-}). \tag{9.441}$$

Thus we have

$$F(\theta)\sin\theta = \frac{8}{\pi^2}\int ds \int dt\theta(1 - 2\epsilon')\theta(2\epsilon')\text{Tr}(T_x^{(\epsilon',s)})\int e^{-\nu t}\cos 2st\sin\nu\theta d\nu$$

$$+ \frac{8}{\pi^2}\int ds \int dt\theta(2 - 2\epsilon')\theta(2\epsilon' - 1)\text{Tr}(T_x^{(1-\epsilon',s)})\cos 2st$$

$$\times \int e^{-\nu t}\sin\nu\theta d\nu - \int_0^\infty \text{Tr}(T_x^{k+} + T_x^{k-})\sin\nu\theta d\nu.$$

In the first integral,

$$k = n + 1 - \epsilon' \quad \nu = 2k - 1 = 2n + 1 - \tau \quad d\nu = -d\tau.$$

Similarly in the second

$$k = n + 2 - \epsilon' \quad \nu = 2n + 1 + \tau \quad \tau = 2 - 2\epsilon'; \quad d\nu = d\tau.$$

Thus we have

$F(\theta)\sin\theta$

$$= \frac{8}{\pi^2} \int_0^\infty ds \int_0^1 d\tau \mathrm{Tr}\left(\mathrm{T}_x^{\left(\frac{\tau}{2},s\right)}\right) \int_0^\infty dt \cos 2st$$

$$\times \sum_{n=0}^\infty [e^{-(2n+1-\tau)t}\sin(2n+1-\tau)\theta + e^{-(2n+1+\tau)t}\sin(2n+1+\tau)\theta]$$

$$-\frac{2}{\pi}\int \mathrm{Tr}(\mathrm{T}_x^{k+} + \mathrm{T}_x^{k-})\sin\nu\theta d\nu. \tag{9.442}$$

The summation can be easily carried out recalling that

$$\sum e^{-(2n+1\mp\tau)t}\sin(2n+1\mp\tau)\theta = \mathrm{Im}\, e^{-t(1\mp\tau)+i(1\mp\tau)\theta} \times \sum e^{-2nt+2in\theta}$$

$$\tag{9.443}$$

and using the standard summation formula for the geometric series. Thus

$$F(\theta)\sin\theta = \frac{8}{\pi^2}\int_0^\infty ds \int_0^1 d\tau \mathrm{Tr}(\mathrm{T}_x^{\left(\frac{\tau}{2},s\right)})$$

$$\times \left[\sin(1+\tau)\theta \int_0^\infty dt \frac{\cos 2st \cosh(1-\tau)t}{\cosh 2t - \cos 2\theta}\right.$$

$$\left. + \sin(1-\tau)\theta \int_0^\infty dt \frac{\cos 2st \cosh(1+\tau)t}{\cosh 2t - \cos 2\theta}\right]$$

$$-\frac{2}{\pi}\int_0^\infty d\nu \sin\nu\theta \mathrm{Tr}(\mathrm{T}_x^{k+} + \mathrm{T}_x^{k-}). \tag{9.444}$$

The integrals can be written in the form,

$$\frac{1}{2}\int_{-\infty}^\infty e^{2ist}\frac{\cosh(1\mp\tau)tdt}{\cosh 2t - \cos 2\theta}$$

which can be evaluated by the sum of the residues at the poles in the upper half-plane located at

$$t = i(\theta + n\pi), \ n = 0, 1, 2, \dots$$

and at

$$t = i(n\pi - \theta), \ n = 1, 2, 3 \dots.$$

It can be easily ascertained that

$$\sin(1+\tau)\theta \int_0^\infty \frac{\cos 2st \cosh(1-\tau)t}{\cosh 2t - \cos 2\theta} dt$$

$$+ \sin(1-\tau)\theta \int_0^\infty \frac{\cos 2st \cosh(1+\tau)t}{\cosh 2t - \cos 2\theta} dt = \frac{\pi}{2} \text{Re} \frac{\cosh\left[\pi\left(s + \frac{i\tau}{2}\right) - 2s\theta\right]}{\cosh \pi\left(s + \frac{i\tau}{2}\right)}.$$

We therefore obtain

$$F(\theta)\sin\theta = \frac{4}{\pi} \int_0^\infty ds \int_0^1 d\tau \text{Tr}(T_x^{\left(\frac{\tau}{2}, s\right)}) \text{Re} \frac{\cosh\left[\pi\left(s + \frac{i\tau}{2}\right) - 2s\theta\right]}{\cosh \pi\left(s + \frac{i\tau}{2}\right)}$$

$$- \frac{4}{\pi} \int dk \sin(2k-1)\theta \text{Tr}[T_x^{k+} + T_x^{k-}]. \qquad (9.445)$$

It now remains to relate $x(e)$ with $F(\theta)\sin\theta$. For this we equate two different calculations for the residue at the pole at $\lambda = -\frac{3}{2}$ of the generalized function[8]

$$(x_3^2 - x_2^2 - x_1^2)_+^\lambda$$

as an analytic function of λ. Let us consider

$$I(\lambda) = \int_{x_0 > 0} (x_3^2 - x_2^2 - x_1^2)_+^\lambda x(u) d\mu(u) \qquad (9.446)$$

where $d\mu(u)$ stands for the invariant measure for the elliptic elements

$$x_3 = \sin\theta \cosh\rho \quad x_2 = \sin\theta \sinh\rho \sin\eta$$

$$x_1 = \sin\theta \sinh\rho \cos\eta; \ 0 \leq \eta \leq 4\pi, \ 0 \leq \rho < \infty; \ 0 \leq \theta \leq \pi$$

where $\theta = \frac{\theta_0}{2}$, ρ, η are defined and $x_0 = \cos\theta$.
It now immediately follows

$$d\mu(u) = \frac{dx_1 dx_2 dx_3}{|x_0|}. \qquad (9.447)$$

Thus

$$I(\lambda) = \int_{x_0 > 0} (x_3^2 - x_2^2 - x_1^2)_+^\lambda \phi(x) dx_1 dx_2 dx_3 \qquad (9.448)$$

$$\phi(x) = \frac{x(u)}{|x_0|}.$$

Setting

$$x_1^2 + x_2^2 = v \quad x_3^2 = u$$

$$v = ut$$

we have

$$I(\lambda) = -\frac{1}{4} \int du\, u^{\lambda + \frac{1}{2}} \int_0^1 (1-t)^\lambda \psi(u, tu) dt \qquad (9.449)$$

where

$$\psi(u, tu) = \int_0^{4\pi} \phi(\sqrt{ut}\cos\eta, \sqrt{ut}\sin\eta, \sqrt{u})d\eta. \qquad (9.450)$$

Since $\phi(x)$ vanishes outside a bounded set the integral $I(\lambda)$ converges in the usual sense for $\mathrm{Re}\,\lambda > -1$. For $\mathrm{Re}\,\lambda < -1$ it is to be understood in the sense of its regularization (analytic continuation):

$$I(\lambda) = \frac{1}{4} \frac{1}{(e^{2\pi i\lambda} - 1)(e^{2\pi i\lambda} + 1)} \int_\infty^{0+} du\, u^{\lambda+\frac{1}{2}} \int_0^{1+} (1-t)^\lambda \psi(u, tu) dt. \qquad (9.451)$$

We now define

$$\Phi(\lambda, u) = -\frac{1}{4} \frac{1}{(e^{2\pi i\lambda} - 1)} \int_0^{1+} (1-t)^\lambda \psi(u, tu) dt. \qquad (9.452)$$

Just as the generalized function $(x_+^\lambda, \phi)\Phi(\lambda, u)$ is regular for all λ except at

$$\lambda = -1, -2, -3, \ldots$$

where it has simple poles.

On the other hand at regular points of $\Phi(\lambda, u)$ the integral

$$I(\lambda) = -\frac{1}{(e^{2\pi i\lambda} + 1)} \int_\infty^{0+} du\, u^{\lambda+\frac{1}{2}} \Phi(\lambda, u) \qquad (9.453)$$

also have poles at

$$\lambda = -\frac{3}{2}, -\frac{5}{2}, \ldots$$

which are once again simple poles. The analytic function $I(\lambda)$ can be written as

$$I(\lambda) = \sum_{\eta=0}^\infty \frac{\frac{1}{n!}\left[\frac{\partial^n}{\partial u^n}\Phi(\lambda, u)\right]_{u=0}}{\lambda + \frac{3}{2} + n} + E(\lambda) \qquad (9.454)$$

where $E(\lambda)$ is an entire function. Thus

$$\mathrm{Res}[I(\lambda)]_{\lambda=-\frac{3}{2}} = \Phi\left(-\frac{3}{2}, 0\right) = -\frac{1}{2}\psi(0, 0).$$

Now

$$\psi(0, 0) = \int_0^{4\pi} \phi(0, 0, 0)d\eta = 4\pi x(e).$$

Thus

$$\text{Res}[I(\lambda)]_{\lambda=-\frac{3}{2}} = -2\pi x(e). \qquad (9.455)$$

We shall calculate the same thing in another way by writing

$$I(\lambda) = \int_0^{\frac{\pi}{2}} \sin^{2\lambda+1}\theta[F(\theta)\sin\theta].$$

We now write $\sin\theta = \theta[1 - u(\theta)]$ where

$$u(0) = \left[\frac{du}{d\theta}\right]_{\theta=0} = 0.$$

Thus

$$I(\lambda) = \int_0^{\frac{\pi}{2}} \theta^{2\lambda+1}G(\theta)d\theta \qquad (9.456a)$$

where

$$G(\theta) = F(\theta)\sin\theta[1 - u(\theta)]^{2\lambda+1}. \qquad (9.456b)$$

Since $G(\theta)$ has a compact support and is regular at $\theta = 0$ we have [8] for $\text{Re}\,\lambda > -\frac{n}{2} - 1$

$$I(\lambda) = \int_0^{\frac{\pi}{2}} \theta^{2\lambda+1}\left[G(\theta) - \sum_{r=0}^{n-1}\frac{G^{(r)}(0)\theta^r}{r!}\right] + \sum_r \frac{G^{(r)}(0)}{r!(2\lambda+2+r)}. \qquad (9.457)$$

Hence

$$\text{Re}\,s[I(\lambda)]_{\lambda=-\frac{3}{2}} = \frac{1}{2}G'(0) = \frac{1}{2}\left\{\frac{d}{d\theta}[\sin\theta F(\theta)]\right\}_{\theta=0}. \qquad (9.458)$$

Equating Eqs. (9.455) and (9.458) we have

$$x(e) = -\frac{1}{4\pi}\left\{\frac{d}{d\theta}[\sin\theta F(\theta)]\right\}_{\theta=0}. \qquad (9.459)$$

Combining Eqs. (9.445) and (9.459) we immediately obtain

$$x(e) = \frac{2}{\pi^2}\int_0^\infty ds \int_0^1 d\tau \text{Tr}(T_x^{\left(\frac{\tau}{2},s\right)}) \times s\text{Re}\tanh\pi\left(s + \frac{i\tau}{2}\right)$$

$$+ \frac{2}{\pi^2}\int_{\frac{1}{2}}^\infty dk\left(k - \frac{1}{2}\right)\text{Tr}(T_x^{k+} + T_x^{k-}). \qquad (9.460)$$

Replacing $x(e)$ by $x_1^\dagger x_2(e)$ and using Eq. (9.375) we have

$$\int \overline{x_1(u)} x_2(u) d\mu(u)$$

$$= \frac{2}{\pi^2} \int_0^\infty ds \int_0^1 d\tau \operatorname{Tr}[\operatorname{T}_{x_1}^{(\frac{\tau}{2},s)^\dagger} \operatorname{T}_{x_2}^{(\frac{\tau}{2},s)}] s \operatorname{Re} \tanh \pi \left(s + \frac{i\tau}{2} \right)$$

$$+ \frac{2}{\pi^2} \int_{\frac{1}{2}}^\infty dk \left(k - \frac{1}{2} \right) \operatorname{Tr}(\operatorname{T}_{x_1}^{k+\dagger} \operatorname{T}_{x_2}^{k+} + \operatorname{T}_{x_1}^{k-\dagger} \operatorname{T}_{x_2}^{k-}). \qquad (9.462)$$

This is the analogue of the Plancherel formula for the ordinary Fourier transform. Harish–Chandra's Plancherel formula for $\mathrm{SL}(2, \mathbb{R})$ corresponds to the point contributions at $\tau = 0, 1$:

$$\int \overline{x_1(u)} x_2(u) d\mu(u)$$

$$= \frac{2}{\pi^2} \sum_{\tau=0,1} \int ds \, s \operatorname{Re} \tanh \pi \left(s + \frac{i\tau}{2} \right) \operatorname{Tr} \left[\operatorname{T}_{x_1}^{\dagger(\frac{\tau}{2},s)} \operatorname{T}_{x_2}^{(\frac{\tau}{2},s)} \right]$$

$$+ \frac{2}{\pi^2} \sum_{k=1,\frac{3}{2}\cdots} \left(k - \frac{1}{2} \right) \operatorname{Tr} \left[\operatorname{T}_{x_1}^{k+\dagger} \operatorname{T}_{x_2}^{k+} + \operatorname{T}_{x_1}^{k-\dagger} \operatorname{T}_{x_2}^{k-} \right].$$

The Plancherel formula for the universal covering group of $\mathrm{SL}(2, \mathbb{R})$ [denoted by $\widetilde{\mathrm{SL}}(2, \mathbb{R})$] was written down by Pukanszky.[9] Following Pukanszky's work the more general problem of Plancherel theorem for semisimple groups was attempted by Herb and Wolf[10] and by Duflo and Vergne[11] who claim to be in agreement with Pukanszky.[9] The edifice of Herb and Wolf's work[10] on general semisimple groups is built on Pukanszky's work on $\widetilde{\mathrm{SL}}(2, \mathbb{R})$ ("we also need this special result to do the general case").

However Pukanszky–Herb–Wolf formula for $\widetilde{\mathrm{SL}}(2, \mathbb{R})$ which in our notation reads,

$$x(e) = \frac{2}{\pi^2} \int_0^\infty ds \int_0^1 d\tau s \operatorname{Re} \tanh \pi(s + i\tau) \operatorname{Tr} [\operatorname{T}_x^{(\tau,s)}]$$

$$+ \frac{2}{\pi^2} \int_{\frac{1}{2}}^\infty dk \left(k - \frac{1}{2} \right) \operatorname{Tr} [\operatorname{T}_x^{k+} + \operatorname{T}_x^{k-}]$$

suffers from a serious flaw. All these authors treat the set of unitarily equivalent representations C_s^ϵ and $C_s^{1-\epsilon}$ as distinct. This flaw in Pukanszky's work has been carried into the later work of Herb and Wolf[10] as well as in that of Duflo and Vergne.

Therefore rectification of this flaw seems to be necessary before the more general problem of semisimple group is attempted.

The general method adopted in this problem was first suggested by Gel'fand[12] and followed for SU(1,1) later by Vilenkin and Klimyk.[13] However the Vilenkin–Klimyk[13] method does not go beyond the integral and half-integral representations of Bargmann. The method is applied here for the first time for the covering group $\widetilde{SL}(2, R)$.

The SU(1, 1) group and its covering group have manifold applications in quantum optics, construction of squeezed states etc. The Barut–Girardello coherent states,[14] for example, have been applied to quantum optics of laser by Agarwal[15] following a suggestion of Klauder.[15] The squeezed states were envisaged to have very little uncertainty in one sector, of course, at the expense of the other. These problems have been considered in some detail by Klauder and Skagerstam.[16]

References

[1] V. Bargmann, *Ann. Math.* **48** (1947) 568.

[2] I. M. Gel'fand, M. I. Graev and N. Ya. Vilenkin, *Generalized Functions*, Vol. 5, (Academic N.Y., 1966), p. 396.

[3] See Ref. 2.

[4] (a) L. Pukanszky, *Trans. Am. Math. Soc.* **100** (1961) 116. (b) S. D. Majumdar, *J. Math. Phys.* **17** (1976) 194. (c) D. Basu and S. D. Majumdar, *J. Math. Phys.* **20** (1979) 459.

[5] I. M. Gel'fand and M. A. Naimark, I. M. Gel'fand, *Collected Papers* (Springer Verlag, Berlin 1988), Vol. II, pp. 41, 182.

[6] V. Bargmann, *Commun. Pure Appl. Math.* **14** (1961) 187; **20** (1967) 1. I. E. Segal Ill. *J. Math.* **6** (1962) 500.

[7] (a) S. Bal, K. V. Shajesh and D. Basu, *J. Math. Phys.* **38** (1997) 3209. (b) D. Basu, S. Bal and K. V. Shajesh, *J. Math. Phys.* **41** (2000) 1. (c) The Plancherel formula for the universal covering group of SL(2, R) revisited [arxiv:0710.2224V3].

[8] I. M. Gel'fand and G. E. Shilov, *Generalized Functions*, Vol. 1 (1964) p. 253. [see also Chap. 6].

[9] L. Pukanszky, *Math. Annalen* **156** (1964) 96.

[10] R. Herb and J. A. Wolf, *Compositio Mathematica* **57**(3) (1986) 271.

[11] M. Duflo and M. Vergne, La Formula de Plancherel de groups de Lie semi-simples reals, *Advance Studies in Pure Math.* **14** (1988) 289.

[12] I. M. Gel'fand, *Collected Papers*, Vol. I, (Proc. Int. Congr. Math. 1954, Amsterdam (1957)), 253–276, *Zbl.* **79**, 326.

[13] N. Ja. Vilenkin and A. U. Klimyk, *Representation of Lie Groups and Special Functions*, Vol. 1 Chap. 6, (Kluwer Academic, Boston, 1991), p. 298.

[14] (a) A. O. Barut and L. Girardello, *Commun. Math. Phys.* **21** (1971) 41.
(b) D. Basu, *J. Math. Phys.* **33**(1) (1992).
[15] G. S. Agarwal, *Phys. Rev. Lett.* **5** (1985) 687.
[16] J. R. Klauder and B. S. Skagerstam, *Coherent States, Application in Physics and Mathematical Physics* (World Scientific, Singapore, 1985).

CHAPTER 10

The Four-Dimensional Lorentz Group

In this chapter we shall consider the representation theory and harmonic analysis for the group of complex unimodular matrices

$$\begin{pmatrix} a & b \\ c & d \end{pmatrix}; \quad ad - bc = 1.$$

As will be shown below this group is locally isomorphic to the group of Lorentz transformations. It is the simplest of the so-called simple Lie groups and clearly shows the difference of harmonic analysis on this group from the traditional harmonic analysis in Euclidean spaces, namely, the Fourier integral. Since the analogues of exponentials $(e^{i\lambda x})$ are unitary representations, we shall describe the Gel'fand–Naimark[1] construction of the unitary representations of the group in some detail. In a sense this group is simpler insofar as it readily generalizes to arbitrary complex semi-simple groups[2] unlike the real group discussed in the last chapter.

10.1 The Four-Dimensional Lorentz Group

It is well known that the laws of classical mechanics do not depend upon the choice of any particular fixed coordinate system and are not altered by going from one inertial frame to another. If ν_x, ν_y, ν_z are the components of the velocity of the second system, relative to the first then, $x = x + \nu_x t, y = y + \nu_y t, z = z + \nu_z t$. A transformation of this form is called a Galilean transformation.

In this classical mechanics proceeds from the assumption that the time t may be taken the same for both the coordinate systems. The theory of relativity rejects this assumption and regards the velocity of light c as a constant. It assigns to each inertial system its own time t. The passage from the inertial system x, y, z whose time is t is accomplished by a linear transformation of the variables x, y, z, t which leaves invariant the quadratic form,

$$x^2 + y^2 + z^2 - c^2 t^2 \tag{10.1}$$

where c is the velocity of light.

The invariance of the form (10.1) is a mathematical expression of the fact that the velocity of light in vacuum is the same for any inertial system.

It will be convenient to use, instead of the variable t a new variable

$$x_4 = ct. \tag{10.2}$$

A general Lorentz transformation can then be described as a linear transformation

$$x_i = \sum_j a_{ij} x_j. \tag{10.3}$$

The Lorentz transformation a is then determined by the condition

$$x_1^2 + x_2^2 + x_3^2 - x_4^2 = x_1'^2 + x_2'^2 + x_3'^2 - x_4'^2. \tag{10.4}$$

In matrix notation

$$x' = ax. \tag{10.5}$$

We now introduce

$$g = \begin{pmatrix} 1 & 0 & 0 & 0 \\ 0 & 1 & 0 & 0 \\ 0 & 0 & 1 & 0 \\ 0 & 0 & 0 & -1 \end{pmatrix}. \tag{10.6}$$

Then $x_1^2 + x_2^2 + x_3^2 - x_4^2 = \tilde{x}gx$, hence condition (10.4) can be written as $\tilde{x}'g\tilde{x}' = \tilde{x}gx$.

From the above it easily follows

$$(\det a)^2 = 1$$

$$\det a = \pm 1.$$

The condition $\det a = 1$ corresponds to the proper Lorentz transformation. We shall consider the proper Lorentz group corresponding to $\det a = 1$.

10.2   The Spinor Description of Proper Lorentz Group

We denote by $\mathrm{SL}(2, \mathrm{C})$ the aggregate of all complex matrices

$$g = \begin{pmatrix} a & b \\ c & d \end{pmatrix} \tag{10.7}$$

of order 2 whose determinant is equal to unity, $\det g = ad - bc = 1$.

Evidently $\mathrm{SL}(2, \mathrm{C})$ is a group under matrix multiplication. It is called complex unimodular group of order 2.

In what follows we shall need the important fact that the elements of the proper Lorentz group $SO(3,1)$ can be given by means of the matrices $g \in SL(2,C)$. This description of the group $SL(2,C)$ can be arrived at in the following way. Let us consider the Hermitian matrix

$$X = \begin{pmatrix} X_{11} & X_{12} \\ X_{21} & X_{22} \end{pmatrix} = \begin{pmatrix} x_4 + x_3 & x_1 + ix_2 \\ x_1 - ix_2 & x_4 - x_3 \end{pmatrix}. \tag{10.8}$$

We shall regard X_{ik} as variables and define a linear transformation of these variables by putting

$$X' = gXg^\dagger \tag{10.9}$$

and

$$\dot{X} = \begin{pmatrix} X_{11} \\ X_{12} \\ X_{21} \\ X_{22} \end{pmatrix} = \begin{pmatrix} x_4 + x_3 \\ x_1 + ix_2 \\ x_1 - ix_2 \\ x_4 - x_3 \end{pmatrix} = Tx \tag{10.10}$$

where

$$x = \begin{pmatrix} x_1 \\ x_2 \\ x_3 \\ x_4 \end{pmatrix}, \quad T = \begin{pmatrix} 0 & 0 & 1 & 1 \\ 1 & i & 0 & 0 \\ 1 & -i & 0 & 0 \\ 0 & 0 & -1 & 1 \end{pmatrix}. \tag{10.11}$$

From Eq. (10.9), $\det X' = \det X$

$$x_4'^2 - x_3'^2 - x_2'^2 - x_1'^2 = x_4^2 - x_3^2 - x_2^2 - x_1^2.$$

Hence the transformation $X \to X'$ is a Lorentz transformation. The transformation Eq. (10.9) can be written as

$$X' = gXg^\dagger = \begin{pmatrix} a & b \\ c & d \end{pmatrix} \begin{pmatrix} X_{11} & X_{12} \\ X_{21} & X_{22} \end{pmatrix} \begin{pmatrix} \bar{a} & \bar{c} \\ \bar{b} & \bar{d} \end{pmatrix}. \tag{10.12}$$

Using the definitions of X_{ik} and the column matrix $\dot{X}$ (see Eq. 10.10) we have from Eq. (10.12),

$$\dot{X}' = \begin{pmatrix} X_{11}' \\ X_{12}' \\ X_{21}' \\ X_{22}' \end{pmatrix} = \begin{pmatrix} |a|^2 & a\bar{b} & \bar{a}b & |b|^2 \\ a\bar{c} & a\bar{d} & b\bar{c} & b\bar{d} \\ \bar{a}c & \bar{b}c & \bar{a}d & \bar{b}d \\ |c|^2 & c\bar{d} & \bar{c}d & |d|^2 \end{pmatrix} \begin{pmatrix} X_{11} \\ X_{12} \\ X_{21} \\ X_{22} \end{pmatrix} = ATx \tag{10.13}$$

where A is the matrix multiplying the column matrix $\dot{X}$.

The Lorentz Transformation is therefore, given by

$$L = T^{-1}AT.$$

Calculating T^{-1} and carrying out the matrix multiplication we finally obtain,

$$L = \frac{1}{2}\begin{pmatrix} a\bar{d} + \bar{b}c + b\bar{c} + \bar{a}d & i(a\bar{d} + \bar{b}c - b\bar{c} - \bar{a}d) \\ i(\bar{b}c + \bar{a}d - b\bar{c}) & \bar{a}d + a\bar{d} - b\bar{c} - \bar{b}c \\ a\bar{b} - c\bar{d} + \bar{a}b - \bar{c}d & i(a\bar{b} - c\bar{d} - \bar{a}b + \bar{c}d) \\ a\bar{b} + c\bar{d} + \bar{a}b + \bar{c}d & i(a\bar{b} + c\bar{d} - \bar{a}b - \bar{c}d) \end{pmatrix}$$

$$\begin{pmatrix} (a\bar{c} + ac - b\bar{d} - \bar{b}d) & (a\bar{c} + \bar{a}c + b\bar{d} + \bar{b}d) \\ i(\bar{a}c - a\bar{c} - \bar{b}d + b\bar{d}) & i(\bar{a}c - a\bar{c} + \bar{b}d - b\bar{d}) \\ |a|^2 - |c|^2 - |b|^2 + |d|^2 & |a|^2 - |c|^2 + |b|^2 - |d|^2 \\ |a|^2 + |c|^2 - |b|^2 - |d|^2 & |a|^2 + |c|^2 + |b|^2 + |d|^2 \end{pmatrix} . \quad (10.14)$$

Thus the pure rotation

$$a_3(\theta) = \begin{pmatrix} \cos\theta & -\sin\theta & 0 & 0 \\ \sin\theta & \cos\theta & 0 & 0 \\ 0 & 0 & 1 & 0 \\ 0 & 0 & 0 & 1 \end{pmatrix}, \tilde{a}_3(\theta) = \begin{pmatrix} e^{i\theta/2} & 0 \\ 0 & e^{-i\theta/2} \end{pmatrix} \quad (10.15)$$

$$a_2(\theta) = \begin{pmatrix} \cos\theta & 0 & \sin\theta & 0 \\ 0 & 1 & 0 & 0 \\ -\sin\theta & 0 & \cos\theta & 0 \\ 0 & 0 & 0 & 1 \end{pmatrix}, \tilde{a}_2(\theta) = \begin{pmatrix} \cos\frac{\theta}{2} & -\sin\frac{\theta}{2} \\ \sin\frac{\theta}{2} & \cos\frac{\theta}{2} \end{pmatrix} \quad (10.16)$$

$$a_1(\theta) = \begin{pmatrix} 1 & 0 & 0 & 0 \\ 0 & \cos\theta & -\sin\theta & 0 \\ 0 & \sin\theta & \cos\theta & 0 \\ 0 & 0 & 0 & 1 \end{pmatrix}, \tilde{a}_1(\theta) = \begin{pmatrix} \cos\frac{\theta}{2} & i\sin\frac{\theta}{2} \\ i\sin\frac{\theta}{2} & \cos\frac{\theta}{2} \end{pmatrix} \quad (10.17)$$

$$a_1' = \left(\frac{da_1}{d\theta}\right)_{\theta=0} = \begin{pmatrix} 0 & 0 & 0 & 0 \\ 0 & 0 & -1 & 0 \\ 0 & 1 & 0 & 0 \\ 0 & 0 & 0 & 0 \end{pmatrix},$$

$$\tilde{a}_1' = \left(\frac{d\tilde{a}_1}{d\theta}\right)_{\theta=0} = \frac{i}{2}\begin{pmatrix} 0 & 1 \\ 1 & 0 \end{pmatrix} \quad (10.18)$$

$$a_2' = \left(\frac{d\tilde{a}_2}{d\theta}\right)_{\theta=0} = \frac{1}{2}\begin{pmatrix} 0 & 1 \\ 1 & 0 \end{pmatrix}, a_3' = \left(\frac{da_3}{d\theta}\right)_{\theta=0} = \frac{i}{2}\begin{pmatrix} 1 & 0 \\ 0 & -1 \end{pmatrix} \quad (10.19)$$

$$J_1 = i\tilde{a}_1' = -\frac{1}{2}\begin{pmatrix} 0 & -1 \\ 1 & 0 \end{pmatrix} = -\frac{1}{2}\sigma_1 \quad (10.20)$$

$$J_2 = i\tilde{a}_2' = \frac{1}{2}\begin{pmatrix} 0 & -i \\ i & 0 \end{pmatrix} = \frac{1}{2}\sigma_2 \quad (10.21)$$

$$J_3 = i\tilde{a}_3' = -\frac{1}{2}\begin{pmatrix} 1 & 0 \\ 0 & -1 \end{pmatrix} = -\frac{1}{2}\sigma_3. \quad (10.22)$$

We now consider the Lorentz boosts

$$F_1 = i\tilde{b}_1' = \frac{i\sigma_1}{2}, \quad \tilde{b}_1(\tau) = \begin{pmatrix} \cosh\left(\frac{\tau}{2}\right) & \sinh\left(\frac{\tau}{2}\right) \\ \sinh\left(\frac{\tau}{2}\right) & \cosh\left(\frac{\tau}{2}\right) \end{pmatrix}$$

$$F_2 = i\tilde{b}_2' = \frac{i\sigma_2}{2}, \quad \tilde{b}_2(\tau) = \begin{pmatrix} \cosh\left(\frac{\tau}{2}\right) & i\sinh\left(\frac{\tau}{2}\right) \\ -i\sinh\left(\frac{\tau}{2}\right) & \cosh\left(\frac{\tau}{2}\right) \end{pmatrix}$$

$$F_3 = i\tilde{b}_3' = \frac{i\sigma_3}{2}, \quad \tilde{b}_3(\tau) = \begin{pmatrix} e^{\tau/2} & 0 \\ 0 & e^{-\tau/2} \end{pmatrix}. \quad (10.23)$$

The Lie algebra of $SL(2, C)$ is given by the commutation relations

$$[J_k, J_l] = i\epsilon_{klm}J_m \quad (10.24a)$$

$$[J_k, F_l] = i\epsilon_{klm}F_m \quad (10.24b)$$

$$[F_k, F_l] = -i\epsilon_{klm}J_m. \quad (10.24c)$$

10.3 Determination of the Infinitesimal Operators of a Representation

In place of the operators $J_k, F_k, k = 1, 2, 3$, it will be convenient for us to seek the following linear combinations of them

$$J_\pm = J_1 \pm iJ_2, \ F_\pm = F_1 \pm iF_2.$$

It is easy to find the commutators of these operators

$$[J_+, J_3] = -J_+, [J_+, F_+] = [J_-, F_-] = [J_3, F_3] = 0$$
$$[J_-, J_3] = -J_-, [J_+, J_-] = 2J_3$$

$$[F_+, F_3] = +J_+, [F_-, F_3] = -J_-$$

$$[F_+, F_-] = -2J_3, [J_-, F_3] = F_-$$

$$[J_+, F_-] = -[J_-, F_+] = 2F_3$$

$$[F_+, J_3] = -F_+, [F_-, J_3] = -F_-. \tag{10.25}$$

A given representation of the group $SL(2, C)$ is also a representation of the group $SU(2)$. Let H^j be the corresponding invariant subspaces so that the space R of the representation $g \to T_g$ is the direct sum of the subspaces H^j. Let $f_{jm}, m = -j, -j+1, \ldots j$ be a canonical basis for H^j. Since H^j is invariant relative to the operators $J_\pm, J_3$ of the $SU(2)$ subgroup we have

$$J_\pm f_{jm} = \sqrt{(J \mp m)(J \pm m + 1)} f_{jm\pm 1} \tag{10.26a}$$

$$J_3 f_{jm} = m f_{jm}. \tag{10.26b}$$

We now proceed to determine the action of $F_\pm, F_3$ on f_{jm}. We first note that $\mp F_\pm, \sqrt{2} F_3$ are the ± 1 and 0 components of a vector operator A^1, i.e.

$$-F_+ = A_1^1 \quad F_3 = \frac{1}{\sqrt{2}} A_0^1$$

$$F_- = A_{-1}^1. \tag{10.27}$$

Hence we have

$$F_+ f_{jm} = -A_1^1 f_{jm} = -\sum_{j'm'} (f_{j'm'}, A_1^1 f_{jm}) f_{j'm'}. \tag{10.28}$$

Now by the Wigner–Eckart theorem of $SU(2)$,

$$(f_{j'm'}, A_1^1 f_{jm}) = C \begin{pmatrix} j & 1 & j' \\ m & 1 & m' \end{pmatrix} \langle j' \parallel A^1 \parallel j \rangle. \tag{10.29}$$

Thus

$$F_+ f_{jm} = \sum_{j'm'} C \begin{pmatrix} j & 1 & j' \\ m & 1 & m' \end{pmatrix} \langle j' \parallel A^1 \parallel j \rangle f_{j'm'}. \tag{10.30}$$

The Clebsch–Gordan coefficient vanishes unless

$$j' = j+1, \; j, j-1 \quad \text{and} \quad m' = m+1. \tag{10.31}$$

Hence

$$F_+ f_{jm} = -C \begin{pmatrix} j & 1 & j-1 \\ m & 1 & m+1 \end{pmatrix} C_{j-1,j} \; f_{j-1,m+1}$$

$$-C \begin{pmatrix} j & 1 & j \\ m & 1 & m+1 \end{pmatrix} C_{jj} \; f_{j,m+1} - C \begin{pmatrix} j & 1 & j+1 \\ m & 1 & m+1 \end{pmatrix} C_{j+1,j} \; f_{j+1,m+1}$$

where

$$C_{j-1,j} = \langle j-1\|A^1\|j\rangle \quad C_{jj} = \langle j\|A^1\|j\rangle$$
$$C_{j+1,j} = \langle j+1\|A^1\|j\rangle. \tag{10.32}$$

Using the tabulated values of the Clebsch–Gordan coefficients from Condon and Shortley[3] and absorbing the purely j-dependent factors appropriately we have

$$\begin{aligned}
F_+ f_{jm} &= -\sqrt{(j-m)(j-m-1)}A_j\, f_{j-1,m+1} \\
&\quad + \sqrt{(j-m)(j+m+1)}B_j\, f_{j,m+1} \\
&\quad - \sqrt{(j+m+1)(j+m+2)}C_{j+1}\, f_{j+1,m+1}. \tag{10.33}
\end{aligned}$$

We now consider

$$\begin{aligned}
F_3 f_{jm} &= \frac{1}{\sqrt{2}}A_0^1\, f_{jm} \\
&= \frac{1}{\sqrt{2}}C\begin{pmatrix} j & 1 & j-1 \\ m & 0 & m \end{pmatrix}\langle j-1\|A^1\|j\rangle\ f_{j-1,m} \\
&\quad + \frac{1}{\sqrt{2}}C\begin{pmatrix} j & 1 & j \\ m & 0 & m \end{pmatrix}\langle j\|A^1\|j\rangle\ f_{jm} \\
&\quad + \frac{1}{\sqrt{2}}C\begin{pmatrix} j & 1 & j+1 \\ m & 0 & m \end{pmatrix}\langle j+1\|A^1\|j\rangle\ f_{j+1,m}. \tag{10.34}
\end{aligned}$$

Using the tabulated values of the Clebsch–Gordan coefficients and making the replacements

$$A_j \to -A_j,\ B_j \to -B_j,\ C_{j+1} \to -C_{j+1}.$$

We have finally

$$\begin{aligned}
F_3 f_{jm} &= \sqrt{(j-m)(j+m)}A_j\, f_{j-1,m} - mB_j f_{jm} \\
&\quad - \sqrt{(j-m+1)(j+m+1)}C_{j+1}f_{j+1,m} \tag{10.35a}
\end{aligned}$$

$$\begin{aligned}
F_+ f_{jm} &= \sqrt{(j-m)(j-m-1)}A_j f_{j-1,m+1} \\
&\quad - B_j\sqrt{(j-m)(j+m+1)}f_{j,m+1} \\
&\quad + \sqrt{(j+m+1)(j+m+2)}C_{j+1}f_{j+1,m+1}. \tag{10.35b}
\end{aligned}$$

We note that there is some arbitrariness in the normalization of the basis f_{jm}. In fact if we dilate all the vectors of the basis in each of the subspace H^j in the same manner, i.e. if we take

$$f'_{jm} = h(j)f_{jm} \qquad (10.36)$$

where $h(j)$ is some number, then the form of the operators J_+, J_-, J_3 in the new basis is the same as before, for they act independently in each H^j, such a change of basis carries the numbers A_j, B_j, C_j into

$$A'_j = \frac{h(j)}{h(j-1)}A_j$$

$$B'_j = B_j$$

$$C'_{j+1} = \frac{h(j)}{h(j+1)}C_{j+1}.$$

Consequently

$$A'_j C'_j = A_j C_j \qquad (10.37)$$

i.e. the product $A'_j C'_j$ is preserved. By a suitable choice of multipliers $h(j)$ the numbers C'_j and A'_j can be made equal.

To this end it suffices to take

$$h(j) = C \prod_{j_0+1}^{j} \sqrt{\frac{C_\nu}{A_\nu}} \qquad (10.38)$$

so that

$$A'_j = C'_j = D_j \quad \text{(say)}.$$

Thus

$$F_3 f_{jm} = \sqrt{(j-m)(j+m)}D_j f_{j-1,m} - mB_j f_{jm}$$
$$-\sqrt{(j-m+1)(j+m+1)}D_{j+1}f_{j+1,m} \qquad (10.39)$$

$$F_+ f_{jm} = \sqrt{(j-m)(j-m-1)}D_j f_{j-1,m+1}$$
$$-B_j\sqrt{(j-m)(j+m+1)}f_{j,m+1}$$
$$+\sqrt{(j+m+1)(j+m-2)}D_{j+1}f_{j+1,m+1}. \qquad (10.40)$$

It remains to determine D_j and B_j. For this purpose we apply the commutation relation

$$F_+F_3 - F_3F_+ = J_+. \qquad (10.41)$$

Acting both sides of the above on f_{jm} we have

$$\sqrt{(j-m)(j+m+1)}f_{j,m+1} = \sqrt{(j-m)(j+m)}D_j F_+ f_{j-1,m}$$
$$-mB_j F_+ f_{jm} - \sqrt{(j-m+1)(j+m+1)}$$

$$D_{j+1}F_+ f_{j+1,m} - \sqrt{(j-m)(j-m-1)}D_j F_3 f_{j-1,m+1}$$

$$-\sqrt{(j-m)(j+m+1)}B_j F_3 f_{\widehat{j,m+1}}$$

$$+\sqrt{(j+m+1)(j+m+2)}D_{j+1}F_3 f_{\widehat{j+1,m+1}}.$$

Acting by F_3, F_+ with the help of Eqs. (10.40) and (10.41) and equating the coefficients of $f_{j,m+1}$ from both the sides

$$(2j-1)D_j^2 - B_j^2 - (2j+3)D_{j+1}^2 = 1. \tag{10.42}$$

Similarly equating the coefficient of $f_{j-1,m+1}$ to zero we have

$$[(j+1)B_j - (j-1)B_{j-1}]D_j = 0. \tag{10.43}$$

Let j_0 be the lowest value of j. Thus $D_{j_0} = 0$ because it multiplies $f_{j_0-1,m}$.

The following two cases are possible:

(a) D_j does not vanish for any of the values $j = j_0 + n$, $n = 1, 2, 3 \ldots$
(b) D_j vanishes for some of the values of $j = j_0 + n$, $n = 1, 2, \ldots$.

Case (a).

$$(j+1)B_j - (j-1)B_{j-1} = 0.$$

We now define

$$\rho_j = j(j+1)B_j$$

$$B_j = \frac{\rho_j}{j(j+1)}$$

thus

$$\frac{\rho_j}{j} - \frac{\rho_{j-1}}{j} = 0, \quad \rho_j = \rho_{j-1} = \text{const.} = ij_0 c \tag{10.44}$$

where c is a complex number

$$B_j = \frac{\rho_j}{j(j+1)} = \frac{ij_0 \cdot c}{j(j+1)}. \tag{10.45}$$

We now proceed to determine D_j:

$$(2j-1)D_j^2 - (2j+3)D_{j+1}^2 = 1 + B_j^2 = 1 - \frac{j_0^2 c^2}{j^2(j+1)^2}. \tag{10.46}$$

If we now multiply both sides of Eq. (10.46) by $(2j+1)$ and define

$$\sigma_j = (2j-1)(2j+1)D_j^2$$

then

$$\sigma_j - \sigma_{j+1} = (2j+1) - j_0^2 c^2 \left[\frac{1}{j^2} - \frac{1}{(j+1)^2}\right].$$

Hence

$$\sigma_{j_0} - \sigma_j = (j^2 - j_0^2) - j_0^2 c \left[\frac{1}{j_0^2} - \frac{1}{j^2} \right] = \frac{(j^2 - j_0^2)(j^2 - c^2)}{j^2}.$$

Since

$$\sigma_{j_0} = (2j_0 - 1)(2j_0 + 1)D_{j_0}^2 = 0$$

because $D_{j_0} = 0$. We have

$$\sigma_j = -\frac{(j^2 - j_0^2)(j^2 - c^2)}{j^2} \qquad (10.47)$$

using the definition of σ_j we immediately obtain

$$D_j = \frac{i}{j} \sqrt{\frac{(j^2 - j_0^2)(j^2 - c^2)}{(4j^2 - 1)}}.$$

We therefore finally obtain

$$F_3 f_{jm} = \sqrt{(j - m)(j + m)} D_j f_{j-1,m} - mB_j f_{jm}$$
$$- \sqrt{(j - m + 1)(j + m + 1)} D_{j+1} f_{j+1,m}$$
$$F_+ f_{jm} = \sqrt{(j - m)(j - m - 1)} D_j f_{j-1,m+1}$$
$$- B_j \sqrt{(j - m)(j + m + 1)} f_{j,m+1}$$
$$+ \sqrt{(j + m + 1)(j + m + 2)} D_{j+1} f_{j+1,m+1} \qquad (10.48)$$
$$F_- f_{jm} = -\sqrt{(j + m)(j + m - 1)} D_j f_{j-1,m-1}$$
$$- B_j \sqrt{(j + m)(j - m + 1)} f_{j,m-1}$$
$$- \sqrt{(j - m + 1)(j - m + 2)} D_{j+1} f_{j+1,m-1}$$

where

$$D_j = \frac{i}{j} \sqrt{\frac{(j^2 - j_0^2)(j^2 - c^2)}{4j^2 - 1}}, \quad B_j = \frac{ij_0 c}{j(j + 1)}. \qquad (10.49)$$

Thus we arrive at the following theorem:

Theorem. *Each irreducible representation* (IR) *of the proper Lorentz group is determined by a pair of numbers* (j_0, c) *where* j_0 *is an integral or half-integral non-negative number and* c *is a complex number. The* (IR) *corresponding to a given pair* (j_0, c) *is then, with a suitable choice of basis* f_{jm}

in the space of the representation, given by the above formulae together with

$$J_{\pm} f_{jm} = \sqrt{(j \mp m)(j \pm m + 1)} f_{j,m\pm1}. \qquad (10.50)$$

If $c^2 = (j_0 + n)^2$ for some natural number n then the representation is finite-dimensional and the possible values of m and j are $m = -j, -j+1 \ldots, j, j = j_0, j_0 + 1, \ldots, j_0 + n$. If however $c \neq (j_0 + n)$ for any natural number n the representation is infinite-dimensional and the possible values of the indices m and j are

$$m = -j, -j + 1, \ldots, j$$

$$j = j_0, j_0 + 1, j_0 + 2, \ldots.$$

10.4 Unitary Representation

Let us now elucidate the conditions under which a given representation $g \to T_g$ is unitary. In a unitary representation F_3 is Hermitian, i.e.

$$(F_3 f_{jm}, f_{jm}) = (f_{jm}, F_3 f_{jm})$$

which yields

$$-j_0 \bar{c} = j_0 c. \qquad (10.51)$$

This is possible in the following two cases:

(a) c pure imaginary, $j_0 = $ arbitrary
(b) $j_0 = 0, c$ arbitrary.

Similarly

$$(F_3 f_{jm}, f_{j-1,m}) = (f_{jm}, F_3 f_{j-1,m})$$

yields

$$\overline{D}_j = -D_j. \qquad (10.52)$$

Thus

$$\frac{1}{j} \sqrt{\frac{(j^2 - j_0^2)(j^2 - c^2)}{4j^2 - 1}} = \text{real number}.$$

Hence the expression $j^2 - c^2$ must be positive. Obviously this is possible when c^2 is real, i.e. when c is real or purely imaginary. In the second case $-c^2 > 0$. Thus we have

(i) $j_0 = \left| \frac{m}{2} \right|$, $c = is$, where m is an integer and $-\infty < s < \infty$.

When c is real we must have $j_0 = 0$ (by virtue of what has been said about $j_0 c = -j_0 \bar{c}$). Hence $\frac{j^2 - c^2}{4j^2 - 1} \geq 0$, for $c^2 \geq 0$ for all $j = 0, 1, 2 \ldots$ only if $0 \leq c^2 \leq 1$. In order to see this, it is sufficient to put $j = 1$. The theorem is thereby proved. We therefore obtain two families of irreducible unitary representations.

(a) Principal series of representations:

$$j_0 = \text{non-negative integer or half-integer}, \quad c = is, \quad \infty < s < -\infty. \tag{10.53a}$$

(b) Supplementary series of representation:

$$j_0 = 0, \quad -1 < c < 1. \tag{10.53b}$$

The formula (10.53a), (10.53b) for unitary representations were first obtained by Gel'fand and Harish-Chandra.[4]

10.5 Principal Series of Representations[5]

In this section we describe an important class of infinite dimensional representation of $SL(2, C)$, namely, the principal series of representations. The discussion presented below is a reproduction of Gel'fand and Naimark's celebrated paper on $SL(2, C)$. The purpose of this section is the construction of the principal series of representations. Several subgroups of $SL(2, C)$ will play an important role for this construction. The operators T_g of a representation will be realized as operators in a function space. Elements of this space are functions on the cosets of $SL(2, C)$ by some subgroups.

10.5.1 *The subgroup* K

Let us denote by K the set of all complex matrices k of the form

$$k = \begin{pmatrix} k_{11} & k_{12} \\ 0 & k_{22} \end{pmatrix}, \quad k_{11} k_{22} = 1. \tag{10.54}$$

Evidently K is a subgroup of $SL(2, C)$.

Thus a matrix k is determined by two independent complex parameters, e.g. by k_{22} and k_{12}. A left translation $k \to k'k$ is a linear transformation of these parameters. The left invariant differential $d\omega = k^{-1}dk$.

So that

$$d\omega_{12} = k_{22}dk_{12} - k_{12}dk_{22} \tag{10.55a}$$

$$d\omega_{22} = k_{11}dk_{22}. \tag{10.55b}$$

Hence

$$d\mu_l(k) = d^2\omega_{12}d^2\omega_{22}$$

$$= \left|\frac{\partial(\omega_{12}, \omega_{22})}{\partial(k_{12}, k_{22})}\right|^2 d^2k_{12}d^2k_{22} \tag{10.56}$$

where $d^2z = dx\,dy$, etc.

Since the Jacobian in Eq. (56) is one we have

$$d\mu_l(k) = d^2k_{12}d^2k_{22}. \tag{10.57}$$

Similarly from

$$d\omega' = dk \cdot k^{-1}$$

we have

$$d\mu_r(k) = |k_{22}|^{-4}d^2k_{12}d^2k_{22}. \tag{10.58}$$

Let us write

$$\beta(k) = \frac{d\mu_l(k)}{d\mu_r(k)}$$

$$= |k_{22}|^4 \tag{10.59}$$

which is called Radon–Nicodym derivative.

10.5.2 *The subgroup* **H**

We denote by H the set of all matrices

$$h = \begin{pmatrix} h_{11} & 0 \\ h_{21} & h_{22} \end{pmatrix}, \quad h_{11}h_{22} = 1. \tag{10.60}$$

Evidently H is a subgroup of $SL(2, C)$.

The subgroup H is related to the subgroup K as follows. Let us take the matrix

$$s = \begin{pmatrix} 0 & 1 \\ 1 & 0 \end{pmatrix}. \tag{10.61}$$

Then the automorphism $g \to s^{-1}gs$ transforms H into K and K into H.

Hence the right and left invariant measure takes the form

$$d\mu_l(h) = |h_{22}|^{-4}d^2h_{12}d^2h_{22}$$

$$d\mu_r(h) = d^2h_{22}d^2h_{12} \tag{10.62}$$

so that the Radon–Nicodym derivative becomes

$$\frac{d\mu_l(h)}{d\mu_r(h)} = |h_{22}|^{-4}. \tag{10.63}$$

10.5.3   *The subgroup* Z

Let us denote by Z the set of matrices Z such that

$$z = \begin{pmatrix} 1 & 0 \\ z & 1 \end{pmatrix}. \tag{10.64}$$

Evidently Z is a subgroup of H, the subgroup Z is commutative and is isomorphic to additive group of complex numbers. The invariant measure on Z is

$$d\mu_l(z) = d\mu_r(z) = d\mu(z) = d^2z. \tag{10.65}$$

10.5.4   *The subgroup* $\sum$

We denote by $\sum$ the set of matrices ζ such that

$$\zeta = \begin{pmatrix} 1 & \zeta \\ 0 & 1 \end{pmatrix}. \tag{10.66}$$

The invariant measure on $\sum$ is given by

$$d\mu_l(\zeta) = d\mu_r(\zeta) = d\mu(\zeta) = d^2\zeta. \tag{10.67}$$

10.5.5   *The subgroup* D

Let us denote by D the set of all diagonal matrices of the form

$$\delta = \begin{pmatrix} \lambda^{-1} & 0 \\ 0 & \lambda \end{pmatrix}. \tag{10.68}$$

Evidently $D = K \cap H$. The group D is isomorphic to the multiplicative group of complex numbers. The invariant measure on D is given by,

$$d\mu(\delta) = \frac{d\sigma\,d\tau}{|\lambda|^2} \quad \text{where } \lambda = \sigma + i\tau, \quad \text{(Chapter 7)}. \tag{10.69}$$

10.5.6   *Some relations between* SL*(2,* C*) and the subgroups* H, K, $\sum$, Z, D

We will need some theorems about the presentation of an element in SL(2, C) as a product of the elements of the subgroups defined above. These will be useful in the study of coset spaces SL(2, C) by these subgroups.

(i) Each elements $k \in K$ can be uniquely presented in the form

$$k = \delta\zeta \tag{10.70a}$$

and in the form

$$k = \zeta\delta. \qquad (10.70b)$$

The Eq. (10.70a) immediately gives

$$k_{12} = \lambda^{-1}\zeta k_{22} = \lambda \qquad (10.71)$$

which determines λ and ζ.

Similarly (10.70b) yields

$$k_{12} = \lambda\zeta, \quad k_{22} = \lambda \qquad (10.72)$$

which uniquely determine λ and ζ.

Similarly each element $h \in H$ can be uniquely presented in the form

$$h = \delta z. \qquad (10.73)$$

Each element $g \in SL(2, \mathbb{C})$ with $g_{22} \neq 0$ can be uniquely presented in the form,

$$g = kz \qquad (10.74)$$

and in the form

$$g = zk. \qquad (10.75)$$

The Eq. (10.74) yields

$$g_{12} = k_{12}, \quad g_{21} = k_{22}z$$
$$h_{22} = k_{22} \qquad (10.76a)$$

which requires

$$z = \frac{g_{21}}{g_{22}} \qquad (10.76b)$$

which is meaningful only if $g_{22} \neq 0$.

Similarly for $g_{22} \neq 0$, $g \in SL(2, \mathbb{C})$ can also be written in the form

$$g = \zeta h. \qquad (10.77)$$

10.5.7 Cosets of SL(2, C) by $\sum$ and K

We denote by $\widetilde{h}$ the right cosets of $SL(2, \mathbb{C})$ by $\sum$, i.e. sets of elements $g \in SL(2, \mathbb{C})$ of the form ζg_0 where ζ runs through the subgroup $\sum$ for a fixed g_0. The set of all these cosets and the corresponding homogeneous space will be denoted by $\widetilde{H}$. The right multiplication by another fixed element $g \in SL(2, \mathbb{C})$ sends each cosets $\widetilde{h}$ into another coset $\widetilde{h}'$ and therefore yields the transformation of the space $\widetilde{H}$.

So to each element g there corresponds the transformation,

$$\widetilde{h} \rightarrow \widetilde{h}' = \widetilde{h}_g \tag{10.78}$$

of the space $\widetilde{\mathrm{H}}$.

For $g_{22} \neq 0$, the equality $g = \zeta h$ means that g and h belong to the same coset $\widetilde{h}$. In view of the uniqueness of the representation $g = \zeta h$ the coset $\widetilde{h}$ contains only one element of H. Thus $\widetilde{\mathrm{H}}$ can be identified with H and a transformation in $\widetilde{\mathrm{H}}$ induces a transformation in H. Thus the transformation $\widetilde{h} \rightarrow \widetilde{h}'$ induces the following transformation.

$$h \rightarrow h' = h_g. \tag{10.79}$$

The above equality means that the matrices $h' = h_g$ and hg belong to the same coset, i.e. we have

$$hg = \zeta h'$$

$$\begin{pmatrix} h_{11} & 0 \\ h_{21} & h_{22} \end{pmatrix} \begin{pmatrix} g_{11} & g_{12} \\ g_{21} & g_{22} \end{pmatrix} = \begin{pmatrix} 1 & \zeta \\ 0 & 1 \end{pmatrix} \begin{pmatrix} h'_{11} & 0 \\ h'_{21} & h'_{22} \end{pmatrix}. \tag{10.80}$$

Thus

$$\left. \begin{array}{l} h'_{21} = g_{11}h_{21} + g_{21}h_{22} \\ h'_{12} = g_{12}h_{21} + g_{22}h_{22} \end{array} \right\}. \tag{10.81}$$

10.5.8   *Cosets by* **K**

We denote by $\widetilde{z}$ the right cosets of $\mathrm{SL}(2,\mathrm{C})$ by K, i.e. the set $\mathrm{K}g_0$ such that an element g of this set is of the form kg_0 where k runs through K and g_0 is fixed. If $g_{22} \neq 0$ we have $g = kz$. This equality means that the matrices g and z lie in the same coset $\widetilde{z}$. The decomposition $g = kz$ is unique and, therefore, the class $\widetilde{z}$ contains only one element $z \in Z$. The right multiplication by g yields the transformation $\widetilde{z} \rightarrow \widetilde{z}' = \widetilde{z}g$ of the space $\widetilde{\mathrm{Z}}$. We denote this transformation by

$$\widetilde{z} \rightarrow \widetilde{z}'$$

$$z \rightarrow z_g = z'. \tag{10.82}$$

Since z and $z_g = z'$ belong to the same coset by K we have

$$zg = kz'$$

$$\begin{pmatrix} 1 & 0 \\ z & 1 \end{pmatrix} \begin{pmatrix} g_{11} & g_{12} \\ g_{21} & g_{22} \end{pmatrix} = \begin{pmatrix} k_{11} & k_{12} \\ 0 & k_{22} \end{pmatrix} \begin{pmatrix} 1 & 0 \\ z' & 1 \end{pmatrix} \tag{10.83}$$

which yields

$$\left. \begin{array}{l} k_{22}z' = g_{11}z + g_{21} \\ k_{22} = g_{12}z + g_{22} \end{array} \right\}. \tag{10.84}$$

Thus

$$z' = z_g = \frac{g_{11}z + g_{21}}{g_{12}z + g_{22}}. \tag{10.85}$$

10.5.9 The principal series of irreducible representation of the group SL(2, C)

Now we turn to the construction of the irreducible representation of the group SL(2, C). To do this we first consider a particular reducible representation of the group called the quasi regular representation.

Let us denote by B(H) the Hilbert space of square integrable functions on H (w.r.t. the right invariant measure). The inner product is defined by

$$(f_1, f_2) = \int \overline{f_1(h)} f_2(h) d\mu_r(h). \tag{10.86}$$

Let us now put

$$T_g f(h) = f(h_g) \tag{10.87}$$

where $h_g = h'$ is given by Eq. (10.81).

Although the group action is not a right translation we shall show that the scalar product is invariant under this transformation.

$$(T_g f_1, T_g f_2) = \int \overline{f_1(h_g)} f_2(h_g) d\mu_r(h). \tag{10.88}$$

Now setting

$$h_g = h'$$

$$\zeta h' = hg$$

$$\zeta \cdot dh' + d\zeta \cdot h' = dh \cdot g. \tag{10.89}$$

Multiplying Eq. (10.89) on the right by h'^{-1} and on the left by ζ^{-1} we have

$$dh' \cdot h'^{-1} + \zeta^{-1} \cdot d\zeta = \zeta^{-1} dh \cdot gh'^{-1}$$

$$= \zeta^{-1} dh \cdot g\, g^{-1} h^{-1} \zeta$$

$$= \zeta^{-1} dh \cdot h^{-1} \zeta$$

$$dh' \cdot h'^{-1} = -\zeta^{-1} \cdot d\zeta + \zeta^{-1} dh \cdot h^{-1} \zeta. \tag{10.90}$$

Thus

$$dw' = \zeta^{-1}dw\zeta - \zeta^{-1}d\zeta \tag{10.91}$$

where $dw(dw')$ is the right invariant differential.

We put

$$du = \zeta^{-1}dw\zeta \tag{10.92}$$

and regard dw_{21} and dw_{22} as the independent (complex) invariants.

Now

$$d^2u_{21}d^2u_{22} = \left|\det\begin{vmatrix}1 & \zeta \\ 0 & 1\end{vmatrix}\right|^2 d^2w_{21}d^2w_{22}$$

$$= d\mu_r(h). \tag{10.93}$$

Thus we have

$$d^2u_{21}d^2u_{22} = d\mu_r(h)$$
$$dw' = du - \zeta^{-1}d\zeta$$
$$dw'_{21} = du_{21}$$
$$dw'_{22} = du_{22}.$$

Hence finally

$$d\mu_r(h') = d\mu_r(h) \tag{10.94}$$

and we have

$$(T_g f_1, T_g f_2) = \int \overline{f_1(h')} f_2(h') d\mu_r(h')$$

$$= (f_1, f_2). \tag{10.95}$$

It shows that the quasi regular representation (10.87) is unitary. We now proceed to decompose the quasi regular representation into irreducible representations. Let us consider

$$\|f\|^2 = \int |f(h)|^2 d\mu_r(h)$$

$$= \int |f(h)|^2 \beta(h) d\mu_l(h). \tag{10.96}$$

It can be easily shown that under the decomposition

$$h = \delta z \tag{10.97}$$

the left invariant measure decomposes as

$$d\mu_l(h) = d\mu_l(\delta z)$$
$$= d\mu(\delta)d\mu(z). \tag{10.98}$$

Since

$$\beta(\delta z) = \beta(\delta)\beta(z)$$
$$= \beta(\delta) \quad (\because \quad \beta(z) = 1) \tag{10.99}$$

we finally obtain

$$\|f\|^2 = \int |f(\delta z)\beta^{\frac{1}{2}}(\delta)|^2 d\mu(\delta)d\mu(z). \tag{10.100}$$

Let X be the group of characters of the group D. Let $\chi(\delta)$ be the character. Let us consider the "Fourier transform" of the function $f(\delta z)\beta^{\frac{1}{2}}(\delta)$ by considering $\chi(\delta)$ as the kernel of the transform

$$f_\chi(z) = \int f(\delta z)\beta^{\frac{1}{2}}(\delta)\chi(\delta)d\mu(\delta). \tag{10.101}$$

We now explicitly determine $\chi(\delta)$ by using the unitarity and the group multiplication law

$$\delta = \begin{pmatrix} \lambda^{-1} & 0 \\ 0 & \lambda \end{pmatrix}.$$

$\chi(\delta) = \chi(\lambda)$ must satisfy

$$\chi(\lambda_1)\chi(\lambda_2) = \chi(\lambda_1\lambda_2).$$

From this it follows

$$\chi^2(0) = \chi(0) \therefore \chi(0) = 1 \quad \text{(rejecting the null solution)}$$
$$\chi(r_1 e^{i\phi_1})\chi(r_2 e^{i\phi_2}) = \chi(r_1 r_2 e^{i(\phi_1+\phi_2)}).$$

Setting

$$r_1 = r_2 = 1.$$

Similarly

$$\chi(e^{i\phi_1})\chi(e^{i\phi_2}) = \chi(e^{i(\phi_1+\phi_2)}) \tag{10.102a}$$
$$\chi(r_1)\chi(r_2) = \chi(r_1 r_2). \tag{10.102b}$$

Solution of the first

$$\chi(e^{i\phi}) = e^{-i\mu\phi}$$
$$\chi(e^{i\phi}) = \chi(e^{i(\phi+2\pi)}) \quad \text{(Single valuedness)}$$

requires $\mu = m$ where m is an integer. Similarly

$$\chi(r) = r^p.$$

Now unitarity demands

$$\overline{\chi(r)}\chi(r) = 1$$

$$p + \bar{p} = 0, \quad \text{i.e.} \quad p = -i\rho.$$

Thus

$$\chi(\delta) = r^{-i\rho}e^{im\phi}$$

$$= |\delta|^{-i\rho}\left(\frac{\bar{\delta}}{|\delta|}\right)^{-m}.$$

Finally

$$\chi(\delta) = |\delta|^{m-i\rho}\bar{\delta}^{-m}. \tag{10.103}$$

Setting

$$|\delta| = e^t, \quad -\infty < t < \infty$$

$$\chi(\delta) = e^{-i\rho t}e^{-im\phi} \tag{10.104}$$

which is the Fourier transform-series kernel.

Thus

$$f_\chi(z) = \frac{1}{2\pi}\int f(\delta z)\beta^{\frac{1}{2}}(\delta)e^{-it\rho}e^{-im\phi}d\mu(\delta)$$

so that

$$f(\delta z)\beta^{\frac{1}{2}}(\delta) = \frac{1}{2\pi}\sum e^{-im\phi}\int e^{-it\rho}f_\chi(z)d\rho. \tag{10.105}$$

By Plancherel theorem

$$\int |f_\chi(z)|^2 d\mu(\chi) = \int |f(\delta z)\beta^{\frac{1}{2}}(\delta)|^2 d\mu(\delta). \tag{10.106}$$

In this particular case of course

$$\int |f_\chi(z)|^2 d\mu(\chi) = \sum_m \int |f_{\rho m}(z)|^2 d\rho. \tag{10.107}$$

We therefore obtain

$$|f|^2 = \int \left|f(\delta z)\beta^{\frac{1}{2}}(\delta)\right|^2 d\mu(\delta)d\mu(z)$$

$$= \int |f_\chi(z)|^2 d\mu(\chi)d\mu(z)$$

$$= \int\int |f_\chi(z)|^2 d\mu(z)d\mu(\chi). \tag{10.108}$$

Let B(Z) be the Hilbert space of square integrable functions of Z with inner product given by

$$(f_1, f_2) = \int \overline{f_1(z)} f_2(z) d\mu(z). \qquad (10.109)$$

The formula (10.108) thus means that the B(H) is the continuous direct sum of the Hilbert spaces. $B^\chi(z) = B(z)$.

Let us now consider the transformation of $f_\chi(z)$ when we come from $f(h)$ to $f(h_g)$:

$$h' = h_g; \quad h = \delta z; \quad h' = \delta' z'. \qquad (10.110)$$

Further h' and hg belong to the same coset; hence

$$\zeta h' = hg = \delta z g. \qquad (10.111)$$

Also

$$hg = \zeta h' = \zeta \delta' z'.$$

Thus

$$\delta z g = \zeta \delta' z'$$
$$zg = \delta^{-1} \zeta^{-1} \zeta \delta' z'$$
$$= \delta^{-1} \zeta \delta \delta^{-1} \delta' z'.$$

It is easy to check

$$\delta^{-1} \zeta \delta = \begin{pmatrix} 1 & \delta^2 \zeta \\ 0 & 1 \end{pmatrix} = \begin{pmatrix} 1 & \zeta' \\ 0 & 1 \end{pmatrix} \in \Sigma.$$

Further we denote

$$\delta^{-1} \delta' = \delta_1$$

so that

$$zg = \zeta' \delta_1 z' = kz' \qquad (10.112)$$

because

$$k = \zeta' \delta_1 \in K$$

with

$$k_{12} = \zeta' \delta_1$$
$$k_{11} = k_{22}^{-1} = \delta_1^{-1}.$$

The equality (10.112) means that zg and z' belong to the same coset by K, i.e.

$$z' = z_g; \quad zg = kz'$$

$$h_g = h' = \delta' z' = \delta \delta_1 z_g. \tag{10.113}$$

But then we can write

$$T_g^\chi f_\chi(z) = f_\chi'(z)$$

$$= \int f(h_g)\beta^{\frac{1}{2}}(\delta)\chi(\delta)d\mu(\delta)$$

$$= \int f(\delta\delta_1 z_g)\beta^{\frac{1}{2}}(\delta)\chi(\delta)d\mu(\delta)$$

$$= \int f(\delta z_g)\beta^{\frac{1}{2}}(\delta\delta_1^{-1})\chi(\delta\delta_1^{-1})d\mu(\delta)$$

$$\text{(using right invariance)}$$

$$= \int f(\delta z_g)\beta^{\frac{1}{2}}(\delta)\beta^{-\frac{1}{2}}(\delta_1)\chi(\delta)\overline{\chi(\delta_1)}d\mu(\delta)$$

$$= \beta^{\frac{1}{2}}(\delta_1)\overline{\chi(\delta_1)}\int f(\delta z_g)\beta^{\frac{1}{2}}(\delta)\chi(\delta)d\mu(\delta).$$

The above integral is evidently

$$f_\chi(z_g)$$

so that

$$T_g^\chi f_\chi(z) = \beta^{-\frac{1}{2}}(\delta_1)\overline{\chi(\delta_1)}f_\chi(z_g) \tag{10.114}$$

$$\overline{\chi(\delta_1)} = |\delta_1|^{m+i\rho}\delta_1^{-m}. \tag{10.115}$$

Now

$$z_g = z'$$

implies

$$zg = kz' = \zeta'\delta_1 z'$$

$$= \begin{pmatrix} \delta_1^{-1} + \zeta'\delta_1 z' & \zeta'\delta_1 \\ \delta_1 z' & \delta_1 \end{pmatrix}.$$

Carrying out the matrix multiplication appearing on the left and equating the elements we have

$$\delta_1 = g_{12}z + g_{22} \tag{10.116}$$

$$\delta_1 z' = g_{11}z + g_{21}$$

$$z' = \frac{g_{11}z + g_{21}}{g_{12}z + g_{22}}. \tag{10.117}$$

Substituting Eqs. (10.115)–(10.117) in Eq. (10.114) we have

$$T_g^\chi f_\chi(z) = |g_{12}z + g_{22}|^{m+i\rho-2}(g_{12}z + g_{22})^{-m} f\left(\frac{g_{11}z + g_{21}}{g_{12}z + g_{22}}\right). \tag{10.118}$$

Thus a representation T_g^χ is determined by two parameters ρ and m where ρ is an arbitrary real number and m is an arbitrary integer.

Theorem. *All representations of the principle series are irreducible.*

Proof. Let us set in the formula (10.118)

$$g = z_0 = \begin{pmatrix} 1 & 0 \\ z_0 & 1 \end{pmatrix}$$

then,

$$T_{z_0}^\chi = f(z + z_0). \tag{10.119}$$

We further put

$$g = \delta = \begin{pmatrix} \lambda^{-1} & 0 \\ 0 & \lambda \end{pmatrix}$$

$$T_\delta^\chi f_\chi(z) = |\lambda|^{m+i\rho-2}(\lambda)^{-m} f(z\lambda^{-2}). \tag{10.120}$$

We shall prove that each bounded operator A which commutes with all the operators T_g^χ is a multiple of the identity.

Let us consider instead of the function $f(z)$ its Fourier transform

$$\phi(p) = \phi(p_1 + ip_2)$$

$$= \frac{1}{2\pi} \int f(x + iy) e^{-i(xp_1 + yp_2)} dx dy$$

$$= \frac{1}{2\pi} \int f(z) e^{-i\text{Re}(z\bar{p})} d^2 z.$$

Then

$$(\phi_1, \phi_2) = \int \overline{\phi_1(p)}\phi_2(p) d^2 p.$$

Let us consider the Fourier transform of the operations $T_{z_0}^\chi, T_\delta^\chi, A$.

$$\hat{T}_{z_0}^\chi \phi(p) = \int f(z + z_0) e^{-i\text{Re}(z\bar{p})} d^2 z$$

$$= e^{i\text{Re}(z_0\bar{p})} \int f(z') e^{-i\text{Re}(z'\bar{p})} d^2 z'$$

$$= e^{i\text{Re}(z_0\bar{p})} \phi(p).$$

Further

$$\hat{T}_\delta^\chi \phi(p) = |\lambda|^{m+ip-2}\lambda^{-m}\frac{1}{2\pi}\int f(z\lambda^{-2}) \times e^{-\mathrm{Re}(z\bar{p})}d^2z.$$

Setting

$$z\lambda^{-2} = z_1$$

$$d^2z = \left|\frac{dz}{dz_1}\right|^2 d^2z_1$$

$$= |\lambda|^4 d^2z_1.$$

Thus

$$\hat{T}_\delta^\chi \phi(p) = |\lambda|^{m+ip-2}\lambda^{-m}\frac{1}{2\pi}|\lambda|^4 \times \int f(z_1)e^{-i\mathrm{Re}[z_1(\lambda^2\bar{p})]}d^2z_1$$

$$= |\lambda|^{m+ip+2}\lambda^{-m}\phi(\bar{\lambda}^2 p).$$

The operator $\hat{A}$ commutes with $e^{i\mathrm{Re}(z_0\bar{p})}$ and hence it must be an operator of multiplication

$$\hat{A}\phi(\eta) = \omega(\eta)\phi(\eta).$$

Let us set

$$\hat{T}_\delta^\chi \phi(p) = \mathrm{F}(p)$$

$$\hat{A}\hat{T}_\delta^\chi \phi(p) = \hat{A}\mathrm{F}(p) = \omega(p)\mathrm{F}(p)$$

$$= \omega(p)|\lambda|^{m+ip+2}\lambda^{-m}\phi(\lambda^{-2}p).$$

Now

$$\hat{T}_\delta^\chi \hat{A}\phi(p) = |\lambda|^{m+ip+2}\lambda^{-m}\phi(p\bar{\lambda}^2)$$

$$= |\lambda|^{m+ip+2}\lambda^{-m}\omega(p\bar{\lambda}^2)\phi(p\bar{\lambda}^2).$$

Thus

$$|\hat{A}, \hat{T}_\lambda^\chi| = 0$$

therefore yields $\omega(p) = \omega(p\bar{\lambda}^2)$. Thus $\omega(p) = \mathrm{const.} = c\mathrm{I}$ and the representation is operator irreducible.

Gel'fand and coworkers have shown that for complex groups operator irreducibility ensures true irreducibility.

10.6   Supplementary Series Representations

We have proved that for any real ρ the formula

$$T_g f(z) = |g_{12}z + g_{22}|^{m+ip-2}(g_{12}z + g_{22})^{-m}f\left(\frac{g_{11}z + g_{21}}{g_{12}z + g_{22}}\right) \qquad (10.121)$$

defines a unitary representation of the Lorentz group $(SL(2, \mathbb{C}))$. The representations are unitary w.r.t. the scalar product

$$(f_1, f_2) = \int \overline{f_1(z)} f_2(z) d^2 z. \tag{10.122}$$

The question arises whether it is possible to construct a Hilbert space, i.e. define a scalar product in such a way that the above formula defines a unitary representation in this space for other values of ρ. We shall see such a definition is in fact possible, resulting in another family of unitary representations, namely, the supplementary series.

In finite dimensional space the general form of scalar product is a positive definite Hermitian quadratic form

$$(x, y) = \sum g_{pq} \bar{x}_p y_q \tag{10.123}$$

a special case of which is the traditional Hermitian bilinear form,

$$(x, y) = \sum \overline{x}_p y_p. \tag{10.124}$$

It is evident that the scalar product can be regarded as the analogue of expression; the analogue will be the nonlocal bilinear form

$$(f_1, f_2) = \int K(z_1, z_2) \overline{f_1(z_1)} f_2(z_2) d^2 z_1 d^2 z_2. \tag{10.125}$$

For the representation to be unitary, it is necessary that the scalar product remains invariant under the group

$$(T_g f_1, T_g f_2) = (f_1, f_2). \tag{10.126}$$

Let us now define

$$\alpha(g) = |g_{22}|^{m+i\rho-2} g_{22}^{-m}. \tag{10.127}$$

Then

$$\alpha(zg) = |g_{12} z + g_{22}| (g_{12} z + g_{22})^{-m}. \tag{10.128}$$

So that

$$T_g f(z) = \alpha(zg) f(z_g)$$

$$(T_g, f_1, T_g f_2) = \int K(z_1, z_2) \overline{\alpha(z_1 g)} f_1(z_{1g}) \alpha(z_2 g) f(z_{2g}) d^2 z_1 \tag{10.129}$$

$$(f_1, f_2) = \int K(z_1', z_2') \overline{f_1(z_1')} f_2(z_2') d^2 z_1' d^2 z_2'. \tag{10.130}$$

Let us now set in the above integral

$$z_1' = z_{1g} = \frac{g_{11}z_1 + g_{21}}{g_{12}z_1 + g_{22}}.$$

From this it immediately follows

$$d^2z_1' = \left|\frac{dz_1'}{dz_1}\right|^2 d^2z_1 = |g_{12}z_1 + g_{22}|^{-4}d^2z_1, d^2z_2' = |g_{12}z_2 +_{22}|^{-4}d^2z_2.$$

Hence from Eq. (10.129) and (10.130) we have

$$\overline{\alpha(z_1g)}\alpha(z_2g)K(z_1, z_2) = |g_{12}z_1 + g_{22}|^{-4} \times |g_{12}z_2 + g_{22}|^{-4}K(z_{1g}, z_{2g}).$$

Thus

$$K\left(\frac{g_{11}z_1 + g_{21}}{g_{12}z_1 + g_{22}}, \frac{g_{11}z_2 + g_{21}}{g_{12}z_2 + g_{22}}\right)$$
$$= |g_{12}z_1 + g_{22}|^{m-i\bar{\rho}+2}|g_{12}z_2 + g_{22}|^{m+i\rho+2}$$
$$\times \overline{(g_{12}z_1 + g_{22})}^{-m}(g_{12}z_2 + g_{22})^{-m}K(z_1, z_2). \qquad (10.131)$$

Let us now set

$$g = \begin{pmatrix} 1 & 0 \\ z_0 & 1 \end{pmatrix}.$$

Equation (10.131) immediately yields

$$K(z_1 + z_0, z_2 + z_0) = K(z_1, z_2). \qquad (10.132)$$

The translational invariance (10.132) requires

$$K(z_1, z_2) = K_1(z_1 - z_2).$$

Thus it can be written as

$$K_1\left(\frac{z_1 - z_2}{(g_{12}z_1 + g_{22})(g_{12}z_2 + g_{22})}\right)$$
$$= |g_{12}z_1 + g_{22}|^{m-i\rho+2}|g_{12}z_2 + g_{22}|^{m+i\rho+2}$$
$$\times \overline{(g_{12}z_1 + g_{22})}^{-m}(g_{12}z_2 + g_{22})^{-m}K_1(z_1 - z_2).$$

Setting $z_2 = 0$ and choosing g_{12} so that $g_{12}z_1 + g_{22} = 1$. We have

$$K_1\left(\frac{z_1}{g_{22}}\right) = K_1(z_1)|g_{22}|^{m+i\rho+2}g_{22}^{-m}. \qquad (10.133)$$

On the other hand, setting $z_1 = 0$ and choosing g_{12} such that $g_{12}z_2+g_{22} = 1$.

We have

$$K_1\left(\frac{-z_2}{g_{22}}\right) = |g_{22}|^{m-i\bar{\rho}+2}\overline{g_{22}}^{-m}\,K_1(-z_2).$$

Setting $z_2 = -z_1$

$$K_1\left(\frac{z_1}{g_{22}}\right) = K_1(z_1)|g_{22}|^{m-i\bar{\rho}+2} \times \overline{g_{22}}^{-m}. \qquad (10.134)$$

From Eqs. (10.133) and (10.134) we have

$$|g_{22}|^{i\rho}g_{22}^{-m} = |g_{22}|^{-i\bar{\rho}}\overline{g_{22}}^{-m}. \qquad (10.135)$$

Setting $g_{22} = e^{i\theta}$, so that $|g_{22}| = 1$ we have $e^{-im\theta} = e^{im\theta}$, which is possible only if $m = 0$.

Thus

$$|g_{22}|^{i\rho} = |g_{22}|^{-i\bar{\rho}}$$

$$\bar{\rho} = -\rho \quad \text{so that } \frac{\rho}{2} = ic, \rho = +2ic.$$

Now finally setting $z_1 = g_{22} = z$

$$K_1(1) = K_1(z)|z|^{2-2c}.$$

Thus

$$K_1(z_1 - z_2) = C|z_1 - z_2|^{-2+2c}, \quad C = K_1(1)$$

and the finite element of the group

$$T_g f(z) = |g_{12}z + g_{22}|^{-2-2c} f\left(\frac{g_{11}z + g_{21}}{g_{12}z + g_{22}}\right) \qquad (10.136)$$

is unitary under the scalar product

$$(f_1, f_2) = C \iint |z_1 - z_2|^{-2+2c}\overline{f_1(z_1)}f_2(z_2)d^2z_1 d^2z_2.$$

The integral converges in the usual sense for $c < \frac{1}{2}$. For $\frac{1}{2} < c < 1$ the integral is to be understood in the sense of its regularization.

10.6.1 The character problem and the Plancherel formula for SL(2, C)

As in Chapter 9 we start from the invariant definition of character of Gel'fand and Naimark. We recapitulate the basic facts of the theory of character and develop the Plancherel formula as a completeness condition of the character.

We denote by X the set of infinitely differentiable functions $x(g)$ on the group which are equal to zero outside a bounded set.

If $g \to T_g$ is a representation of the group G we set as before

$$T_x = \int d\mu(g)x(g)T_g \tag{10.137}$$

where $d\mu(g)$ is the left and right invariant measure (assumed coincident) on G and the integration extends over the entire group manifold.

The product $T_{x_1}T_{x_2}$ can be written in the form,

$$T_{x_1}T_{x_2} = \int d\mu(g)x(g)T_g$$

where

$$x(g) = \int x_1(g)x_2(g_1^{-1}g)d\mu(g_1). \tag{10.138}$$

The function $x(g)$ defined by Eq. (10.138) will be called the product of the functions x_1, x_2 and denoted by $x_1x_2(g)$. Let us suppose that $g \to T_g$ is a unitary representation of the group G realized in the Hilbert space H of the functions $f(z)$ with the scalar product

$$(f,g) = \int \overline{f(z)}g(z)d\lambda(z) \tag{10.139}$$

where $d\lambda(z)$ is the measure in H.

Then the operator T_x is an integral operator in H with a kernel,

$$T_x f(z) = \int K(z,z_1)f(z_1)d\lambda(z_1).$$

It then follows that $K(z,z_1)$ is a positive definite Hilbert–Schmidt kernel, satisfying

$$\int |K(z,z_1)|^2 d\lambda(z)d\lambda(z_1) < \infty.$$

Such a kernel has a trace $T_r(T_x) = \int K(z,z)d\lambda(z)$.

Using the definition one can prove that $T_r(T_x)$ can be written in the form,

$$T_r(T_x) = \int x(g)\pi(g)d\mu(g). \tag{10.140}$$

The function $\pi(g)$ is the character of the representation $g \to T_g$. It should be noted that in this definition the matrix representation of the group does not appear and as well be shown later it makes a complete synthesis of the finite and infinite dimensional irreducible unitary representation.

10.6.2 *The principal series of representations*[4]

The principal series of representations will be realized in the Hilbert space of functions (non-analytic) $L^2(Z)$. The finite element of the group is given by,

$$T_g f(z) = |g_{12}z + g_{22}|^{m+i\rho-2}(g_{12}z + g_{22})^{-m} f\left(\frac{g_{11}z + g_{21}}{g_{12}z + g_{22}}\right). \quad (10.141)$$

If we define

$$\chi(g) = |g_{22}|^{m-i\rho}\overline{g_{22}}^{-m}. \quad (10.142)$$

Then

$$\overline{\chi(zg)} = |g_{22} + g_{12}z|^{m+i\rho}(g_{22} + g_{12}z)^{-m} \quad (10.143)$$

$$T_g f(z) = \overline{\chi(zg)}\beta^{\frac{1}{2}}(zg)f(z_g). \quad (10.144)$$

The representations (10.141) or (10.144) are unitary under scalar product

$$(f, g) = \int \overline{f(z)}g(z)d\mu(z) \quad (10.145)$$

where $d\mu(z) = d^2z$.

We now construct the operator of the group ring

$$T_x = \int d\mu(g)x(g)T_g$$

where $x(g)$ is an arbitrary test function on the group with bounded support. Now,

$$T_x f(z) = \int x(g)T_g f(z)d\mu(g)$$

$$= \int d\mu(g)x(g)\overline{\chi(zg)}\beta^{\frac{1}{2}}(zg)f(z_g) \quad (10.146)$$

z_g is defined by

$$kz_g = zg, \quad z_g = \frac{g_{11}z + g_{21}}{g_{12}z + g_{22}} = \frac{(zg)_{21}}{(zg)_{22}}.$$

So that

$$T_x f(z) = \int d\mu(g)x(g)\overline{\chi(zg)}\beta^{-\frac{1}{2}}(zg)f\left(\frac{(zg)_{21}}{(zg)_{22}}\right). \quad (10.147)$$

Let us now make a left translation

$$g \to z^{-1}g$$

$$T_x f(z) = \int d\mu(g)x(z^{-1}g)\overline{\chi(g)}\beta^{-\frac{1}{2}}(g)f\left(\frac{g_{21}}{g_{22}}\right). \quad (10.148)$$

Setting $g = kz_1$ and noting $\chi(kz_1) = \chi(k), \beta(kz_1) = \beta(k)$ we have

$$\mathbf{T}_x f(z) = \int x(z^{-1}kz_1)\overline{\chi(k)}\beta^{-\frac{1}{2}}(k)f(z_1)d\mu(g). \tag{10.149}$$

It can be verified that under the decomposition $g = kz_1$, the invariant measure decomposes as

$$d\mu(g) = d\mu_1(k)d\mu(z_1). \tag{10.150}$$

Proof. It can be easily seen

$$g^{-1}dg = z_1^{-1}[k^{-1}dk + dz_1z_1^{-1}]z_1$$

which can be written in the form,

$$g^{-1}dg = z_1^{-1}d\mu z_1 \tag{10.151}$$

where

$$d\mu = du + d\vartheta \tag{10.152}$$

$$du = k^{-1}dk, \quad d\vartheta = dz_1z_1^{-1}. \tag{10.153}$$

The independent elements of $d\vartheta$ are taken to du_{12}, du_{22} so that

$$d\mu_l(k) = d^2u_{12}d^2u_{22} \tag{10.154}$$

and

$$d^2\vartheta = d\mu(z_1).$$

If we write

$$d\omega = g^{-1}dg$$

we have

$$d\omega_{22} = d\mu_{22} - z_1 d\mu_{12}$$

$$= du_{22} - z_1 du_{12}$$

$$d\omega_{12} = d\mu_{12} = du_{12}$$

$$d\omega_{21} = 2z_1 du_{22} - z_1^2 du_{12} + dz_1.$$

Thus

$$d\mu(g) = d^2\omega_{22}d^2\omega_{12}d^2\omega_{21}$$

$$= \left| \frac{\partial(\omega_{12}, \omega_{22}, \omega_{21})}{\partial(\mu_{12}, \mu_{22}, \mu_{21})} \right|^2 d^2\mu_{12}d^2\mu_{22}d^2\mu_{21}$$

$$= \left| \frac{\partial(\omega_{12}, \omega_{22}, \omega_{21})}{\partial(u_{12}, u_{22}, u_{21})} \right|^2 d^2u_{12}d^2u_{22}d^2u_{21}. \tag{10.155}$$

Since the Jacobian is the triangular determinant

$$\begin{vmatrix} 1 & -z_1 & -z_1^2 \\ 0 & 1 & 2z_1 \\ 0 & 0 & 1 \end{vmatrix} = 1$$

we have

$$d\mu(g) = d^2 u_{12} d^2 u_{22} d^2 z_1$$

$$= d\mu_1(k) d\mu(z_1). \tag{10.156}$$

We therefore obtain

$$T_u f(z) = \int x(z^{-1} k z_1) \overline{\chi(k)} \beta^{-\frac{1}{2}}(k) d\mu_1(k) f(z_1) d\mu(z_1)$$

$$= \int K(z, z_1) f(z_1) d\mu(z_1) \tag{10.157}$$

where

$$K(z, z_1) = \int x(z^{-1} k z_1) \overline{\chi(k)} \beta^{-\frac{1}{2}}(k) d\mu_l(k) \tag{10.158}$$

is the integral kernel of the group ring.

Hence following Gel'fand and Naimark

$$T_r(T_x) = \int K(z, z) d\mu(z)$$

$$= \int x(z^{-1} k z) \overline{\chi(k)} \beta^{-\frac{1}{2}}(k) d\mu_l(k) d\mu(z). \tag{10.159}$$

Let us suppose that

$$\xi_1 = \begin{pmatrix} x_{11} \\ x_{12} \end{pmatrix} \quad \text{and} \quad \xi_2 = \begin{pmatrix} x_{21} \\ x_{22} \end{pmatrix}$$

are the eigenvectors of the matrix $\tilde{g}$ with eigenvalues δ_1 and δ_2 respectively, assumed distinct. Therefore ξ_1 and ξ_2 are linearly independent so that

$$\det x = \det \begin{pmatrix} x_{11} & x_{12} \\ x_{21} & x_{22} \end{pmatrix} \neq 0.$$

We normalize the eigenvector such that $\det x = 1$

$$g = x^{-1} \delta x \tag{10.160}$$

where $x \in \mathrm{SL}(2, \mathbb{C})$. Hence x can be decomposed as

$$x = k_1 z. \tag{10.161}$$

Thus

$$g = z^{-1} k_1^{-1} \delta k_1 z.$$

Since

$$k_1^{-1} \delta k_1 \in K.$$

We have

$$g = z^{-1} k z \qquad (10.162)$$

with

$$k_{pp} = \delta_p, \quad p = 1, 2. \qquad (10.163)$$

If the eigenvalues δ_p are distinct then for a given ordering of them, the matrices k, z are determined uniquely by the matrix g.

It follows that for a given choice of δ_p the parameters z and k_{12} are uniquely determined.

We note there are exactly two representations of the matrix g by means of the formula (10.162) corresponding to two distinct possibilities.

$$k_{11} = \delta_1, \qquad k_{22} = \delta_2 = \frac{1}{\delta_1}$$

$$k_{11} = \delta_2 = \frac{1}{\delta_1}, \quad k_{22} = \delta_1. \qquad (10.164)$$

Let us now remove from K the elements with $k_{11} = k_{22} = 1$. This operation cuts the group K into two connected disjoint components. Neither of these components contains two matrices which differ only by permutation of the two diagonal elements. In correspondence with this partition the integral on the r.h.s. of Eq. (10.159) is represented as a sum of two integrals.

$$T_r(T_x) = \int_z d\mu(z) \int_{K_1} d\mu_1(k) \overline{\chi(k)} \beta^{-\frac{1}{2}}(k) x(z^{-1} k z)$$

$$+ \int_z d\mu(z) \int_{K_2} d\mu_1(k) \overline{\chi(k)} \beta^{-\frac{1}{2}}(k) x(z^{-1} k g). \qquad (10.165)$$

As z runs over the subgroup Z and k runs over the components K_1 or K_2 the matrix g runs over the group $SL(2, C)$, we shall now prove that in K_1 or K_2

$$d\mu_1(k) d\mu(z) = \frac{|k_{22}|^2}{|k_{22} - k_{11}|^2} d\mu(g)$$

$$= \frac{|\delta_2|^2 d\mu(g)}{|\delta_2 - \delta_1|^2}. \qquad (10.166)$$

To prove this we start from the left invariant differential

$$dw = g^{-1} \cdot dg \tag{10.167}$$

where $g \in \mathrm{SL}(2, \mathbb{C})$ and dg is the matrix of the differentials g_{pq}:

$$dg = \begin{pmatrix} da & db \\ dc & dd \end{pmatrix}. \tag{10.168}$$

The elements dw are invariant under the left translation. Choosing $dw_{12}, dw_{21}, dw_{22}$ as the independent invariant differentials we arrive at the left invariant measure on $\mathrm{SL}(2, \mathbb{C})$

$$d\mu(g) = d^2 w_{12} d^2 w_{21} d^2 w_{22}. \tag{10.169}$$

To prove formula (10.166) we write Eq. (10.162) in the form

$$zg = kz \tag{10.170}$$

$$z \cdot dg + dz \cdot g = dk \cdot z + k \cdot dz$$

$$k^{-1} z \cdot dg + k^{-1} dz \cdot g = k^{-1} dkz + dz$$

$$k^{-1} z \, dg = [k^{-1} \cdot dk + dz \cdot z^{-1} - k^{-1} dz \cdot z^{-1} k] z.$$

Thus

$$g^{-1} dg = dw = z^{-1} d\mu z \tag{10.171}$$

where

$$d\mu = k^{-1} dk + dz \cdot z^{-1} - k^{-1} dz \cdot z^{-1} k.$$

From Eq. (10.171) we obtain

$$dw_{12} = d\mu_{12}$$

$$dw_{22} = d\mu_{22} - z d\mu_{12} \tag{10.172}$$

$$dw_{21} = d\mu_{21} + 2z d\mu_{22} - z^2 d\mu_{12}.$$

It now follows that

$$d\mu(g) = \left| \frac{\partial(w_{12}, w_{22}, w_{21})}{\partial(\mu_{12}, \mu_{22}, \mu_{21})} \right|^2 d^2 \mu_{12} d^2 \mu_{22} d^2 \mu_{21}.$$

The Jacobian is a triangular determinant with one along the main diagonal

$$\begin{vmatrix} 1 & -z & -z^2 \\ 0 & 1 & 2z \\ 0 & 0 & 1 \end{vmatrix} = 1.$$

Thus

$$d\mu(g) = d^2\mu_{12}d^2\mu_{22}d^2\mu_{21}. \tag{10.173}$$

Now

$$d\mu = du + d\vartheta \tag{10.174}$$

where

$$du = k^{-1}dk, \ d\vartheta = dz \cdot z^{-1} - k^{-1}dz \cdot z^{-1}k \tag{10.175}$$

du is a triangular determinant with two independent (complex) elements which are chosen as du_{12}, du_{22}.

On the other hand $d\vartheta$ has only one independent element, (say) $d\vartheta_{21}$. Thus

$$d\mu_{21} = du_{21} + d\vartheta_{12}$$

$$d\mu_{22} = du_{22} + d\vartheta_{22}$$

$$d\mu_{21} = d\vartheta_{21}.$$

Since du and $d\vartheta$ are independent it follows that $\frac{\partial u_{pq}}{\partial \vartheta_y} = 0$. Thus

$$d\mu(g) = d^2\mu_{12}d^2\mu_{22}d^2\mu_{21}$$

$$= \left| \frac{\partial(\mu_{12}, \mu_{22}, \mu_{21})}{\partial(u_{12}, u_{22}, \vartheta_{21})} \right|^2 d^2u_{12}d^2u_{22}d^2\vartheta_{21}. \tag{10.176}$$

Since the Jacobian is a triangular determinant with along the main diagonal

$$\begin{vmatrix} 1 & 0 & 0 \\ 0 & 1 & 0 \\ \dfrac{\partial\vartheta_{12}}{\partial\vartheta_{21}} & \dfrac{\partial\vartheta_{22}}{\partial\vartheta_{21}} & 1 \end{vmatrix}.$$

We have

$$d\mu(g) = d^2u_{12}d^2u_{22}d^2\vartheta_{21}$$

$$= d\mu_l(k)d^2\vartheta_{21}. \tag{10.177}$$

Now writing k as

$$k = \zeta^{-1}\delta\zeta \tag{10.178}$$

so that

$$k_{pp} = \delta_p, \quad p = 1, 2$$

$$\zeta = \frac{k_{12}}{(\delta_1 - \delta_2)}. \tag{10.179}$$

We have

$$d\vartheta = dz \cdot z^{-1} - \zeta^{-1}\delta^{-1}\zeta dz \cdot z^{-1}\zeta^{-1}\delta\zeta$$

$$= \zeta^{-1}[dp]\zeta \tag{10.180}$$

$$dp = d\lambda - \delta^{-1}d\lambda\delta \tag{10.181}$$

where

$$d\lambda = \zeta dz \cdot z^{-1}\zeta^{-1} \tag{10.182}$$

$$= \zeta d\sigma \zeta^{-1}. \tag{10.183}$$

Equation (10.180) yields

$$d\vartheta_{21} = dp_{21} = d\lambda_{21} - \frac{\delta_1}{\delta_2}d\lambda_{21}$$

$$= d\lambda_{21}\frac{(\delta_2 - \delta_1)}{\delta_2}$$

$$\therefore \quad d^2\vartheta_{21} = d^2 p_{21} = \frac{|\delta_2 - \delta_1|^2}{|\delta_2|^2}d^2\lambda_{21}. \tag{10.184}$$

Finally

$$d\lambda_{21} = d\sigma_{21}.$$

Thus

$$d^2\lambda_{21} = d^2\sigma_{21} = d\mu(z).$$

We have therefore

$$d^2\vartheta_{21} = \frac{|\delta_2 - \delta_1|^2}{|\delta_2|^2}d\mu(z). \tag{10.185}$$

Using Eqs. (10.177) and (10.185) we therefore obtain

$$d\mu_1(k)d\mu(z) = \frac{|\delta_2|^2 d\mu(g)}{|\delta_2 - \delta_1|^2}$$

which is Eq. (10.166).

We now recall that in K_1, $k_{22} = \delta_2$, and in K_2, $k_{22} = \delta_1$ so that calling the two distinct eigenvalues as $\lambda_g, \lambda_g^{-1}$ we have

$$T_r(T_x^{(m,p)}) = \int x(g)\pi^{(m,p)}(g)d\mu(g)$$

where

$$\pi^{(m,p)}(g) = \frac{\left[|\lambda_g|^{ip}\left(\frac{\lambda_g}{|\lambda_g|}\right)^{-m} + |\lambda_g|^{-ip}\left(\frac{\lambda_g}{|\lambda_g|}\right)^{m}\right]}{|\lambda_g - \lambda_g^{-1}|^2}. \tag{10.186}$$

10.6.3   Character of the supplementary of representations

The finite element of the group for the supplementary series of representations is given by

$$T_g f(z) = |g_{12}z + g_{22}|^{-2-2c} f\left(\frac{g_{11}z + g_{21}}{g_{12}z + g_{22}}\right). \tag{10.187}$$

These representations are unitary under the scalar product

$$(f_1, f_2) = \iint d^2z_1 d^2z_2 |z_1 - z_2|^{-2+2c} \overline{f_1(z_1)} f_2(z_2). \tag{10.188}$$

We now investigate the condition under which the norm $\|f\|$ is positive definite. It should be pointed out that the scalar product converges in the usual sense for $0 < c < 1$. For $-1 < c < 0$ it is to be understood in the sense of its regularization.

Since the functions $f(z)$ vanish outside some bounded set their Fourier transforms will be locally summable functions:

$$\phi(\omega) = \frac{1}{2\pi} \int f(z) e^{+i\mathrm{Re}(\bar{z}\omega)} d^2z \tag{10.189}$$

$$\int |\phi(\omega)| d^2\omega < \infty.$$

We consider first $0 < c < \frac{1}{2}$. Then

$$\int |z_1 - z_2|^{-2+2c} f(z_1) d^2z_1$$

$$= \int |z|^{-2+2c} f(z + z_2) d^2z$$

$$= \int d^2z |z|^{-2+2c} \frac{1}{2\pi} \int \overline{\phi(\omega)} e^{i\mathrm{Re}[(z+z_2)\bar{\omega}]} d^2\omega$$

$$= \frac{1}{2\pi} \int d^2\omega \overline{\phi(\omega)} e^{i\mathrm{Re}(z_2\bar{\omega})} \times \int e^{i\mathrm{Re}\, z\bar{\omega}} |z|^{-2+2c} d^2z$$

$$= \frac{1}{2} \int d^2\omega \overline{\phi(\omega)} e^{i\mathrm{Re}(z_2\bar{\omega})} \int J_0(r|\omega|) r^{2c-1} dr$$

$$= \frac{1}{2} \int d^2\omega \overline{\phi(\omega)} e^{i\mathrm{Re}(z_2\bar{\omega})} |\omega|^{-2c} \times \int J_0(t) t^{2c-1} dt.$$

The above integral converges in the usual sense for $c < \frac{1}{2}$. For $\frac{1}{2} < c < 1$ it is to be understood or in the sense of its regularization (at ∞). Thus

$$\int J_0(t) t^{2c-1} dt = 2^{2c-1} \frac{\Gamma(c)}{\Gamma(1-c)}.$$

Thus

$$(f, f) = \frac{c2^{2c-1}}{2} \frac{\Gamma(c)}{\Gamma(1-c)} \int |\omega|^{-2c} \overline{\varphi(\omega)}$$

$$\int f(z_2) e^{i\operatorname{Re}(z_2\bar{\omega})} d^2 z_2 = \pi c 2^{2c-1} \frac{\Gamma(c)}{\Gamma(1-c)} \int \phi(\omega)|^2 |\omega|^{2c} d^2\omega. \tag{10.190}$$

This is positive definite for $0 < c < 1$.

10.6.4 The character of the supplementary series of representations

We construct the operator of the group ring

$$\mathbf{T}_x = \int d\mu(g) x(g) \mathbf{T}_g.$$

Thus

$$f(z)\mathbf{T}_x = \int d\mu(g) x(g) \mathbf{T}_g f(z)$$

$$\mathbf{T}_x = \int d\mu(g) x(g) |g_{12}z + g_{22}|^{-2-2c} f\left(\frac{g_{11}z + g_{12}}{g_{12}z + g_{22}}\right)$$

$$= \int d\mu(g) x(g) |(zg)_{22}|^{-2-2c} f\left(\frac{(zg)_{21}}{(zg)_{22}}\right).$$

We now make the left translation $g \to z^{-1}g$

$$\therefore \mathbf{T}_x \int d\mu(g) x(z^{-1}g) |g_{22}|^{-2-2c} f\left(\frac{g_{21}}{g_{22}}\right) \tag{10.191}$$

Let us now write $g \in \mathrm{SL}(2, \mathrm{C}), g = kz_2$ so that $k_{22}z_2 = g_{21}, \ k_{22} = g_{22}$.
Now under the above decomposition the measure decomposes as

$$d\mu(g) = d\mu(kz_2) = d\mu_l(k) d\mu(z_2).$$

Thus

$$\mathbf{T}_x f(g) = \int d\mu_l(k) d\mu(z_2) x(z^{-1}kz_2) |k_{22}|^{-2-2c} f(z_2)$$

$$= \int \mathbf{K}(z, z_2) f(z_2) d\mu(z_2) \tag{10.192}$$

where

$$\mathbf{K}(z, z_2) = \int x(z^{-1}kz_2) |k_{22}|^{-2-2c} d\mu_l(k). \tag{10.193}$$

It should be pointed out that $\mathbf{K}(z, z_2)$ is not the integral kernel of the group ring because $\mathbf{T}_x$ now is as operator in the Hilbert space $\mathbf{H}_c$ with the non-local scalar product

$$(f_1, f_2) = C \int |z_1 - z_2|^{-2+2c} \overline{f_1(z_1)} f_2(z_2) d^2 z_1 d^2 z_2.$$

To write the action of $\mathbf{T}_x$ on $f(z)$ in the form consistent with the scalar product, we introduce the Fourier transforms.

$$K(z, z_2) = \frac{1}{2\pi} \int K(z, \omega) e^{-i\mathrm{Re}(\bar{z}_2 \omega)} d^2 \omega$$

$$f(z_2) = \frac{1}{2\pi} \int \phi(\omega') e^{i\mathrm{Re}(\overline{z_2}\omega')} d^2 \omega'. \tag{10.194}$$

Thus

$$\int K(z, z_2) f(z_2) = \int K(z, \omega) \phi(\omega) d^2 \omega.$$

We now define

$$|\omega|^{2c} K(z, \omega) = K_1(z, \omega) 2^{2c-1} \frac{\Gamma(c)}{\Gamma(1-c)}$$

$$= K_1(z, \omega) \int J_0(t) t^{2c-1} dt = K_1(z, \omega) \rho^{2c} \int_0^\infty J_0(\rho r) r^{2c-1} dr.$$

In other words

$$K(z, \omega) = K_1(z, \omega) \int_0^\infty J_0(\rho r) r^{2c-1} dr. \tag{10.195}$$

Now

$$J_0(\rho r) = \frac{1}{2\pi} \int_0^{2\pi} e^{i(\rho r \cos(\theta - \phi))} d\theta$$

$$= \frac{1}{2\pi} \int e^{i\mathrm{Re}(\xi\bar{\omega})} d\theta.$$

Hence we have

$$K(z, \omega) = K_1(z, \omega) \frac{1}{2\pi} \int \int |\xi|^{-2+2c} e^{i\mathrm{Re}(\xi\bar{\omega})} d^2 \xi.$$

Substituting this we obtain

$$\int K(z, z_2) f(z_2) = \int K_1(z, \omega) |\xi|^{-2+2c} e^{i\mathrm{Re}(\xi\bar{\omega})} \varphi(\omega) d^2 \omega.$$

Now we introduce

$$Q(z, z_1) = \frac{1}{(2\pi)} \int K_1(z, \omega) e^{+i\mathrm{Re}(\bar{z}_1 \omega)} d^2 \omega$$

$$f(z_2) = \frac{1}{(2\pi)} \int \varphi(\omega) e^{+i\mathrm{Re}(\bar{z}_2 \omega)} d^2 \omega. \tag{10.196}$$

We now note that

$$\int e^{i\mathrm{Re}[\omega(\bar{\xi}+\bar{z}_2-\bar{z}_1)]}|\xi|^{-2+2c}d^2\xi d^2\omega = |z_1 - z_2|^{-2+2c}. \qquad (10.197)$$

Thus

$$\mathbf{T}_x f(z) = \int\int Q(z,z_1)|z_1 - z_2|^{-2+2c}f(z_2)d^2z_1 d^2z_2. \qquad (10.198)$$

Comparing the above equation we obtain

$$\mathbf{K}(z,z_2) = \int Q(z,z_1)|z_1 - z_2|^{-2+2c}d^2z_1. \qquad (10.199)$$

Defining the Fourier transforms as

$$Q(z,z_1) = \frac{1}{(2\pi)^2}\int L(\omega,\omega_1)e^{i\mathrm{Re}(z\bar{\omega}-z_1\bar{\omega}_1)}d^2\omega d^2\omega_1$$

$$f(z_2) = \frac{1}{2\pi}\int \phi(\omega')e^{-\mathrm{Re}(z_2\bar{\omega}')}d^2\omega'.$$

So that

$$\mathbf{T}_x\phi(\omega) = \int L(\omega,\omega_1)\phi(\omega_1)\int |\xi|^{-2+2c}e^{i\mathrm{Re}(\xi\bar{\omega_1})}d^2\xi$$

carrying out the integration we have

$$\mathbf{T}_x\phi(\omega) = 2\pi\frac{\Gamma(c)2^{2c-1}}{\Gamma(1-c)}\int d^2\omega_1 L(\omega,\omega_1)|\omega_1|^{-2c}\phi(\omega_1) \qquad (10.200a)$$

$$\mathbf{T}_r(\mathbf{T}_x) = (2\pi)\frac{\Gamma(c)2^{2c-1}}{\Gamma(1-c)}\int d^2\omega L(\omega,\omega)|\omega|^{-2c} \qquad (10.200b)$$

$$= \int Q(z,z_1)|z_1 - z|^{-2+2c}d^2z d^2z_1. \qquad (10.201)$$

Using the Eq. (10.199)

$$\mathbf{T}_r(\mathbf{T}_x) = \int \mathbf{K}(z,z)d^2z$$

$$= \int x(z^{-1}kz)|k_{22}|^{-2-2c}d\mu_l(k)d\mu(z). \qquad (10.202)$$

Proceeding as in the case of the principal series we obtain

$$\mathbf{T}_r(\mathbf{T}_x) = \int x(g)\pi^s(g)\mu(g)$$

where

$$\pi^s(g) = \frac{(|\lambda_g|^{-2c} + |\lambda_g|^{2c})}{|\lambda_g - \lambda_g^{-1}|^2}. \qquad (10.203)$$

10.6.5   *The Plancherel formula for* **SL(2, C)**

Let $x(g)$ be an integrable function on $G = SL(2, C)$

$$I = \int |x(g)|^2 d\mu(g) < \infty.$$

We introduce the three (complex) parameters

$$p_1 = \frac{g_{11}}{g_{12}}, \quad p_2 = \frac{1}{g_{12}}, \quad p_3 = \frac{g_{22}}{g_{12}} \tag{10.204}$$

then we have

$$\int |x(g)|^2 d\mu(g) = \int |\tilde{x}(p_1, p_2, p_3)|^2 |p_2|^{-6} d\mu(p_1) d\mu(p_2) d\mu(p_3) \tag{10.205}$$

where

$$d\mu(p) = d^2 p = dp_x dp_y, \quad \text{etc.}$$

The above formula means that the function $|\tilde{x}(p_1, p_2, p_3)|^2 |p_2|^{-6}$ is integrable in p_1, p_2, p_3. We assume $|\tilde{x}(p.)|^2 |p_2|^{-4}$ also to be integrable so that we can take the Fourier transform of this function,

$$\chi(\zeta_1, \zeta_2, \zeta_3) = \frac{1}{(2\pi)^3} \int \tilde{x}(p_1, p_2, p_3) |p_2|^{-4}$$

$$\times e^{-iRe(\bar{p}_1 \zeta_1 - \bar{p}_2 \zeta_2 + \bar{p}_3 \zeta_3)} d\mu(p_1) d\mu(p_2) d\mu(p_3). \tag{10.206}$$

We now start from

$$K(z_1, z_2; n, \rho) = \int x(z_1^{-1} k z_2) |\delta_2|^{n+i-2} \delta_2^{-n} d\mu_l(k)$$

where

$$k = \delta \zeta, \quad d\mu_l(k) = d\mu(\delta) d\mu(\zeta)$$

$$\delta = \begin{pmatrix} \delta_1 & 0 \\ 0 & \delta_2 \end{pmatrix} = \begin{pmatrix} \lambda^{-1} & 0 \\ 0 & \lambda \end{pmatrix}$$

$$\zeta = \begin{pmatrix} 1 & \zeta \\ 0 & 1 \end{pmatrix}$$

so that

$$K(z_1, z_2; n, \rho) = \int |\lambda|^{n+i\rho} \lambda^{-n} d\mu(\delta) \times \frac{1}{|\lambda|^2} \int x(z_1^{-1} \delta \zeta z_2) d\mu_l(\zeta).$$

If we set

$$\phi(z_1, z_2; \lambda) = \frac{1}{|\lambda|^2} \int d\mu(\zeta) x(z_1^{-1} \delta \zeta z_2)$$

$$K(z_1, z_2; n, \rho) = \int |\lambda|^{n+i\rho} \lambda^{-n} \phi(z_1, z_2; \lambda) d\mu_l(\delta). \tag{10.207}$$

If we now set

$$g = z_1^{-1} \delta \zeta z_2$$

the parameter p_1, p_2, p_3 are given by

$$\left. \begin{array}{l} p_1 = z_2 + \zeta^{-1} \\[2mm] p_2 = \dfrac{\lambda}{\zeta} \\[2mm] p_3 = -z_1 + \dfrac{\lambda}{\zeta}. \end{array} \right\} \tag{10.208}$$

We now introduce the Fourier transform,

$$\phi(\omega_1, \omega_2; \lambda) = \frac{1}{(2\pi)^2} \int \phi(z_1, z_2; \lambda) |p_2|^{-4} e^{i\mathrm{Re}(\overline{\omega_1} z_1 - \overline{\omega_2} z_2)} d\mu(z_1) d\mu(z_2).$$

Introducing p_1, p_2, p_3 we have

$$\phi(\omega_1, \omega_2; \lambda) = \frac{1}{(2\pi)^2} \int \tilde{x}(p_1, p_2, p_3)$$
$$\times |p_2|^{-4} e^{i\mathrm{Re}(\overline{\omega_1} z_1 - \overline{\omega_2} z_2)} d\mu(z_1) d\mu(z_2) d\mu(\zeta).$$

Since

$$\left. \begin{array}{l} z_1 = -p_3 + \dfrac{\lambda^2}{\zeta} = -p_3 + \lambda p_2 \\[3mm] z_2 = p_1 - \dfrac{1}{\zeta} = p_1 - \dfrac{p_2}{\lambda} \end{array} \right\} \tag{10.209}$$

$$\phi(\omega_1, \omega_2; \lambda) = \frac{1}{(2\pi)^2} \int \tilde{x}(p_1, p_2, p_3) \times |p_2|^{-4}$$
$$\times e^{-\mathrm{Re}(\bar{\omega}_2 p_1 - (\bar{\omega}_1 \lambda + \frac{\bar{\omega}_2}{\lambda}) p_2 + \bar{\omega}_1 p_3)} d\mu(p_1) d\mu(p_2) d\mu(p_3). \tag{10.210}$$

We therefore obtain

$$\phi(\omega_1, \omega_2; \lambda)$$
$$= \frac{1}{(2\pi)^3} \int \tilde{x}(p_1, p_2; p_3) |p_2|^{-4} e^{i\mathrm{Re}(\bar{p}_1 \zeta_1 - \bar{p}_2 \zeta_2 + \bar{p}_3 \zeta_3)} d\mu(p_1) d\mu(p_2) d\mu(p_3) \tag{10.211}$$

with

$$\zeta_1 = \omega_2, \quad \zeta_2 = \omega_1 \bar{\lambda} + \frac{\bar{\omega}^2}{\lambda}, \quad \zeta_3 = \omega_1$$

$$\phi = \chi(\zeta_1, \zeta_2, \zeta_3)$$

$$\frac{1}{2\pi} \phi(\omega_1, \omega_2, \lambda) = \chi\left(\omega_2, \omega_1 \bar{\lambda} + \frac{\omega_2}{\lambda}, \omega_1\right)$$

$$\left. \begin{array}{l} \zeta_2 = \omega_1 \bar{\lambda}, + \dfrac{\omega_2}{\bar{\lambda}} \\[2mm] \bar{\zeta}_2 = \overline{\omega_1} \lambda, + \dfrac{\bar{\omega}_2}{\lambda}. \end{array} \right] \tag{10.212}$$

Now

$$\chi(\zeta_1, \zeta_2, \zeta_3) = \frac{1}{(2\pi)^3} \int e^{-\mathrm{Re}(\bar{p}_1 \zeta_1 - \bar{p}_2 \zeta_2 + \bar{p}_3 \zeta_3)} |p_2|^{-4}$$
$$\times \tilde{x}(p_1, p_2, p_3) d\mu(p_1) d\mu(p_2) d\mu(p_3)$$

so that

$$\frac{\partial^2 \chi}{\partial \zeta_2 \partial \bar{\zeta}_2} = -\frac{1}{4} \frac{1}{(2\pi)^3} \int e^{-i\mathrm{Re}(\bar{p}_1 \zeta_1 - \bar{p}_2 \zeta_2 + \bar{p}_3 \zeta_3)} |p_2|^{-2}$$
$$\times \tilde{x}(p_1, p_2, p_3) d\mu(p_1) d\mu(p_2) d\mu(p_3).$$

From the usual Plancherel formula for Fourier transform we have

$$-\int \chi(\zeta) \frac{\partial^2 \chi(\zeta)}{\partial \zeta_1 \partial \bar{\zeta}_2} d\mu(\zeta_1) d\mu(\zeta_2) d\mu(\zeta_3)$$
$$= \frac{1}{4} \frac{1}{(2\pi)^3} \int |x(p_1, p_2, p_3)|^2 |p_2|^{-6} d\mu(p_1) d\mu(p_2) d\mu(p_3)$$
$$= \frac{1}{4} \int |x(g)|^2 \, d\mu(g). \tag{10.213}$$

We also have

$$\zeta_2 = \omega_1 \bar{\lambda} + \frac{\omega_2}{\bar{\lambda}}, \quad \bar{\zeta}_2 = \bar{\omega}_1 \lambda + \frac{\bar{\omega}_2}{\lambda}$$

$$\left| \frac{\partial \zeta_2}{\partial \bar{\lambda}} \right|^2 = \left| \omega_1 - \frac{\omega_2}{\bar{\lambda}^2} \right|^2$$

$$d^2 \zeta_2 = \left| \frac{\partial \zeta_2}{\partial \bar{\lambda}} \right|^2 d^2 \lambda = \left| \omega_1 - \frac{\omega_2}{\bar{\lambda}^2} \right|^2 d^2 \lambda \tag{10.214}$$

$$\zeta_1 = \omega_2, \ \zeta_3 = \omega_1$$

$$\left. \begin{array}{c} d\mu(\zeta_1) = d\mu(\omega_2) \\ d\mu(\zeta_3) = d\mu(\omega_1) \end{array} \right] \tag{10.215}$$

$$\frac{\partial}{\partial \lambda} = \frac{\partial \zeta_2}{\partial \bar{\lambda}} \frac{\partial}{\partial \zeta_2} = \left(\omega_1 - \frac{\omega_2}{\bar{\lambda}_2} \right) \frac{\partial}{\partial \zeta_2} \tag{10.216}$$

$$\frac{\partial^2}{\partial \zeta_1 \partial \bar{\zeta}_2} = \frac{1}{\left| \omega_1 - \frac{\omega_2}{\bar{\lambda}_2} \right|^2} \frac{\partial^2}{\partial \lambda \partial \bar{\lambda}}. \tag{10.217}$$

Substituting Eqs. (10.214)–(10.217) we have

$$\frac{1}{4} \int |x(g)|^2 d\mu(g) = -\frac{1}{(2\pi)^2} \int \phi(\omega_1, \omega_2; \lambda)$$

$$\times \frac{\partial^2 \overline{\phi(\omega_1, \omega_2; \lambda)}}{\partial \lambda \partial \bar{\lambda}} d\mu(\omega_1) d\mu(\omega_2) d\mu(\lambda).$$

Once again using traditional Plancherel formula

$$\int \phi(\omega_1, \omega_2; \lambda) \times \frac{\partial^2}{\partial \lambda \partial \bar{\lambda}} \overline{\phi(\omega_1, \omega_2; \lambda)} d\mu(\omega_1) d\mu(\omega_2)$$

$$= \int \phi(z_1, z_2; \lambda) \times \frac{\partial^2}{\partial \lambda \partial \bar{\lambda}} \overline{\phi(z_1, z_2; \lambda)} d\mu(z_1) d\mu(z_2)$$

$$\int |x(g)|^2 d\mu(g) = -\frac{4}{(2\pi)^2} \times \frac{1}{4} \int \phi(z_1, z_2; \lambda)$$

$$\times \left(\frac{\partial^2}{\partial t^2} + \frac{\partial^2}{\partial \theta^2} \right) \overline{\phi(z_1, z_2; \lambda)} d\mu(z_1) d\mu(z_2) d\mu(\lambda). \tag{10.218}$$

Setting $\lambda = e^t \, e^{i\theta}$

$$\mathrm{K}(z_1, x_2; n, \rho) = \int_{-\infty}^{\infty} \int_0^{2\pi} dt \, d\theta e^{+itp} e^{-in\theta} \phi(z_1, z_2; t, \theta)$$

it has been shown by Gel'fand and Naimark that for fixed z_1, z_2, the function $\phi(z_1, z_2; \lambda) = \phi(z_1, z_2; t, \theta)$ vanishes outside a region $|t| < c$ so that

$$-(n^2 + \rho^2) \mathrm{K}(z_1, z_2; n, \rho)$$

$$= \int_{-\infty}^{\infty} dt \int_0^{2\pi} d\theta e^{it\rho} e^{-in\theta} \left(\frac{\partial^2}{\partial t^2} + \frac{\partial^2}{\partial \theta^2} \right) \phi(z_1, z_2; t, \theta). \tag{10.219}$$

By the usual Plancherel theorem have

$$\frac{1}{(2\pi)^2} \sum_{-\infty}^{\infty} \int_0^{\infty} d\rho |K(z_1, z_2; n, \rho)|^2 (n^2 + \rho^2)$$

$$= -\frac{1}{(2\pi)^2} \int \phi(z_1, z_2; t, \theta) \overline{\nabla \phi(z_1, z_2; t\theta)} dt \, d\theta \qquad (10.220)$$

where we have used

$$K(z_1, z_2; n, \rho) = K(z_1, z_2; n, -\rho)$$

$$\nabla = \frac{\partial^2}{\partial t^2} + \frac{\partial^2}{\partial \theta^2}. \qquad (10.221)$$

It is left as an exercise for the reader to verify,

$$\text{Tr}[T_x^{(n,\rho)\dagger} T_x^{(n,\rho)}] = \int |K(z_1, z_2; -n, -\rho)|^2 d\mu(z_1) d\mu(z_2) \qquad (10.222)$$

$$\int d\mu(g)|x(g)|^2 = \frac{1}{2\pi^2} \sum_{-\infty}^{\infty} \int_0^{\infty} |K(z_1, z_2; n, \rho)|^2 (n^2 + \rho^2) d\rho d\mu(z_1) d\mu(z_2)$$

$$= \frac{1}{2\pi^2} \sum_{-\infty}^{\infty} \int_0^{\infty} \text{Tr}[T_x^{(n,\rho)\dagger} T_x^{(n,\rho)}](n^2 + \rho^2) d\rho. \qquad (10.223)$$

10.6.6   *The* SU(1, 1) *content of* SL(2, C)[6]

It will be seen that the $SU(2)$ and $SU(1,1)$ bases of $SL(2,C)$ have nearly the same form and are functions of the same variable $x = z\bar{z}$. For obtaining the reduction we can, therefore, expand the $SU(2)$ bases $\Psi_{jm}^{SU(2)}$ in a series of functions $\Psi_{km}^{SU(1,1)}$. This is easily done by breaking up $\Psi_{jm}^{SU(2)}$ into a series of the appropriate type by means of Burchnall–Chaundy formula. The series thus obtained yields the continuous and discrete spectra of the $SU(1,1)$ representations after the Sommerfeld–Watson transformation just as the Clebsch–Gordan problem discussed in Chapter 9. The infinitesimal operators of the group can be expressed as differential operators of the from,

$$F_3 = iz\frac{\partial}{\partial z} + i\bar{z}\frac{\partial}{\partial \bar{z}} - i(c-1)$$

$$F_+ = i\frac{\partial}{\partial z} - i\bar{z}^2\frac{\partial}{\partial \bar{z}} + i\bar{z}(c - j_0 - 1)$$

$$F_- = -iz^2\frac{\partial}{\partial z} + i\frac{\partial}{\partial \bar{z}} + iz(c + j_0 - 1)$$

$$J_3 = -z\frac{\partial}{\partial z} + \bar{z}\frac{\partial}{\partial \bar{z}} + j_0$$

$$J_+ = -\frac{\partial}{\partial \bar{z}} - \bar{z}^2\frac{\partial}{\partial \bar{z}} + \bar{z}(c - j_0 - 1)$$

$$J_- = z^2\frac{\partial}{\partial z} + \frac{\partial}{\partial \bar{z}} - z(c + j_0 - 1) \tag{10.224}$$

where $z = \xi_1/\xi_2$, $\bar{z} = \overline{\xi_1}/\overline{\xi_2}$ and (ξ_1, ξ_2) and $(\overline{\xi_1}, \overline{\xi_2})$ are spinors transforming according to the fundamental representation and the complex conjugate representation of $SL(2, \mathbb{C})$ respectively. The compact $SU(2)$ basis resembles the coupling of two angular momenta and are given by

$$g_{jm} = z^{j_0-m}(1 + z\bar{z})^{c-j-1}F(-j - m, j_0 - j; -2j; 1 + z\bar{z})$$

$$= z^{j_0-m}\psi_{jm}. \tag{10.225}$$

The $SU(1,1)$ bases are likewise formally identical to the coupling of two $SU(1,1)$ irreps. and are given by

$$\phi_{km} = z^{j_0-m}(1 - x)^{c+k-1}F(k + m, -j_0 + k; m - j_0 + 1; x) \tag{10.226}$$

for $x = z\bar{z} < 1$ and $m - j_0 \geq 0$.

Similarly

$$\phi_{km} = z^{j_0-m}u_{km} = z^{j_0-m}(1 - x)^{c+k-1}F(k - m, j_0 + k; 1 + j_0 - m; x) \tag{10.227}$$

for $x < 1$, $j_0 - m \geq 0$, and

$$u_{km} = x^{-k-j_0}(1 - x)^{c+k-1}F\left(k + m, j_0 + k; j_0 + m + 1; \frac{1}{x}\right)$$

$$\text{for } x > 1, \ j_0 + m \geq 0 \tag{10.228}$$

$$u_{km} = (-x)^{m-k}(1 - x)^{c+k-1}F\left(k - m, k - j_0; 1 - j_0 - m; \frac{1}{x}\right)$$

$$x > 1, \ j_0 + m \leq 0. \tag{10.229}$$

10.6.7 *Reduction of the principal series of representation*

For the determination of the complete spectrum of $SU(1,1)$ representations, a knowledge of the discrete part of the spectrum seems to be necessary. This can be easily done by applying the operator F_3 and using the recurrence relations of the hypergeometric function. If

$$f_{km} = a_k \phi_{km}$$

$$F_3 f_{k,m} = -\frac{i}{2}[(m+k)(m-k)]^{\frac{1}{2}} \alpha_k (c-k) \frac{a_k}{a_{k-1}} f_{k-1,m}$$

$$+ \frac{imc j_0}{k(1-k)} f_{k,m-2i}[(m+k-1)(m-k+1)]^{\frac{1}{2}}(c+k-1)$$

$$\frac{a_k}{a_{k+1}} f_{k+1,m} \tag{10.230}$$

where

$$\alpha_k = \frac{j_0^2 - k^2}{k^2(4k^2 - 1)}. \tag{10.231}$$

The condition

$$(F_3 f_{k+1,m}, f_{km}) = (f_{k+1,m}, F_3 f_{km})$$

then gives

$$\left| \frac{a_{k+1}}{a_k} \right|^2 = \frac{|c-k|^2 (j_0^2 - k^2)}{4k^2(4k^2 - 1)}. \tag{10.232}$$

For this to be positive k must have the set of values

$$k = j_0, j_0 - 1, j_0 - 2 \cdots - \frac{1}{2} \text{ or zero.} \tag{10.233}$$

To determine the continuous part of the spectrum of k values we expand the function ψ_{jm} in powers of $x = z\bar{z}$ and use the Burchnall–Chaundy formula.

$$x^r = \sum_{n=0}^{\infty} (-)^n \frac{(\alpha)_n (\beta)_n}{(\gamma + n - 1)_n n!} {}_3F_2 \begin{bmatrix} -r, n+\gamma-1, -n \\ \alpha, \beta \end{bmatrix}$$

$$\times (1-x)^n F(\alpha+n, \beta+n, \gamma+2n, 1-x). \tag{10.234}$$

For $x < 1$, $m \geq j_0$, $\alpha = m+1-c$, $\beta = 1-j_0-c$, $\gamma = 2(1-c)$ the process yields.

$$\psi_{jm} = \frac{2^{c-j-1} e^{i\pi(m-j_0)}}{\Gamma(1+m-c)\Gamma(1-j_0-c)}$$

$$\times \sum_{n=0}^{\infty} (-)^n \frac{\Gamma(m+1-c+n)\Gamma(1-c+n-j_0)(1-2c+n)}{\Gamma(1-2c+2n)n!}$$

$$S_{jm\,j_0}^n (1-x)^n F(1-c+n-m, 1-c+n+j_0; 2(1-c+n); (1-x)) \tag{10.235}$$

where

$$S^n_{jm\,j_0} = \sum_{t=0}^{j-m} \frac{(p)_t(q)_t}{(r)_t t!} 2^t \, {}_3F_2 \left[\begin{matrix} -n, 1-2c+n, 1-c-t+j \\ (1+m-c), (1-j_0-c) \end{matrix} ; \frac{1}{2} \right]$$

$$p = -j+m, \quad q = -j_0-j, \quad r = -2j. \tag{10.236}$$

We now express the sum as a contour integral in the complex k–plane by constructing the meromorphic function

$$\chi(k) = \frac{\Gamma(m+k)\Gamma(k-j_0)\Gamma(1-k-c)\Gamma(k-c)}{\Gamma(2k-1)} S^k_{jm\,j_0} a_{km} \tag{10.237}$$

where

$$S^k_{jm\,j_0} = \sum_{t=0}^{j-m} \frac{(p)_t(q)_t}{(r)_t t!} 2^t \, {}_3F_2 \left[\begin{matrix} k-c, 1-k-c, -c-t+j+1 \\ 1+m-c, 1-j_0-c \end{matrix} ; \frac{1}{2} \right]$$

$$a_{km} = (1-x)^{c+k-1} F(k-m, j_0+k, 2k, 1-x). \tag{10.238}$$

The various terms in the sum are easily recognized as the residues of $\chi(k)$ at $k = n+1-c, n = 0,1,2\ldots$ which are the positions of the simple poles of $\Gamma(1-c-k)$.

In addition, $\chi(k)$ have simple poles at

$$k = c-n, k = j_0 - n, k = -m-n, n = 0,1,2\ldots. \tag{10.239}$$

Since $c = is$ (pure imaginary) for the principal series the singularities at $c-n, -m-n, j_0-n$ (for $n > 0$, and $n > j_0$) lie on the left half of the complex k-plane. The singularities at $j_0 - n$, for $n = j_0 - 1, j_0 - 2, \ldots$ lie the region of interest Re $k > 0$.

Because of the occurrence of $\Gamma(2k-1)$ no singularities occur at $k = \frac{1}{2}$ or $k = 0$. Since $\chi(k)$ goes to zero rapidly as $|k|$ tends to infinity in the region Re $k > 0$. We choose a contour C consisting of the infinite semicircle S on the right and the line Re $k = \frac{1}{2}$. The singularities enclosed by the contour are the simple poles at $k = n+1-c$, $n = 0,1,2\ldots$ and at $k = j_0 - l$, $(l = 0,1,2\ldots j_0 - \frac{1}{2}$ or $j_0)$. Therefore by Cauchy's residue theorem,

$$\frac{1}{2\pi i} \oint \chi(k)dk = \sum_{n=0}^{\infty} \text{Res}[\chi(k)]_{k=n+1-c} + \sum_l \text{Res}[\chi(k)]_{k=j_0-l}. \tag{10.240}$$

Since the first term on the r.h.s equals ψ_{jm} by our previous analysis and since the integral vanishes on S we have,

$$\psi_{jm} = \frac{1}{2\pi i} \int_{\frac{1}{2}-i\infty}^{\frac{1}{2}+i\infty} \chi(k)dk + \sum_i \text{Res}[\chi(k)]_{k=j_0-l}. \tag{10.241}$$

Folding the integral about the real axis and evaluating the residues of the last term, we have

$$\psi_{jm} = A_{mj_0c}\left[\frac{1}{2\pi i}\int_{\frac{1}{2}}^{\frac{1}{2}+i\infty} dk \frac{\Gamma(m+k)\Gamma(m-k+1)\Gamma(1-k-j_0)}{\Gamma(2k-1)}\right.$$

$$\left.\frac{\Gamma(k-c)\Gamma(1-k-c)}{\Gamma(1-2k)}S_{jmj_0}^k u_{km}\right]$$

$$+\sum_{1/\frac{3}{2}}^{j_0}\frac{1}{\Gamma(1+j_0-k)}\frac{\Gamma(1-c-k)S_{jmj_0}^k}{\Gamma(2k-1)}\times\frac{\Gamma(2k)\Gamma(m-j_0)}{\Gamma(k-j_0)}\times u_{km}$$

$$\tag{10.242}$$

where

$$A_{mj_0c} = \frac{2^{c-j-1}e^{i\pi(m-j_0)}\Gamma(m-j_0+1)}{\Gamma(m+1-c)\Gamma(1-j_0-c)}.\tag{10.243}$$

We can also write down the expansion for $x > 1, j_0 + m < 0$

$$\psi_{jm} = B_{mj_0c}\left[\frac{1}{2\pi i}\frac{1}{\Gamma(1-j_0-m)}\right.$$

$$\times\int_{\frac{1}{2}}^{\frac{1}{2}+i\infty} dk \frac{\Gamma(k-m)\Gamma(1-k-m)\Gamma(k-k_0)\Gamma(1-k-k_0)}{\Gamma(2k-1)}$$

$$\left.\times\frac{\Gamma(k-c)\Gamma(1-k-c)}{\Gamma(1-2k)}S_{j,-mj_0}^k u_{km}e^{i\pi(j_0-c)}\right]$$

$$+\sum_{k=1/\frac{3}{2}}^{j_0}\frac{\Gamma(k-m)\Gamma(1-k-c)\Gamma(k-c)}{\Gamma(2k-1)\Gamma(1-k+j_0)}S_{j,-mj_0}^k a_{km}\tag{10.244}$$

where

$$B_{mj_0c} = \frac{2^{\sigma-j-1}\exp[i\pi(j-j_0)]}{\Gamma(1-c-m)\Gamma(1-j_0-c)}$$

and for $j_0 + m > 0$ the discrete spectrum terminates at $k = m$. The above formulas show that in addition to the principal series of SU(1,1) in the first case the reduction yields only the positive discrete series while in the second case only the negative discrete series.

10.6.8   *Reduction of the supplementary series*

For the supplementary series $j_0 = 0$ and c is a real number lying in the interval $0 < c < 1$. Since $j_0 = 0$ the spectrum of the SU(1,1) representations does not contain any discrete part. Using the method of the previous section we have for $x < 1$,

$$\psi_{jm} = 2^{\sigma-j-1}(\mp)^m \sum_{n=0}^{\infty} (-)^n \frac{\Gamma(|m|+1-c+n)\Gamma(1-c+n)\Gamma(1-2c+n)}{\Gamma(1-2c+2n)n!}$$

$$S_{j|m|0}^n (1-x)^n F(-m+1-c+n, 1-c+n, 2(1-c+n), 1-x).$$

$$(10.245)$$

Where the positive or negative sign is to be taken according as $m \geq 0$. The series on the r.h.s can be regarded as the sum of the residues at $k = n+1-c$ of the analytic function $\chi(k)$ with $j_0 = 0$. Besides these remaining all singularities lie on the left half of the complex k-plane. If $0 < c < \frac{1}{2}$ the only singularities lying on the semi-infinite plane $\operatorname{Re} k > \frac{1}{2}$ and are enclosed by the contour C of the previous section are simple poles at $k = n+1-c$, $n = 0, 1, 2, \ldots$. Since the integral on the semicircular part again vanishes, we have

$$\psi_{jm} = \frac{2^{\sigma-j-1}(-)^m}{\Gamma(1-c)\Gamma(m+1-c)\Gamma(m+1)} \frac{1}{2\pi} \int_{\frac{1}{2}}^{\frac{1}{2}+i\infty} \frac{\Gamma(m+k)\Gamma(m+1-k)}{\Gamma(1-2k)}$$

$$\times \frac{\Gamma(k)\Gamma(1-k)\Gamma(k-c)\Gamma(1-k-c)}{\Gamma(2k-1)} u_{km} S_{jm0}^k dk. \qquad (10.246)$$

If on the other hand $\frac{1}{2} < c < 1$ then the pole at $k = 1-c$ nearest to the imaginary axis occurs between the points 0 and $\frac{1}{2}$ and so lies outside C. For this range however, one pole of $\Gamma(k-c)$, namely, the one on the extreme left at $k = c$ lies inside C. Because of the relation

$$u_{km} = \frac{\Gamma(m+1)\Gamma(1-2k)}{\Gamma(m+1-k)\Gamma(1-k)}(1-x)^{c+k-1} F(k-m, k, 2k, 1-x)$$

$$+ \frac{\Gamma(m+1)\Gamma(2k-1)}{\Gamma(m+k)\Gamma(k)}(1-x)^{c-k} F(1-k-m, 1-k, 2-2k, 1-x)$$

between the solution of the hypergeometric equation the two terms combine to give $[u_{km}]_{k=c}$ and we have

$$\psi_{jm} = 2^{c-j-i}(-)^m \left[\frac{\Gamma(m+c)\Gamma(c)}{\Gamma(m+1)\Gamma(2c-1)} S_{jm0} u_{cm} \right.$$

$$+ \frac{1}{\Gamma(m+1-c)\Gamma(1-c)\Gamma(m+1)} \frac{1}{2\pi i}$$

$$\times \int_{\frac{1}{2}}^{\frac{1}{2}+i\infty} \frac{\Gamma(m+k)\Gamma(m+1-k)\Gamma(k)\Gamma(1-k)}{\Gamma(1-2k)}$$

$$\times \frac{\Gamma(k-c)\Gamma(1-k-c)}{\Gamma(2k-1)} S_{jm0}^k u_{km} dk \Bigg] . \tag{10.247}$$

The extra term on the r.h.s corresponds to a representation D^c of the exceptional series of $SU(1,1)$. Thus the irreducible unitary representation belonging to the supplementary series of $SL(2,C)$. decomposes into IR's of principal series only of $SU(1,1)$ if $0 < c < \frac{1}{2}$ and into IR's of the principal series and one member of the exceptional series of $SU(1,1)$ if $\frac{1}{2} < c < 1$.

References

[1] I. M. Gel'fand and M. A. Naimark, *I. M. Gelfand Collected Papers* (Springer Verlag, Berlin, 1988) Vol. II (Paper on SL(2, C)).

[2] Harish Chandra, *Trans. Amer. Math. Soc.* **70** (1951) 28; **74** (1953) 185; **76** (1954) 26; **76** (1954) 234.

[3] E. U. Condon and G. H. Shortley, *The Theory of Atomic Spectra* (Cambridge University Press, 1953).

[4] I. M. Gel'fand, R. A. Minlos and Z. Ya. Shapiro, *The Theory of Rotation and Lorentz Groups and Their Applications*; Harish-Chandra, *Proc. Roy. Soc.* (London) **A 189** (1947) 272.

[5] I. M. Gel'fand and M. A. Naimark, Ref. 1.

[6] D. Basu and S. D. Majumdar, *J. Math. Phys.* **17** (1976) 186; N. Mukunda, **9** (1968) 50; **9** (1968) 417.

CHAPTER 11

The Heisenberg–Weyl Group and the Bargmann–Segal Spaces

The group theoretic content of Heisenberg's commutation relation between position and momentum operators was discovered by Weyl about 87 years ago. The quantum mechanical or the Heisenberg–Weyl group is a non-compact non-Abelian group with a non-trivial centre and in one dimension consists of three generators. It is the simplest nilpotent group and plays the same fundamental role in Kirillov's general representation theory[1] of nilpotent groups as $SU(2)$ does in the theory of semisimple groups. However detailed descriptions of the group are not abundant in mathematical physics literature.

In this chapter we use the coherent states introduced by Schrödinger[2] and subsequently by Klauder,[3] Glauber[4] and Sudarshan[5] as a possible basis for the Heisenberg–Weyl group. This basis has enabled us to reconstruct group theoretically what Bargmann[6] and Segal[7] did analytically on the Hilbert spaces of analytic functions. The resulting procedure leads to a derivation of the integral transform pair of Bargmann in a purely group theoretic setting. The theory can be subsequently extended to many other groups. For example the procedure connects various carrier spaces of $SU(1, 1)$ by integral transform pairs.[8]

11.1 The Heisenberg–Weyl Group

The Heisenberg Weyl group (HWG) consists of three infinitesimal operators (generators) satisfying

$$[Q, P] = i H, \quad [Q, H] = [P, H] = 0. \tag{11.1}$$

The above is called the Heisenberg–Weyl algebra. The exponentiation of the generators will yield the HWG. Since H commutes with both Q and P it is itself the center of the algebra.

A finite element of the group can be written as

$$T_{g(x,y,\omega)} = e^{i(xQ + yP + \omega H)}. \tag{11.2}$$

335

Since H commutes with both Q and P, we have

$$e^{i(xQ+yP+\omega H)} = e^{i(xQ+yP)}e^{i\omega H}.$$

We shall now use

$$e^A e^B = e^{A+B+\frac{1}{2}[A,B]}$$

which is valid if

$$[[A,B],A] = [[A,B],B] = 0$$

whence

$$T_{g(x,y,\omega)} = e^{ixQ}e^{iyP}e^{i(\omega+\frac{xy}{2})H} = e^{iyP}e^{ixQ}e^{i(\omega-\frac{xy}{2})H}.$$

From the above we can obtain the group multiplication law

$$T_{g(x_1,y_1,\omega_1)}T_{g(x_2,y_2,\omega_2)} = T_{g(x_1+x_2,y_1+y_2,\omega_1+\omega_2+\frac{[y_1x_2-x_1y_2]}{2})}. \tag{11.3}$$

The left and right invariant measure is the three-dimensional volume element

$$d\mu(g) = dxdyd\omega. \tag{11.4}$$

We now consider the functions $f(g) = f(x,y,\omega)$ defined on the group satisfying

$$\int |f(g)|^2 d\mu(g) < \infty. \tag{11.5}$$

We now introduce the finite element of the group defined as,

$$T_{g'}f(g) = f(g'^{-1}g) \tag{11.6}$$

where

$$g' = g(x',y',\omega'), \quad g'^{-1} = g(-x,-y,-\omega).$$

The representation (11.6) is unitary under the scalar product

$$(f,h) = \int \overline{f(g)}h(g)d\mu(g).$$

We now write

$$T_{g(x,y,\omega)} = T_{g(0,y,\omega-\frac{xy}{2})}T_{g(x,0,0)} \tag{11.7}$$

and consider the transformation induced by the finite element (11.6) on functions

$$f_c(y,u) = f_c\left(y, \omega - \frac{xy}{2}\right). \tag{11.8}$$

Since $g'^{-1}g = g(x - x', y - y', \omega - \omega' + \frac{yx' - xy'}{2})$ we have

$$T_{g'} f_c(y, u) = f_c\left(y - y', u - \omega' + x'y - \frac{x'y'}{2}\right). \tag{11.9}$$

To decompose the function space into the eigenspaces of irreducible representations, we note that

$$H = i\frac{\partial}{\partial u}.$$

We, therefore, introduce the Fourier transform,

$$f_c(y, u) = \frac{1}{\sqrt{2\pi}} \int \phi_\lambda(y) e^{-i\lambda u} du \tag{11.10}$$

so that

$$T_{g'} f_c(y, u) = f_c\left(y - y', u - \omega' + x'y - \frac{x'y'}{2}\right)$$
$$= \int \phi_\lambda(y - y') e^{-i\lambda\left(u - \omega' + x'y - \frac{x'y'}{2}\right)}. \tag{11.11}$$

This immediately yields

$$T_{g'} \phi_\lambda(y) = \phi_\lambda(y - y') e^{i\lambda\left(\omega' - x'y + \frac{x'y'}{2}\right)}. \tag{11.12}$$

The Hermitian generators of the group are given by

$$Q = -\lambda y, \quad P = i\frac{\partial}{\partial y}, \quad H = \lambda. \tag{11.13}$$

For quantum mechanics $\lambda = \hbar$. Henceforth we set $\lambda = \hbar = 1$.

The complete orthonormal coordinate basis is given by,

$$f_q = \delta(q + y), \quad (\text{setting } \lambda = 1). \tag{11.14a}$$

The momentum p-basis is given by

$$f_p = \frac{1}{\sqrt{2\pi}} e^{-ipy}. \tag{11.14b}$$

The coherent state basis introduced by Schrödinger, Klauder, Glauber and Sudarshan is the eigenfunction of the annihilation operator

$$a = (Q + iP)/\sqrt{2}$$

satisfying $af_\alpha = \alpha f_\alpha$, $\alpha \in C$, $\alpha = \alpha_1 + i\alpha_2$ and is given by

$$f_\alpha = \pi^{-\frac{1}{4}} e^{-(\alpha_1^2 + i\alpha_1\alpha_2)} e^{-(y^2 + 2\sqrt{2}\alpha y)/2}. \tag{11.14c}$$

The choice of phase ensures that the overlap between two coherent states is given by Glauber's formula

$$(f_\beta, f_\alpha) = e^{[\overline{\beta}\alpha - \frac{|\alpha|^2}{2} - \frac{|\beta|^2}{2}]}. \tag{11.15}$$

The completeness of the coordinate and coherent state bases is given by

$$\int \overline{f_q(x)} f_q(y) dq = \frac{1}{\pi} \int \overline{f_\alpha(x)} f_\alpha(y) d^2\alpha = \delta(x - y). \tag{11.16}$$

It is now easy to obtain the matrix elements of the group in any basis

$$D_{mn}(g) = (u_m, T_g u_n).$$

We list below the matrix elements in the coordinate and coherent state bases:

$$D_{\beta\alpha}(g) = \exp \left[\overline{\beta}\alpha - \frac{|\alpha|^2}{2} - \frac{|\beta|^2}{2} - i\omega + \left\{ \frac{1}{2}(x^2 + y^2) \right. \right.$$
$$\left. \left. + \sqrt{2}\,\overline{\beta}(y - ix) - \sqrt{2}\alpha(y + ix) \right\} \right] \tag{11.17a}$$

$$D_{qq'}(g) = \delta(y + q - q') \exp i \left[\omega + \frac{1}{2} + \frac{(q + q')x}{2} + \frac{xy}{2} \right] \tag{11.17b}$$

$$D_{\beta q}(g) = \pi^{-\frac{1}{4}} e^{-|\beta|^2/2} \exp \left[\left(\omega - \frac{xy}{2} \right) - \frac{y^2}{2} \right.$$
$$\left. - \frac{1}{2}(\overline{\beta}^2 + q^2) + \sqrt{2}\,\overline{\beta}(q - y) + q(ix + y) \right]. \tag{11.17c}$$

For $g = e$, i.e. $x = y = \omega = 0$

$$D_{\beta q}(e) = \pi^{-\frac{1}{4}} \exp \left[-\frac{(\overline{\beta}^2 + q^2)}{2} + \sqrt{2}\,\overline{\beta}q - \frac{|\beta|^2}{2} \right]. \tag{11.18}$$

In the Schrödinger realization the eigenvalue q of the operator Q is regarded as the configuration space coordinate. The Schrödinger wave function is defined as

$$\psi(q) = (f_q, \psi)$$

where ψ is an arbitrary state with a finite norm.

$$\hat{T}_g \psi(q) = (f_q, T_g \psi) = \int D_{qq'}(g) \psi(q') dq'.$$

Using Eq. (11.17b) we obtain

$$\hat{T}_g \psi(q) = e^{i(\omega + qx + \frac{xy}{2})} \psi(q + y). \tag{11.19}$$

This yields

$$Q = q, \quad P = -i\frac{d}{dq}, \quad H = 1 \tag{11.20}$$

the familiar quantum mechanical operators.

The coherent state wave function is analogously defined as,

$$\phi(\beta) = (f_\beta, \psi) \tag{11.21}$$

$$\hat{T}_g \phi(\beta) = \frac{1}{\pi} \int D_{\beta\alpha}(g)\phi(\alpha)d^2\alpha. \tag{11.22}$$

Setting $g = e$, $T_g = I$

$$\phi(\beta) = \frac{1}{\pi} \int e^{\left(\overline{\beta}\alpha - \frac{|\beta|^2}{2} - \frac{|\alpha|^2}{2}\right)} \phi(\alpha)d^2\alpha.$$

We now define

$$\phi(\alpha) = e^{-\frac{|\alpha|^2}{2}} f(\overline{\alpha})$$

where a possible α-dependence has been suppressed. Thus

$$f(\overline{\beta}) = \int e^{\overline{\beta}\alpha} f(\overline{\alpha})d\mu(\alpha)$$

$$d\mu(\alpha) = \frac{e^{-|\alpha|^2}}{\pi} d^2\alpha. \tag{11.23}$$

The scalar product in the Hilbert space is given by

$$(f_1, f_2) = \int \overline{f_1(\overline{\beta})} f_2(\overline{\beta})d\mu(\beta). \tag{11.24}$$

The form of the scalar product follows from,

$$(\psi_1, \psi_2) = \frac{1}{\pi} \int (\psi_1, f_\beta)(f_\beta, \psi_2)d^2\beta$$

$$= \frac{1}{\pi} \int \overline{\phi_1(\beta)} \phi_2(\beta)d^2\beta.$$

Setting $\overline{\beta} = z$, $\overline{\alpha} = \xi$, Eqs. (11.23) and (11.24) can be written in the form

$$f(z) = \int e^{z\overline{\xi}} f(\xi)d\mu(\xi) \tag{11.25}$$

$$(f_1, f_2) = \int \overline{f_1(z)} f_2(z)d\mu(z). \tag{11.26}$$

We shall now prove the following theorem:

Theorem. *The function $f(z)$ satisfying Eq. (11.25) and having a finite norm according to the scalar product (11.26) is an entire function analytic in z.*

Proof. Since $\frac{\partial f}{\partial \bar{z}} = \int \frac{\partial}{\partial \bar{z}}(e^{z\bar{\xi}})f(\xi)d\mu(\xi) = 0$ the function $f(z)$ is analytic in z.

Now,

$$f(z) = \sum \frac{c_n}{n!} z^n \tag{11.27}$$

where $c_n = \int \bar{\xi}^n f(\xi)d\mu(\xi)$.

We shall show that the radius of convergence of the power series (11.27) is infinite by requiring

$$(f, f) = \|f\|^2 < \infty.$$

Thus the series $\sum_n \frac{|c_n|^2}{n!}$ must be absolutely convergent, i.e.

$$\left|\frac{c_{n+1}}{c_n}\right|^2 \frac{1}{n} \leq 1 \quad \text{i.e.} \quad \left|\frac{c_{n+1}}{c_n}\right| \leq \sqrt{n}$$

for sufficiently large n.

If we now denote the nth term of the power series (11.27) by u_n, we have

$$\lim_{n\to\infty}\left|\frac{u_{n+1}}{u_n}\right| = \lim_{n\to\infty}\left|\frac{c_{n+1}}{c_n}\right|\frac{|z|}{n} \leq \frac{|z|}{\sqrt{n}}\to 0$$

no matter how large $|z|$ is. Hence the radius of convergence of the power series (11.27) is infinite and the analytic function represented by it is an entire function.

Substituting the explicit form of $D_{\beta\alpha}(g)$, setting $\bar{\beta} = z$ etc. and using Eq. (11.25) we have

$$T_g e^{-|z|^2/2} f(z) = e^{-\frac{1}{2}[(x^2+y^2)+\sqrt{2}z(y-ix)+|z|^2]+i\omega} f\left(z + \frac{1}{\sqrt{2}}(y+ix)\right).$$

Hence setting

$$V_g = e^{\frac{1}{2}|z|^2} T_g e^{-\frac{1}{2}|z|^2}$$

we have

$$V_g f(z) = e^{i\omega-\frac{1}{2}[(x^2+y^2)+\sqrt{2}z(y-ix)]} f\left(z + \frac{1}{\sqrt{2}}(y+ix)\right). \tag{11.28}$$

This is the finite element of the HWG in the space of the entire analytic functions $f(z)$. The representation is unitary under the scalar product (11.26) which satisfies,

$$(zf, g) = \left(f, \frac{\partial g}{\partial z}\right). \tag{11.29}$$

A complete orthonormal set in this Hilbert space is given by

$$u_n(z) = \frac{z^n}{\sqrt{n!}}, \quad n = 0, 1, 2, \ldots$$

so that

$$e_z(\xi) = \sum_n \overline{u_n(z)} u_n(\xi) = e^{\bar{z}\xi} \tag{11.30}$$

in agreement with Eq. (11.25).

The position and momentum operators in the realization of (11.28) are given by

$$Q = \frac{1}{\sqrt{2}}\left(z + \frac{\partial}{\partial z}\right)$$

$$P = \frac{i}{\sqrt{2}}\left(z - \frac{\partial}{\partial z}\right)$$

so that

$$a = \frac{1}{\sqrt{2}}(Q + iP) = \frac{\partial}{\partial z} \tag{11.31}$$

$$a^\dagger = z.$$

We are now in a position to derive Bargmann's integral transform and its inversion. We start from

$$T_g \phi(\beta) = (f_\beta, \hat{T}_g \psi) = \int D_{\beta q}(g)\psi(q)dq$$

$$T_g \psi(q) = \frac{1}{\pi} \int \overline{D_{\beta q}(g^{-1})}\phi(\beta)d^2\beta. \tag{11.32}$$

We set $g = e$, i.e. $x = 0 = y = \omega$, $\bar{\beta} = z$, and use Eq. (11.18) and $\phi(\beta) = e^{-|z|^2/2}f(z)$. Thus

$$f(z) = \int A(z, q)\psi(q)dq$$

$$\psi(q) = \int \overline{A(z, q)}f(z)d\mu(z) \tag{11.33}$$

where

$$A(z, q) = \pi^{-\frac{1}{4}} e^{-\frac{(z^2 + q^2)}{2} + \sqrt{2} z q}.$$

The generalization to n-dimensions is quite trivial.

11.2   Representations of SU(1, 1) in the Bargmann–Segal Space and Class of Associated Integral Transforms

11.2.1   *Metaplectic representation*

We shall now use the Hilbert space $B(C)$ as the carrier space of the irreducible unitary representations of the discrete series of $SU(1, 1)$. The finite element of the group in this realization yields a parametrized continuum of integral transforms mapping $B(C)$ onto itself. Each value of the group parameters yields an integral transform pair.

The Lie algebra of $SU(1, 1)$ can be satisfied by the following formal solution of the commutation relations which are explicitly Hermitian.

$$J_1 = \frac{1}{4} \left(z^2 + \frac{d^2}{dz^2} \right)$$

$$J_2 = -\frac{i}{4} \left(z^2 - \frac{d^2}{dz^2} \right) \tag{11.34}$$

$$J_3 = \frac{1}{2} \left(z\frac{d}{dz} + \frac{1}{2} \right).$$

The representation generated by the Hermitian operators Eq. (11.34) is the direct sum

$$D = D^+_{\frac{1}{4}} \oplus D^+_{\frac{3}{4}} \tag{11.35}$$

called the metaplectic representation which is a special representation of $SU(1, 1)$ having many simplifying features.

Under the action of the operators (11.34) the Bargmann–Segal space $B(C)$ decomposes into two invariant subspaces $B_O(C)$ and $B_E(C)$ consisting of odd and even entire analytic functions respectively. The subspace of even entire analytic functions corresponds to $D^+_{\frac{1}{4}}$ and that of odd functions to $D^+_{\frac{3}{4}}$.

A finite element of the group is obtained by exponentiating the operators (11.34). For this purpose we introduce the Euler angle parametrization

$$\begin{pmatrix} \alpha & \beta \\ \overline{\beta} & \overline{\alpha} \end{pmatrix} = \begin{pmatrix} e^{i\phi/2} & 0 \\ 0 & e^{-i\phi/2} \end{pmatrix} \begin{pmatrix} \cosh \frac{\tau}{2} & -\sinh \frac{\tau}{2} \\ -\sinh \frac{\tau}{2} & \cosh \frac{\tau}{2} \end{pmatrix} \begin{pmatrix} e^{i\theta/2} & 0 \\ 0 & e^{-i\theta/2} \end{pmatrix}$$

$$0 \leq \phi \leq 2\pi, \ 0 \leq \theta \leq 4\pi, \ 0 \leq \tau < \infty \tag{11.36}$$

so that

$$\alpha = e^{i(\phi+\theta)/2} \cosh \frac{\tau}{2}$$

$$\beta = -e^{-i(\phi-\theta)/2} \sinh \frac{\tau}{2}.$$

(11.37)

The arbitrary element of the group according to this parametrization is given by

$$T_u = e^{i\phi J_3} e^{i\tau J_1} e^{i\theta J_3}.$$

(11.38)

We shall show that the action of the operator T_u on an arbitrary element $f(z) \in B(C)$ is an integral transform.

Since J_3 is a first order operator, the action of $\exp(i\theta J_3)$ is simple

$$f_\theta(z) = e^{i\theta J_3} f(z) = e^{i\theta/4} f(e^{i\theta/2} z).$$

(11.39)

To obtain the action of the operator $\exp(i\tau J_1)$ we first proceed to obtain a Baker–Campbell–Hausdorff formula for exponentials of operators of the type

$$e^{\alpha \frac{d^2}{dz^2} + \beta z^2}$$

by using a theorem due to Wilcox.[9] Let P and Q be any two operators satisfying the commutation relation

$$[P, Q] = cI$$

(11.40)

where c is a complex number. Thus P and Q may be the annihilation and creation operators, momentum and coordinate operators etc. Then Wilcox's theorem states that

$$e^{\alpha P^2 + \beta Q^2 + \gamma QP} = [Je^{c\gamma}]^{-\frac{1}{2}} N[e^{AP^2 + BQ^2 + GQP}]$$

(11.41)

where N stands for the normal ordering operator which, acting on $f(P, Q)$ moves all the P's to the right of the Q's and

$$\alpha^{-1}A = \beta^{-1}B = [\lambda J]^{-1} \sinh \lambda$$

$$G = c^{-1}(J^{-1} - 1)$$

(11.42)

$$J = \cosh \lambda - \rho\gamma \sinh \lambda; \quad \lambda = c[\gamma^2 - 4\alpha\beta]^{\frac{1}{2}}.$$

To apply this formula to our operator we first set $\alpha = \beta = 0$; then $A = B = 0$; $\lambda = c\gamma$, $J = e^{-c\gamma}$ so that

$$\gamma = \frac{1}{c} \ln(cG + 1).$$

(11.43)

This immediately yields

$$N[e^{GQP}] = e^{\frac{1}{c}\ln(cG+1)QP}. \tag{11.44}$$

If we now set $\gamma = 0$ in Wilcox's formula we obtain

$$e^{\alpha P^2+\beta Q^2} = J^{-\frac{1}{2}}e^{BQ^2}e^{\frac{1}{c}\ln(cG+1)QP} \times e^{AP^2}.$$

Setting $\alpha = \beta = \frac{i\tau}{4}$, $Q = z$, $P = \frac{d}{dz}$ so that $c = 1$, we finally obtain

$$e^{i\tau J_1} = \left(\cosh\frac{\tau}{2}\right)^{-\frac{1}{2}} \exp\left(\frac{i}{2}\tanh\frac{\tau}{2}z^2\right)$$
$$\times \exp\left(\ln\operatorname{sech}\frac{\tau}{2}z\frac{d}{dz}\right)\exp\left(\frac{i}{2}\tanh\frac{\tau}{2}\frac{d^2}{dz^2}\right). \tag{11.45}$$

To determine the action of the above operator on $f(e^{i\theta/2}z)$ we use the fundamental property of the principal vector as given by Eq. (11.31):

$$f(e^{i\theta/2}z) = \int e^{ze^{i\theta/2}\bar{\xi}}f(\xi)d\mu(\xi). \tag{11.46}$$

We first operate Eq. (11.46) with the second order operator appearing on the extreme right. Thus

$$\exp\left(\frac{i}{2}\tanh\frac{\tau}{2}\frac{d^2}{dz^2}\right)f(e^{i\theta/2}z)$$
$$= \int \exp\left(\frac{i}{2}e^{i\theta}\tanh\frac{\tau}{2}\bar{\xi}^2 + ze^{i\theta/2}\bar{\xi}\right)f(\xi)d\mu(\xi).$$

Applying the remaining factors successively we have

$$e^{i\tau J_1}f(e^{i\theta/2}z) = \left(\cosh\frac{\tau}{2}\right)^{-\frac{1}{2}}\int \exp\left[\frac{i}{2}\tanh\frac{\tau}{2}(z^2 + e^{i\theta}\bar{\xi}^2)\right.$$
$$\left. + \operatorname{sech}\frac{\tau}{2}e^{i\theta/2}z\bar{\xi}\right]f(\xi)d\mu(\xi). \tag{11.47}$$

Applying once again the operator $e^{i\phi J_3}$ on both sides of Eq. (11.47), using Eq. (11.37) and setting

$$[T_u f](z) = g_u(z)$$

we obtain,

$$g_u(z) = (\bar{\alpha})^{-\frac{1}{2}}\int \exp\left(-\frac{i}{2\bar{\alpha}}[\beta z^2 + \bar{\beta}\bar{\xi}^2 + 2iz\bar{\xi}]\right)f(\xi)d\mu(\xi) \tag{11.48}$$

which is an integral transform mapping B(C) onto itself.[10]

The formula for the inversion of the transform follow immediately by noting

$$f(\xi) = [T_{u^{-1}}g_u](\xi).\tag{11.49}$$

Since

$$u^{-1} = \begin{pmatrix} \overline{\alpha} & -\beta \\ -\overline{\beta} & \alpha \end{pmatrix}\tag{11.50}$$

we immediately obtain from (11.48) and (11.50)[10]

$$f(\xi) = (\alpha)^{-\frac{1}{2}} \int \exp\left[\frac{i}{2\alpha}(\beta\xi^2 + \overline{\beta}\,\overline{z}^2 - 2i\xi\overline{z})\right] g_u(z)d\mu(z).\tag{11.51}$$

Equations (11.48) and (11.51) constitute an integral transform pair for each allowed value of the group parameters α and β.

We now consider some simple special cases. The transform pair for

$$u = \begin{pmatrix} \sqrt{2} & 1 \\ 1 & \sqrt{2} \end{pmatrix}\tag{11.52}$$

is given by

$$g(z) = 2^{-\frac{1}{4}} \int \exp\left[\frac{-i}{2\sqrt{2}}(z^2 + \overline{\xi}^2 + 2iz\overline{\xi})\right] f(\xi)d\mu(\xi)$$

$$f(\xi) = 2^{-\frac{1}{4}} \int \exp\left[\frac{i}{2\sqrt{2}}(\xi^2 + \overline{z}^2 - 2i\xi\,\overline{z})\right] g(z)d\mu(z).\tag{11.53}$$

Similarly for

$$u = \begin{pmatrix} \sqrt{2} & i \\ -i & \sqrt{2} \end{pmatrix}\tag{11.54}$$

the transform pair is given by

$$g(z) = 2^{-\frac{1}{4}} \int \exp\left[\frac{1}{2\sqrt{2}}(z^2 - \overline{\xi}^2 + 2z\overline{\xi})\right] f(\xi)d\mu(\xi)$$

$$f(\xi) = 2^{-\frac{1}{4}} \int \exp\left[\frac{1}{2\sqrt{2}}(\overline{z}^2 - \xi^2 + 2\xi\,\overline{z})\right] g(z)d\mu(z).\tag{11.55}$$

The Plancherel formula for the transform pair is obtained from the unitarity of the representation,

$$(f_1, f_2) = (T_u f_1, T_u f_2) = (g_{1u}, g_{2u})$$

which yields

$$\int \overline{f_1(\xi)} f_2(\xi)d\mu(\xi) = \int \overline{g_{1u}(z)}g_{2u}(z)d\mu(z).$$

To establish the connection of the integral transform with the canonical transformation of Moshinsky and Quesne we start from,

$$T_u z f = (\overline{\alpha})^{-\frac{1}{2}} \int \exp\left[-\frac{i}{2\overline{\alpha}}(\beta z^2 + \overline{\beta}\,\overline{\xi}^2 + 2iz\overline{\xi})\right] \times \xi f(\xi) d\mu(\xi).$$

Using the analyticity of $f(\xi)$ and integrating by parts we obtain

$$T_u z f = \alpha z T_u f - i\overline{\beta}\frac{\partial}{\partial z}T_u f. \tag{11.56}$$

This is equivalent to the operator equality

$$T_u z T_u^{-1} = \alpha z - i\overline{\beta}\frac{\partial}{\partial z}. \tag{11.57}$$

In a similar manner

$$T_u \frac{\partial}{\partial z} T_u^{-1} = i\beta z + \overline{\alpha}\frac{\partial}{\partial z}. \tag{11.58}$$

If we now introduce the Fock–Bargmann representation of the coordinate and momentum operators,

$$Q = \frac{1}{\sqrt{2}}\left(z + \frac{\partial}{\partial z}\right)$$

$$P = \frac{i}{\sqrt{2}}\left(z - \frac{\partial}{\partial z}\right).$$

and use Eqs. (9.33) and (9.34) connecting $SU(1,1)$ and $SL(2, R)$ parameters, we obtain

$$Q' = T_u Q T_u^{-1} = dQ + bP$$
$$P' = T_u P T_u^{-1} = cQ + aP. \tag{11.59}$$

To get the mapping of the integral transform pair (11.48) and (11.51) in $L^2(R)$ space we introduce the Bargmann transform

$$v_g(x) = \pi^{-\frac{1}{4}} \int \exp\left[-\frac{1}{2}(\overline{z}^2 + x^2) + \sqrt{2}\overline{z}x\right] g_u(z) d\mu(z). \tag{11.60}$$

In the left-hand side we have replaced the subscript $u \in SU(1,1)$ by its $SL(2, R)$ image g because the final result takes a simple form in terms of the $SL(2, R)$ parameters. Substituting Eq. (11.48) in Eq. (11.60) we obtain

$$v_g(x) = \pi^{-\frac{1}{4}}(\overline{\alpha})^{-\frac{1}{2}} \int \exp\left[-\frac{i\overline{\beta}\,\overline{\xi}^2}{2\alpha} - \frac{x^2}{2}\right] I(\overline{\xi}, x) f(\xi) d\mu(\xi) \tag{11.61}$$

where

$$I(\overline{\xi}, x) = \int \exp \left[\frac{1}{2} \gamma z^2 + az + \frac{1}{2} \overline{\delta} \overline{z}^2 + \overline{b} \overline{z} \right] d\mu(z) \quad \text{with}$$

$$\gamma = -\frac{i\beta}{\overline{\alpha}}, \quad a = \overline{\xi}/\overline{\alpha}, \quad \overline{\delta} = -1, \quad \overline{b} = \sqrt{2} x. \quad (11.62)$$

The above integral has been evaluated by Bargmann and the result is

$$I(\overline{\xi}, x) = \left[\frac{\overline{\alpha}}{\overline{\alpha} - i\beta} \right]^{\frac{1}{2}} \exp \left(\left[-\overline{\xi}^2 / 2\overline{\alpha} - i\beta x^2 + \sqrt{2} \, \overline{\xi} x \right] / (\overline{\alpha} - i\beta) \right). \quad (11.63)$$

If we replace $f(\xi)$ by the Bargmann transform,

$$f(\xi) = \pi^{-\frac{1}{4}} \int e^{-\frac{1}{2}(\xi^2 + v^2) + \sqrt{2} v \xi} u(y) dy$$

we have

$$v_g(x) = \frac{e^{i \, \text{sgn} \, b(\pi/4)}}{\sqrt{2\pi |b|}} \int \exp[(dx^2 - 2xy + ay^2)/2ib] \times u(y) dy. \quad (11.64)$$

The inversion formula for the transform follows that the corresponding formula in $B(C)$, namely, Eq. (11.51) or from the requirement

$$u(y) = [T_{g^{-1}} v_g](y)$$

and is given by

$$u(y) = \frac{e^{-i \, \text{sgn} \, b(\pi/4)}}{\sqrt{2\pi |b|}} \int e^{-\frac{1}{2ib}(dx^2 - 2xy + ay^2)} v_g(x) dx. \quad (11.65)$$

Equations (11.64) and (11.65) constitute the one-dimensional version of the Moshinsky–Quesne[11] transform.

11.2.2   *The discrete representations* $\mathbf{D}_k^+$

To obtain arbitrary representations of the discrete class we consider in place of Eq. (11.34) the following set of Hermitian generators

$$J_1 = \frac{1}{4} \left(z_1^2 + z_2^2 + \frac{\partial}{\partial z_1^2} + \frac{\partial^2}{\partial z_2^2} \right)$$

$$J_2 = -\frac{i}{4} \left(z_1^2 + z_2^2 - \frac{\partial^2}{\partial z_1^2} - \frac{\partial^2}{\partial z_2^2} \right) \quad (11.66)$$

$$J_3 = \frac{1}{2} \left(z_1 \frac{\partial}{\partial z_1} + z_2 \frac{\partial}{\partial z_2} + 1 \right).$$

The representation D of SU(1, 1) generated by the above operators is reducible and is a direct sum of the irreducible unitary representations belonging to the positive discrete series of representations,

$$D = \sum_{\substack{\oplus \\ k=\frac{1}{2},1....}} D_k^+.$$

If we now introduce the Hermitian operator

$$K = \frac{1}{2} - \frac{i}{2}\left(z_1\frac{\partial}{\partial z_2} - z_2\frac{\partial}{\partial z_1}\right) \tag{11.67}$$

the Casimir operator becomes a function of K

$$J_1^2 + J_2^2 - J_3^2 = K(1 - K). \tag{11.68}$$

11.2.3 *The reduced Bargmann space*

The subspace $B_k(\mathbb{C})$ of the representation space $B(\mathbb{C}_2)$ in which the operator K is a number, will be called the reduced Bargmann space. The form (11.68) of the operator K suggests that we introduce the polar coordinates

$$z_1 = z\cos\phi, \quad z_2 = z\sin\phi \tag{11.69}$$

where the radius z and the angle ϕ are both complex numbers

$$z_1 = |z|e^{i\arg z}, \quad 0 \le \arg z \le \pi$$
$$\phi = \phi_1 + i\phi_2, \ 0 \le \phi_1 \le 2\pi, \ -\infty < \phi_2 < \infty. \tag{11.70}$$

The operator K and the generators J_1, J_2, J_3 are now given by

$$K = \frac{1}{2} - \frac{i}{2}\frac{\partial}{\partial\phi} \tag{11.71}$$

$$J_1 = \frac{1}{4}\left[z^2 + \frac{\partial^2}{\partial z^2} + \frac{1}{z}\frac{\partial}{\partial z} - \frac{(2K-1)^2}{z^2}\right]$$

$$J_2 = -\frac{i}{4}\left[z^2 - \frac{\partial^2}{\partial z^2} - \frac{1}{z}\frac{\partial}{\partial z} + \frac{(2K-1)^2}{z^2}\right] \tag{11.72}$$

$$J_3 = \frac{1}{2}\left[z\frac{\partial}{\partial z} + 1\right].$$

Since K is diagonal in the subspace $B_k(\mathbb{C})$ of the UIR's D_k^+, we have

$$f(z_1, z_2) = e^{i(2k-1)\phi}f(z) \tag{11.73}$$

where $f(z)$ is an analytic function regular in the upper half-plane $0 \le \arg z \le \pi$.

To obtain the scalar product in $B_k(C)$ we start from the scalar product in $B(C_2)$,

$$(f, g) = \int \overline{f(z_1, z_2)} g(z_1, z_2) d\mu(z_1) d\mu(z_2) \tag{11.74}$$

where integral extends over C_2. Using the transformation (11.69) and Eq. (7.25) of Chapter 7, we obtain

$$dz_1^2 dz_2^2 = |z|^2 d^2 z d\phi_1 d\phi_2 \tag{11.75a}$$

$$e^{-|z_1|^2 - |z_2|^2} = e^{-|z|^2 \cosh 2\phi_2}. \tag{11.75b}$$

Thus

$$(f, g) = \frac{2}{\pi} \delta_{kk'} \int_{\mathrm{Im}\, z > 0} \overline{f(z)} g(z) |z|^2 d^2 z$$
$$\times \int_{-\infty}^{\infty} \exp[-(2k-1)2\phi_2 - |z|^2 \cosh 2\phi_2] d\phi_2. \tag{11.76}$$

Using the standard integral representation of the modified Bessel function of the second kind

$$K_n(x) = \frac{1}{2} \int_{-\infty}^{\infty} dv\, e^{-vn - x \cosh v} \tag{11.77}$$

we obtain the scalar product in $B_k(C)$:

$$(f, g) = \int_{\mathrm{Im}\, z > 0} \overline{f(z)} g(z) d\lambda(z) \tag{11.78}$$

where

$$d\lambda(z) = \frac{2}{\pi} |z|^2 K_{2k-1}(|z|^2) d^2 z. \tag{11.79}$$

We now introduce the principal vectors e_z that are bounded linear functions satisfying

$$f(z) = (e_z, f). \tag{11.80}$$

To find explicit form of e_z in $B_k(C)$ we start from the two-dimensional version of Eq. (11.80), i.e.

$$f(z_1, z_2) = (e_{z_1, z_2}, f) = \int e^{z_1 \overline{\xi_1} + z_2 \overline{\xi_2}} f(\xi_1, \xi_2) d\mu(\xi_1) d\mu(\xi_2). \tag{11.81}$$

We now introduce the polar coordinates in both (z_1, z_2) and (ξ_1, ξ_2) and restrict ourselves to functions of the form (11.73). Thus

$$e^{i(2k-1)\phi} f(z) = \frac{1}{\pi^2} \int_{\operatorname{Im}\xi>0} \exp[z\overline{\xi}\cos(\phi - \overline{\psi}) + i(2k - 1)\psi$$

$$-|\xi|^2 \cosh 2\psi_2] f(\xi) |\xi|^2 d^2\xi d\psi_1 d\psi_2 \qquad (11.82)$$

where ξ and ψ are defined by

$$\xi_1 = \xi \cos\psi, \quad \xi_2 = \xi \sin\psi$$
$$\xi = |\xi| e^{i \arg \xi}, \quad 0 \le \arg \xi \le \pi$$
$$\psi = \psi_1 + i\psi_2, \quad 0 \le \psi_1 \le 2\pi, \quad -\infty < \psi_2 < \infty$$
$$f(\xi_1, \xi_2) = e^{i(2k-1)\psi} f(\xi).$$

We now note that Eq. (11.82) can be written in the form

$$e^{i(2k-1)\phi} f(z) = \frac{1}{\pi^2} \int_{\operatorname{Im}\xi>0} d^2\xi |\xi|^2 f(\xi) \mathrm{I}(z, \overline{\xi}, \xi, \phi) \qquad (11.83)$$

where

$$\mathrm{I}(z, \overline{\xi}, \xi, \phi) = \int_{-\infty}^{\infty} d\psi_2 \int_0^{2\pi} d\psi_1 \exp[z\overline{\xi}\cos(\phi - \overline{\psi}) + i(2k - 1)\psi] e^{-|\xi| \cosh 2\psi_2}.$$
$$(11.84)$$

By a simple change of variables the above integral can be recast in the form,

$$\exp\left[i(2k - 1)\left(\phi + \frac{\pi}{2}\right)\right] \times \frac{1}{2} \int d\alpha_2 \exp[-(2k - 1)\alpha_2 - |\xi|^2 \cosh \alpha_2]$$

$$\times \int_0^{2\pi} d\alpha_1 \exp[z\overline{\xi}\sin\alpha - i(2k - 1)\alpha] \qquad (11.85)$$

where α_1 and α_2 are the real and imaginary parts of the complex number

$$\alpha = \alpha_1 + i\alpha_2, \quad 0 \le \alpha_1 \le 2\pi, \quad -\infty < \alpha_2 < \infty.$$

We first write the α_1 integral as integral over a circle s of radius $\rho = e^{-\alpha_2}$ centered at the origin. Thus writing J for α_1 integral we obtain,

$$\mathrm{J} = \frac{1}{i} \int_s du\, e^{[(\eta/2)(u - u^{-1}) - i(2k-1)]} \qquad (11.86a)$$

where

$$\eta = -iz\overline{\xi}. \qquad (11.86b)$$

The integral appearing in the r.h.s. of Eq. (11.86a) is the standard contour integral representation of the Bessel function and we have

$$J = 2\pi e^{-i(2k-1)\frac{\pi}{2}} I_{2k-1}(z\overline{\xi}) \tag{11.87}$$

where $I_\mu(z)$ stands for the modified Bessel function of the first kind.

Using (11.77) the α_2 integration can now be easily carried out and we have

$$I(z,\overline{\xi},\xi,\phi) = 2\pi e^{i(2k-1)\phi} I_{2k-1}(z\overline{\xi}) K_{2k-1}(|\xi|^2). \tag{11.88}$$

Substituting Eq. (11.88) in Eq. (11.83) we have

$$f(z) = \int_{\mathrm{Im}\,\xi > 0} I_{2k-1}(z\overline{\xi}) f(\xi) d\lambda(\xi). \tag{11.89}$$

The principal vector in $B_k(C)$ defined by Eq. (11.80) is therefore given by

$$e_z(\xi) = I_{2k-1}(\overline{z}\xi). \tag{11.90}$$

We shall now show that the elements $f(z)$ of $B_k(C)$ are entire analytic functions whose behavior near the origin is of the form,

$$f(z) \approx \mathrm{const.}\, z^{2k-1}. \tag{11.91}$$

To prove this we start by noting that a complete orthonormal set in $B_k(C)$ is given by

$$u_n(z) = \frac{2^{-k-n-\frac{1}{2}} z^{2k-1+2n}}{[(2k+n-1)!\, n!]^{\frac{1}{2}}}, \quad n = 0,1,2,\ldots. \tag{11.92}$$

The orthonormaity can be easily verified

$$(u_n, u_m) = \frac{2^{-2k-n-m+1}}{[(2k+n-1)!\, n!(2k+m-1)!\, m!]^{\frac{1}{2}}} \int \overline{z}^{2k-1+2n} z^{2k-1+2m} d\lambda(z).$$

Setting $z = re^{i\theta}$, $0 \le \theta \le \pi$, $0 \le r \le \infty$, $r^2 = x$, we obtain

$$(u_n, u_m) = \delta_{mn} \frac{2^{-2k-2n+1}}{(2k+n-1)!\, n!} \int_0^\infty x^{2k+2n} K_{2k-1}(x) dx$$

$$= \delta_{mn}. \tag{11.93}$$

The completeness can be ensured by noting that

$$\sum_n (f, u_n)(u_n, g) = \int d\lambda(z)\overline{f(z)} \int_{\mathrm{Im}\,\xi > 0} \sum u_n(z)\overline{u_n(\xi)} \times g(\xi) d\lambda(\xi). \tag{11.94}$$

Using the explicit form of the orthonormal vectors u_n as given by (11.92) we obtain,

$$\sum_n u_n(z)\overline{u_n(\xi)} = I_{2k-1}(z\overline{\xi}). \tag{11.95}$$

Hence the ξ-integral in (11.94) reads

$$\int_{\operatorname{Im}\xi > 0} I_{2k-1}(z\overline{\xi})g(\xi)d\lambda(\xi) \tag{11.96}$$

which is equal to $g(z)$ Eq. (11.89). Thus

$$\sum (f, u_n)(u_n, g) = (f, g) \tag{11.97}$$

and the orthonormal set (11.92) is complete. Following the same procedure as in Sec. 11.1 it can be easily proved that $f(z)$ is an entire function.

We conclude this subsection by giving the explicit forms of the generators of the group in $B_k(C)$:

$$J_1 = \frac{1}{4}\left[z^2 + \frac{d^2}{dz^2} + \frac{1}{z}\frac{d}{dz} - \frac{(2k-1)^2}{z^2}\right]$$

$$J_2 = -\frac{i}{4}\left[z^2 - \frac{d^2}{dz^2} - \frac{1}{z}\frac{d}{dz} + \frac{(2k-1)^2}{z^2}\right] \tag{11.98}$$

$$J_3 = \frac{1}{2}\left[z\frac{d}{dz} + 1\right].$$

11.2.4   *Finite element of the group and the associated integral transform*

To find the action of the element

$$T_u = e^{i\theta' J_3} e^{i\tau J_2} e^{i\theta J_3}$$

of the group on $f(z) \in B_k(C)$ we start from

$$e^{i\theta J_3} f(z_1, z_2) = e^{i\theta/2} f(z_1 e^{i\theta/2}, z_2 e^{i\theta/2}) \tag{11.99}$$

$$= e^{i\theta/2} \int e^{(z_1\overline{\xi_1} + z_2\overline{\xi_2})} e^{i\theta/2} f(\xi_1, \xi_2) d\mu(\xi_1) d\mu(\xi_2). \tag{11.100}$$

To find the action of $\exp(i\tau J_1)$ on $f_\theta(z_1, z_2)$ we use the Wilcox decomposition

$$\exp\left[(i\tau/4)\left[z_1^2 + z_2^2 + \frac{\partial^2}{\partial z_1^2} + \frac{\partial^2}{\partial z_2^2}\right]\right]$$

$$= \operatorname{sech}\frac{\tau}{2}\exp\left(\frac{i}{2}\tanh\frac{\tau}{2}(z_1^2 + z_2^2)\right)\exp\left[\ln\operatorname{sech}\frac{\tau}{2}\left(z_1\frac{\partial}{\partial z_1} + z_2\frac{\partial}{\partial z_2}\right)\right]$$

$$\times \exp\left[\frac{i}{2}\tanh\frac{\tau}{2}\left(\frac{\partial^2}{\partial z_1^2} + \frac{\partial^2}{\partial z_2^2}\right)\right]. \tag{11.101}$$

We first operate both sides of Eq. (11.100) with the second order operator appearing on the extreme right of Eq. (11.101). Thus

$$\exp\left[\frac{i}{2}\tanh\frac{\tau}{2}\left(\frac{\partial^2}{\partial z_1^2} + \frac{\partial^2}{\partial z_2^2}\right)\right]f(z_1 e^{i\theta/2}, z_2 e^{i\theta/2})$$

$$= \int \exp\left[\frac{i}{2}e^{i\theta}\tanh\frac{\tau}{2}(\overline{\xi_1}^2 + \overline{\xi_2}^2) + (z_1\overline{\xi_1} + z_2\overline{\xi_2})e^{i\theta/2}\right]$$

$$\times f(\xi_1, \xi_2)d\mu(\xi_1)d\mu(\xi_2). \tag{11.102}$$

Following the previous subsection we now introduce the polar coordinates and restrict ourselves to $B_k(C)$. Then the r.h.s. of Eq. (11.102) becomes

$$\frac{1}{\pi^2}\int_{\operatorname{Im}\xi > 0}\exp\left[\frac{i}{2}e^{i\theta}\tanh\frac{\tau}{2}\overline{\xi}^2\right]I(z, \overline{\xi}, \xi, \phi, \theta)f(\xi)|\xi|^2 d^2\xi \tag{11.103}$$

where

$$I(z, \overline{\xi}, \xi, \phi, \theta) = \int_{-\infty}^{\infty}d\psi_2\int_0^{2\pi}d\psi_1\exp\left[z\overline{\xi}e^{i\theta/2}\cos(\phi - \overline{\psi})\right.$$

$$\left. + i(2k - 1)\psi - |\xi|^2\cosh 2\psi_2\right]. \tag{11.104}$$

The above integral is of the same form as $I(z, \overline{\xi}, \xi, \phi)$ with z replaced by $ze^{i\theta/2}$ so that

$$I(z, \overline{\xi}, \xi, \phi, \theta) = I(ze^{i\theta}, \overline{\xi}, \xi, \phi) = 2\pi e^{i(2k-1)\phi}I_{2k-1}(e^{i\theta/2}z\overline{\xi})K_{2k-1}(|\xi|^2). \tag{11.105}$$

Using the above result and applying the remaining operator factors in Eq. (11.101) successively we obtain

$$\exp(i\tau J_1)f_\theta(z_1, z_2)$$

$$= e^{i(2k-1)\phi}\left(\operatorname{sech}\frac{\tau}{2}\right) \times \int \exp\left[\frac{i}{2}\tanh\frac{\tau}{2}(z^2 + e^{i\theta}\overline{\xi}^2)\right]$$

$$\times I_{2k-1}\left(e^{i\theta/2}\operatorname{sech}\frac{\tau}{2}z\overline{\xi}\right)f(\xi)d\lambda(\xi). \tag{11.106}$$

We now apply $\exp(i\theta' J_3)$ on both sides of the above equation, which yields,

$$e^{i\theta' J_3} e^{i\tau J_1} f_\theta = e^{i(2k-1)\phi} e^{i\theta'/2} \text{sech} \frac{\tau}{2} \int_{\text{Im } \xi > 0} \exp\left[\frac{i}{2} \tanh \frac{\tau}{2} (e^{i\theta' z^2} + e^{i\theta \bar{\xi}^2})\right]$$

$$\times I_{2k-1}\left(e^{i(\theta+\theta')} \text{sech} \frac{\tau}{2} z\bar{\xi}\right) f(\xi) d\lambda(\xi). \tag{11.107}$$

Since in $B_k(C)$ the function $f(z_1, z_2)$ is of the form (11.73), we obtain, after omitting the factor $\exp[i(2k-1)\phi]$ from both sides of Eq. (11.107),

$$g_u(z) = [T_u f](z) = \int e^{-i(\beta z^2 + \bar{\beta} \bar{\xi}^2)/2\bar{\alpha}} I_{2k-1}\left(\frac{z\bar{\xi}}{\bar{\alpha}}\right) \times f(\xi) d\lambda(\xi). \tag{11.108}$$

The inversion formula for this transform once again follows from,

$$f(\xi) = [T_{u^{-1}} g_u](\xi)$$

where u^{-1} is given by Eq. (11.50). This immediately yields

$$f(\xi) = \int e^{i(\beta \xi^2 + \bar{\beta} \bar{z}^2)/2\alpha} I_{2k-1}\left(\frac{\xi \bar{z}}{\alpha}\right) g_u(z) d\lambda(z). \tag{11.109}$$

Equations (11.108) and (11.109) constitute a parametrized continuum of integral transform pairs in $B_k(C)$. Each value of the group parameters yields an integral transform pair.

11.3   The Coherent States of the SU(1, 1) Group and a Class of Associated Integral Transforms

There are two types of coherent states for the SU(1, 1) group, the Barut–Girardello[12] coherent[10] states and the Perelomov[13] coherent states. However the analysis of Barut and Girardello of the basic properties of their coherent state system is not satisfactory. First, the measure in the completeness condition is wrong. Second, the evaluation of the matrix element in their coherent state basis is incorrect leading to an incorrect form of the finite element of the group. Third, it was not recognized that the associated Hilbert space of analytic functions is a subspace of the Bargmann–Segal space invariant under SU(1, 1). This subspace called "the reduced Bargmann space" and denoted by $B_k(C)$ has been introduced in the previous section (11.2.3). The corresponding Hilbert space for the Perelomov coherent states, on the other hand, coincides with Bargmann's canonical carrier space H_k for the positive discrete series, namely, the Hilbert space of functions analytic within the open unit disc.

The coordinate representation of the SU(1, 1) boson operators span an $L_k^2(R^+)$ space with a suitable scalar product. In $L_k^2(R^+)$ the generator of the

E(1) subgroup J_1+J_3 (which is a positive Hermitian operator) is an operator of multiplication just as the position operator Q in the Heisenberg–Weyl group is an operator of multiplication in the Schrödinger representation. This Hilbert space has been introduced by Moshinsky and Quesne[11] and later used by Wolf and coworkers.[14]

We shall now state the main results of this analysis. The integral transform pair connecting H_k and $L_k^2(R^+)$ (coordinate representation of the SU(1, 1) boson operators) is given by

$$g(\xi) = \int_{R^+} A(\xi, r)\psi(r)rdr$$

$$\psi(r) = \int_{|\xi|<1} \overline{A(\xi, r)}g(\xi)d\sigma(\xi)$$

(11.110)

$$A(\xi, r) = \frac{2^{\frac{1}{2}-k}r^{2k-1}e^{\frac{-r^2}{4}\left(\frac{1+\xi}{1-\xi}\right)}}{\sqrt{\Gamma(2k)}(1-\xi)^{2k}}$$

(11.111)

$$d\sigma(\xi) = \frac{(2k-1)}{\pi}(1-|\xi|^2)^{2k-2}d^2\xi.$$

(11.112)

On the other hand the integral transform pair connecting $B_k(C)$ and H_k is given by,

$$g(\xi) = \frac{2^{-k+\frac{1}{2}}}{\sqrt{(2k-1)!}}\int_{\text{Im } z>0} \overline{z}^{2k-1}e^{-\xi \overline{z}^2/2}f(z)d\lambda(z)$$

(11.113)

$$f(z) = \frac{2^{-k+\frac{1}{2}}}{\sqrt{(2k-1)!}}\int_{|\xi|<1} z^{2k-1}e^{-z^2\overline{\xi}/2}g(\xi)d\sigma(\xi)$$

(11.114)

where

$$d\lambda(z) = \frac{2}{\pi}|z|^2 K_{2k-1}(|z|^2)d^2z.$$

(11.115)

11.3.1 *The* SU(1, 1) *coherent states*[10]

Using the parametrization (11.36) and (11.37) the invariant measure on the group is given by

$$du = \sinh \tau d\tau d\phi d\theta, \quad 0 \leq \tau < \infty, \ 0 \leq \theta \leq 4\pi, \ 0 \leq \phi \leq 2\pi. \quad (11.116)$$

In what follows we shall consider only positive discrete class D_k^+. The canonical orthonormal basis for this particular class of unitary irreducible representations can be defined by the following relations

$$J_3 f_m^k = m f_m^k$$

(11.117a)

$$J_\pm f_m^k = \sqrt{(m \pm k)(m \mp k \pm 1)}f_{m\pm1}^k.$$

(11.117b)

The Barut–Girardello coherent states are eigenvectors of J_-

$$J_- f_\sigma^k = \sigma f_\sigma^k \tag{11.118}$$

and can be expressed as a linear combination of the canonical orthonormal basis

$$f_\sigma^k = (2k-1)! \sum_{m=k}^{\infty} \frac{\sigma^{m-k} f_m^k}{[(m-k)!(m+k-1)!]^{\frac{1}{2}}}. \tag{11.119}$$

The overlap between the coherent states can be easily calculated

$$(f_\eta^k, f_\sigma^k) = {}_0F_1(2k; \overline{\eta}\,\sigma). \tag{11.120}$$

We establish now the completeness condition of these coherent states by finding a measure $d\mu(\sigma)$ such that

$$\int (f, f_\sigma^k)(f_\sigma^k, g) d\mu(\sigma) = (f, g) \tag{11.121}$$

where f and g are arbitrary functions having finite norm. Using the expansion in the orthonormal basis we obtain,

$$\int (f, f_\sigma^k)(f_\sigma^k, g) d\mu(\sigma)$$
$$= \sum_{m,m'} \frac{(f, f_m^k)(f_{m'}^k, g) \int \sigma^{m-k} \overline{\sigma}^{m'-k} d\mu(\sigma)}{[(m-k)!(m'-k)!(m+k-1)!(m'+k-1)!]^{\frac{1}{2}}}. \tag{11.122}$$

The r.h.s. of the above equation suggests that $d\mu(\sigma)$ is of the form,

$$d\mu(\sigma) = \mu(|\sigma|)|\sigma|d|\sigma|d\theta = \mu(r)rdrd\theta$$

where $\sigma| = r$, $\arg \sigma = \theta$. We shall consider $\mu(r)$ to be a positive definite quantity because it will appear later as the weight function in a Hilbert space of analytic functions. The integral appearing in the r.h.s. therefore becomes

$$2\pi \delta_{mm'} \int r^{2(m-k)+1} \mu(r) dr. \tag{11.123}$$

We then obtain,

$$\int (f, f_\sigma^k)(f_\sigma^k, g) = (2k-1)!2\pi \sum_{n=0}^{\infty} \frac{(f, f_{k+n}^k)(f_{k+n}^k, g)}{n!(2k+n-1)!} \times \int_0^{\infty} r^{2n+1} \mu(r) dr. \tag{11.124}$$

We therefore require $\mu(r)$ to be such that

$$\int_0^\infty r^{2n+1}\mu(r)dr = \frac{n!(2k+n-1)!}{2\pi(2k-1)!}, \quad n = 0,1,2,\ldots. \tag{11.125}$$

Since n is a positive integer or zero this is a moment problem (not a Mellin transform). If we restrict $\mu(r)$ to be a positive quantity it has a unique solution given by,

$$\mu(r) = \frac{2r^{2k-1}}{\pi(2k-1)!}K_{2k-1}(2r) \tag{11.126}$$

where $K_{2k-1}(2r)$ is a modified Bessel function of the second kind. To verify that this is indeed a solution we substitute it in the l.h.s. of Eq. (11.126) and note that

$$\int_0^\infty K_\nu(\beta r)r^{\mu-1}dr = 2^{\mu-2}\beta^{-\mu}\Gamma\left(\frac{\mu+\nu}{2}\right)\Gamma\left(\frac{\mu-\nu}{2}\right) \tag{11.127}$$

which immediately yields the required result. To prove the uniqueness of the solution we reproduce the argument of Sharma, Mehta, Mukunda and Sudarshan, who considered this moment problem in a different context. We set

$$r^2 = t, \quad \pi(2k-1)!\mu(r) = \phi(t)$$

so that $\phi(t)$ satisfies

$$\int_0^\infty \phi(t)t^n dt = n!(2k+n-1)!. \tag{11.128}$$

A solution of the above equation is given by

$$\phi(t) = 2t^{k-\frac{1}{2}}K_{2k-1}(2\sqrt{t}). \tag{11.129}$$

Following Shohat and Tamarkin, Sharma and coworkers give a sufficient condition under which the moment problem

$$\int_0^\infty \phi(t)t^n dt = \alpha_n, \quad n = 0,1,2,\ldots \tag{11.130}$$

is determined, i.e. $\phi(t)$ is unique as long as it is restricted to be positive. The condition is that the series $\sum(\alpha_n)^{-\frac{1}{2n}}$ is divergent. In the present case

$$\alpha_n = n!(2k+n-1)!$$

so that

$$\lim_{n\to\infty}(\alpha_n)^{-\frac{1}{2n}} \sim \frac{1}{n} \tag{11.131}$$

which is divergent and the solution (11.129) is unique.

The completeness condition satisfied by the coherent states is therefore given by

$$\int (f, f_\sigma^k)(f_\sigma^k, g)d\mu(\sigma) = (f, g) \tag{11.132}$$

where

$$d\mu(\sigma) = \left[\frac{2|\sigma|^{2k-1}}{\pi(2k-1)!}\right] K_{2k-1}(2|\sigma|)d^2\sigma. \tag{11.133}$$

To evaluate the matrix element of the finite transformation in the coherent state basis, we start from the decomposition,

$$\begin{pmatrix} \alpha & \beta \\ \overline{\beta} & \overline{\alpha} \end{pmatrix} = \begin{pmatrix} 1 & \beta/\overline{\alpha} \\ 0 & 1 \end{pmatrix} \begin{pmatrix} 1/\overline{\alpha} & 0 \\ 0 & \overline{\alpha} \end{pmatrix} \begin{pmatrix} 1 & 0 \\ \overline{\beta}/\overline{\alpha} & 1 \end{pmatrix}.$$

A finite element of the group according to this decomposition is given by

$$T_u = e^{(-i\beta/\overline{\alpha})J_+} e^{-(\ln \overline{\alpha}^2)J_3} e^{-(i\overline{\beta}/\overline{\alpha})J_-}. \tag{11.134}$$

This is a non-subgroup decomposition and none of the operator factors is unitary. But the product is unitary whenever J_1, J_2, J_3 are Hermitian. The finite transformation matrix element is given by,

$$D_{\eta\sigma}^k(u) = e^{-i(\beta\overline{\eta}+\overline{\beta}\sigma)/\overline{\alpha}} \sum_{m=k}^{\infty} (\overline{\alpha})^{-2m}(f_\eta^k, f_m^k)(f_m^k, f_\sigma^k)$$

$$= (\overline{\alpha})^{-2k} e^{-i(\beta\overline{\eta}+\overline{\beta}\sigma)/\overline{\alpha}} {}_0F_1(2k; \overline{\eta}\sigma/\overline{\alpha}^2). \tag{11.135}$$

It is interesting to note that the matrix element satisfy the quasi orthogonality condition

$$\int D_{\eta\rho}^k(u)D_{\sigma\tau}^k(u)d\mu(u) = \frac{16\pi^2\delta_{kk'}(f_\eta^k, f_\sigma^k)(f_\tau^k, f_\rho^k)}{(2k-1)}, \quad k, k' = 1, \frac{3}{2}, \dots \tag{11.136}$$

where $d\mu(u)$ is the invariant measure on $SU(1, 1)$. We briefly sketch an independent proof of the quasi orthogonality condition as follows. We construct

$$\int T_u^k Z_{\rho\tau}^k T_{u^{-1}}^k d\mu(u) = \lambda^k I^k \tag{11.137}$$

where the operator $Z_{\rho\tau}^k$ which is independent of the group parameters, is given by

$$(f, Z_{\rho\tau}^k g) = (f, f_\rho^k)(f_\tau^k, g). \tag{11.138}$$

Taking the matrix elements of the operator equality between f_η^k and f_σ^k we obtain

$$\int D_{\eta\rho}^k(u)\overline{D_{\sigma\tau}^k(u)}d\mu(u) = \lambda_{\rho\tau}^k(f_\eta^k, f_\sigma^k).$$
(11.139)

Since $\lambda_{\rho\tau}^k$ is independent of η and σ, we can calculate it by setting $\eta = \sigma = 0$ and explicitly carrying out the integration. Thus

$$\lambda_{\rho\tau}^k = \int |\alpha|^{-4k} e^{-i\overline{\beta}\,\rho/\overline{\alpha}} e^{i\beta\,\overline{\tau}/\alpha} d\mu(u).$$
(11.140)

The integral on the r.h.s. can be calculated by using the Euler angle parametrization and we have

$$\lambda_{\rho\tau}^k = \frac{16\pi^2}{(2k-1)}{}_0F_1(2k; \rho\overline{\tau}) = \frac{16\pi^2}{2k-1}(f_\tau^k, f_\rho^k).$$
(11.141)

Further for $k \neq k'$ using the traditional method the following integral can be shown to be zero

$$\int D_{\eta\rho}^k(u)D_{\sigma\tau}^{k'}(u)d\mu(u) = 0, \quad k \neq k'; \ k, k' = 1, \frac{3}{2}, \ldots.$$

Collecting these results we obtain the quasi orthogonality condition given before.

11.3.2

We shall now consider the Hilbert space of analytic functions associated with these coherent states and show that this Hilbert space coincides with the subspace $B_k(C)$ (reduced Bargmann space) introduced in Sec. 11.2.3. We define

$$\phi(\overline{\sigma}) = (f_\sigma^k, \phi) = \int {}_0F_1(2k; \overline{\sigma}\eta)\phi(\overline{\eta}) \times d\mu(\eta)$$
(11.142a)

where

$$d\mu(\eta) = \frac{2|\eta|^{2k-1}}{\pi(2k-1)!}K_{2k-1}(2|\sigma|)d^2\sigma.$$
(11.142b)

The scalar product in the Hilbert space of the functions $\phi(\overline{\sigma})$ is given by

$$(\phi, \psi) = \int \overline{\phi(\overline{\sigma})}\psi(\overline{\sigma})d\mu(\sigma).$$
(11.143)

Following Sec. 11.1, it can now be shown that the function $\phi(\overline{\sigma})$ is an entire function analytic in $\overline{\sigma}$.

The connection with the reduced Bargmann space $B_k(\mathbb{C})$ follows from the transformation

$$z = (2\overline{\sigma})^{\frac{1}{2}} \quad \text{i.e. } z^2 = 2\overline{\sigma}$$

$$d^2\overline{\sigma} = |z|^2 d^2 z$$

$$\phi(\overline{\sigma}) = 2^{k-\frac{1}{2}}[(2k-1)!]^{\frac{1}{2}} z^{-(2k-1)} f(z). \tag{11.144}$$

From the above equation it follows that as $\overline{\sigma}$ varies over the full complex plane

$$0 < |\overline{\sigma}| < \infty, \quad 0 \leq \arg\overline{\sigma} \leq 2\pi$$

z varies over the upper half-plane,

$$0 < \arg z < \pi, \quad \text{i.e. } \operatorname{Im} z > 0.$$

The transformation (11.144) convert the scalar product (ϕ, ψ) into the scalar product (f, g) in the reduced Bargmann space:

$$(f, g) = \int_{\operatorname{Im} z > 0} \overline{f(z)} g(z) d\lambda(z) \tag{11.145a}$$

$$d\lambda(z) = \frac{2}{\pi} |z|^2 K_{2k-1}(|z|^2) d^2 z. \tag{11.145b}$$

The same transformation in conjunction with the well known formula

$$_0F_1(2k; \overline{\sigma}\,\eta) = 2^{2k-1}(2k-1)!(z\overline{\xi})^{-(2k-1)} I_{2k-1}(z\overline{\xi}). \tag{11.146}$$

Eq. (11.146) yields the principal vector in $B_k(\mathbb{C})$,

$$e_z(\xi) = I_{2k-1}(\overline{z}\xi). \tag{11.147}$$

11.3.3 *The metaplectic representation; the integral transform pair connecting* $\mathbf{B}_O(\mathbb{C})$ $(\mathbf{B}_E(\mathbb{C}))$ *and* $\mathbf{H}_{3/4}$ $(\mathbf{H}_{1/4})$

We first consider the special situation corresponding to metaplectic representation for which the Bargmann–Segal space decomposes into two invariant subspaces $B_O(\mathbb{C})$ and $B_E(\mathbb{C})$ consisting of odd and even entire analytic functions respectively. As we shall see presently $B_O(\mathbb{C})$ and $B_E(\mathbb{C})$ can also be identified as the Hilbert spaces of analytic functions associated with the Barut–Girardello coherent states for $D_{3/4}^+$ and $D_{1/4}^+$ respectively.

The Barut–Girardello coherent states are the eigenfunctions of

$$J_- = \frac{1}{2}\frac{\partial^2}{\partial z^2} \tag{11.148}$$

and these are given by

$$f_\sigma = \frac{\sinh\sqrt{2\sigma}\,z}{(\sinh 2|\sigma|)^{\frac{1}{2}}} \tag{11.149}$$

for $D_{3/4}^+$ and by

$$f_\sigma = \frac{\cosh\sqrt{2\sigma}\,z}{(\cosh 2|\sigma|)^{\frac{1}{2}}} \tag{11.150}$$

for $D_{1/4}^+$. Here $\sigma \in C$ stands for the complex eigenvalue of J_-.

11.3.3.1 $D_{3/4}^+$ *representation*

For this IR the completeness condition is given by

$$(f,g) = \int d\mu(\sigma)(f,f_\sigma)(f_\sigma,g) \tag{11.151a}$$

$$d\mu(\sigma) = \frac{\sinh 2|\sigma|}{\pi|\sigma|}e^{-2|\sigma|}d^2\sigma. \tag{11.151b}$$

The overlap between the coherent states is given by

$$(f_\sigma, f_\rho) = A_\sigma A_\rho \sinh 2\sqrt{\overline{\sigma}\,\rho} \tag{11.152}$$

$$A_\sigma = (\sinh 2|\sigma|)^{-\frac{1}{2}}.$$

We now define

$$(f_\sigma, \psi) = A_\sigma f(\overline{\sigma}) \tag{11.153}$$

where ψ is an arbitrary state with a finite norm.

$$f(\overline{\sigma}) = \int \sinh 2\sqrt{\overline{\sigma}\,\rho}\,f(\overline{\rho})\frac{e^{-2|\rho|}}{\pi|\rho|}d^2\rho. \tag{11.154}$$

If we introduce

$$\sqrt{2\overline{\sigma}} = z, \quad \sqrt{2\overline{\rho}} = z_1$$

so that $0 \le (\arg z, \arg z_1) \le \pi$ we obtain from the above,

$$f(z) = \int_{\mathrm{Im}\,z_1>0} \sinh z\overline{z_1}f(z_1)d\lambda(z_1) \tag{11.155}$$

where we have replaced $f(z^2/2)$ by $f(z)$ and

$$d\lambda(z_1) = \frac{2}{\pi}e^{-|z_1|^2}d^2z_1. \tag{11.156}$$

The scalar product in the Hilbert space of these functions is given by

$$(f, g) = \int_{\text{Im } z > 0} \overline{f(z)} g(z) d\lambda(z). \tag{11.157}$$

It now follows that the function $f(z)$ satisfying Eqs. (11.155)–(11.157) are odd entire functions analytic in the upper half-plane. The function space is thus the odd subspace $B_O(C)$ of the Bargmann–Segal space $B(C)$. The principal vector in $B_O(C)$ is given by

$$e_z(z_1) = \sinh \overline{z} z_1. \tag{11.158}$$

We now consider the Perelomov coherent state and the associated Hilbert space of analytic functions. The Peremolov coherent state is defined by

$$f_a = N_a e^{aJ+} (f_m)_{m=\frac{3}{4}}, \quad a \in C; \ |a| < 1 \tag{11.159}$$

where $f_{3/4} = z$ is the lowest (vacuum) state in the canonical orthonormal basis, and N_a is the normalization factor. The restriction $|a| < 1$ ensures the finiteness of the norm $\|f_a\|$. In $B_O(C)$ these coherent states are given by

$$f_a = N_a z e^{az^2/2} \tag{11.160}$$

where

$$N_a = (1 - |a|^2)^{3/4}. \tag{11.161}$$

The completeness condition of these coherent states is given by,

$$\int (f, f_a)(f_a, g) d\eta(a) \tag{11.162}$$

where

$$d\eta(a) = \frac{1}{2\pi} (1 - |a|^2)^{-2} d^2 a. \tag{11.163}$$

To introduce Bargmann's canonical Hilbert space $H_{3/4}$ of functions analytic in the open unit disc, we define

$$(f_a, \psi) = (1 - |a|^2)^{3/4} g(-\overline{a}). \tag{11.164}$$

We then obtain

$$g(-\overline{a}) = \int_{|b| < 1} (1 - \overline{a}b)^{-\frac{3}{2}} g(-\overline{b}) d\sigma(b) \tag{11.165}$$

where

$$d\sigma(b) = \frac{1}{2\pi} (1 - |b|^2)^{-\frac{1}{2}} d^2 b. \tag{11.166}$$

Setting $\bar{a} = -\xi$ the scalar product in the Hilbert space $H_{3/4}$ of these functions is given by,

$$(g, h) = \int_{|\xi|<1} \overline{g(\xi)} h(\xi) d\sigma(\xi). \tag{11.167}$$

The principal vector in this Hilbert space is given by

$$e_\xi(\zeta) = (1 - \bar{\xi} \zeta)^{-3/2}. \tag{11.168}$$

The Euclidean subgroup $E(1)$ of $SL(2, \mathbb{R})$ consists of matrices of the form

$$\begin{pmatrix} 1 & -b \\ 0 & 1 \end{pmatrix}, \quad -\infty < b < \infty. \tag{11.169}$$

The generator of this subgroup in $B_O(C)$ is the positive Hermitian operator,

$$(J_1 + J_3) = \frac{1}{4} \left(z + \frac{d}{dz} \right)^2. \tag{11.170}$$

In $B_O(C)$ the normalized $E(1)$ basis is given by,

$$f_\rho(z) = \left(\frac{2}{\pi \rho} \right)^{\frac{1}{4}} e^{-\rho} e^{-\frac{z^2}{2}} \sinh 2\sqrt{\rho} z \tag{11.171}$$

where $\rho > 0$ stands for the eigenvalue $J_1 + J_3$. The completeness condition of this basis is given by

$$\int (f, f_\rho)(f_\rho, g) d\rho = (f, g). \tag{11.172}$$

We now introduce the space $L^2_{3/4}(\mathbb{R}^+)$ which consists of functions of the form,

$$\Psi(\rho) = (f_\rho, \psi) = \sqrt{2}\, \psi(\rho). \tag{11.173}$$

Setting $\rho = r^2/4$ and replacing $\psi \left(\frac{r^2}{4} \right)$ by $\psi(r)$, the scalar product in $L^2_{3/4}(\mathbb{R}^+)$ is given by

$$(\psi_1, \psi_2) = \int \overline{\psi_1(r)} \psi_2(r) r dr. \tag{11.174}$$

The integral transforms connecting $L^2_{3/4}(\mathbb{R}^+)$, $H_{3/4}$ and $B_O(C)$ are given by the overlaps

$$(f_\rho, f_a) = 2 \left(\frac{2\rho}{\pi} \right)^{\frac{1}{4}} (1 - |a|^2)^{3/4} (1 + a)^{-3/2} e^{-\rho \left[\frac{(1-a)}{(1+a)} \right]} \tag{11.175a}$$

$$(f_\sigma, f_a) = \frac{(1 - |a|^2)^{3/4}}{[\sinh 2|\sigma|]^{\frac{1}{2}}} (2\bar{\sigma})^{\frac{1}{2}} \exp(a\bar{\sigma}). \tag{11.175b}$$

The integral transform connecting $L^2_{3/4}(\mathbf{R}^+)$ and $H_{3/4}$ follows from,

$$(1 - |a|^2)^{3/4} g(-\bar{a}) = (f_a, \psi) = \int_0^\infty (f_a, f_\rho) \sqrt{2} \psi(\rho) d\rho$$

$$\sqrt{2}\psi(\rho) = (f_\rho, \psi) = \int_{|a|<1} (f_\rho, f_a)(1 - |a|^2)^{3/4} g(-\bar{a}) d\eta(a). \qquad (11.176)$$

Substituting the overlaps, Eq. (11.175) setting $\bar{a} = -\xi$, $\rho = r^2/4$ and replacing $\psi(\frac{r^2}{4})$ by $\psi(r)$ we obtain the integral transform pair mapping $L^2_{3/4}(\mathbf{R}^+)$ onto $H_{3/4}$ consisting of functions analytic in the open unit disc,[8]

$$g(\xi) = \int_0^\infty A(\xi, r)\psi(r) r\, dr \qquad (11.177a)$$

$$\psi(r) = \int \overline{A(\xi, r)} g(\xi) d\sigma(\xi) \qquad (11.177b)$$

$$A(\xi, r) = \frac{2^{+\frac{1}{4}} r^{\frac{1}{2}} e^{-\frac{r^2}{4}\left(\frac{1+\xi}{1-\xi}\right)}}{\sqrt{\pi}(1-\xi)^{\frac{3}{2}}}. \qquad (11.177c)$$

The integral transform pair connecting $H_{3/4}$ and $B_O(\mathbf{C})$ can be obtained in a similar manner

$$f(z) = \int_{|\xi|<1} z e^{-(z^2\bar{\xi})/2} g(\xi) d\sigma(\xi) \qquad (11.178a)$$

$$g(\xi) = \int \bar{z} e^{(-\bar{z}^2\xi)/2} f(z) d\lambda(z). \qquad (11.178b)$$

11.3.3.2 $D^+_{\frac{1}{4}}$ *representation*

For this irreducible representation the completeness condition of the Barat–Girardello coherent states is given by

$$(f, g) = \int (f, f_\sigma)(f_\sigma, g) d\mu(\sigma) \qquad (11.179)$$

$$d\mu(\sigma) = \frac{\cosh 2|\sigma|}{\pi |\sigma|} e^{-2|\sigma|} d^2\sigma. \qquad (11.180)$$

Proceeding as before we can easily verify that the associated Hilbert space of analytic functions satisfying $f(-z) = f(z)$. The scalar product is the same as that in $B_O(\mathbf{C})$ and the principal vector in $B_E(\mathbf{C})$ is given by,

$$e_z(z_1) = \cosh \bar{z} z_1. \qquad (11.181)$$

In $B_E(C)$ the Perelomov coherent state for this representation is given by,

$$f_a = (1 - |a|^2)^{\frac{1}{4}} e^{az^2/2}. \tag{11.182}$$

However it is necessary to point out that the integral in the completeness condition is to be understood in the sense of its regularization as defined in Chapters 6 and 3 and is given by,

$$(f, g) = -\frac{1}{8\pi} \int_\Sigma dt (1-t)^{-2} \int_0^{2\pi} (f, f_a)(f_a, g) d\theta, \ a = \sqrt{t}\, e^{i\theta}, \ \leq \theta \leq 2\pi \tag{11.183}$$

where Σ is a contour that starts from the origin along the positive real axis, encircles the point $+1$ once counterclockwise and returns to the origin along the positive real axis.

The Hilbert space $H_{1/4}$ consisting of functions analytic in the open unit disc can be introduced as before by setting,

$$(f_a, \psi) = (1 - |a|^2)^{\frac{1}{4}} g(-\bar{a}) \tag{11.184}$$

where $g(-\bar{a}) = g(\xi)$ (setting $\bar{a} = -\xi$) is analytic within the unit disc $|\xi| < 1$. The scalar product in $H_{1/4}$ is given by

$$(g, h) = -\frac{1}{8\pi} \int_\Sigma dt (1-t)^{-\frac{3}{2}} \int_0^{2\pi} \overline{g(\xi)} h(\xi) d\theta, \ \xi = \sqrt{t}\, e^{i\theta}, \ 0 \leq t < 1. \tag{11.185}$$

The principal vector in $H_{1/4}$ is given by,

$$e_\xi(\eta) = (1 - \bar{\xi}\eta)^{-\frac{1}{2}}. \tag{11.186}$$

The E(1) basis satisfying $(J_1 + J_3) f_\rho = \rho f_\rho$, $\rho > 0$, is now given by

$$f_\rho = \left(\frac{2}{\pi \rho}\right)^{\frac{1}{4}} e^{-\rho} e^{-z^2/2} \cosh 2\sqrt{\rho}\, z. \tag{11.187}$$

The Hilbert space $L^2_{1/4}(R^+)$ consists of functions

$$\Psi(\rho) = (f_\rho, \Psi) = \sqrt{2}\, \psi(r), \quad \frac{r^2}{4} = \rho \tag{11.188}$$

and is connected to $H_{1/4}$ by the integral transform pair,[8]

$$g(\xi) = \int_0^\infty A(\xi, r) \psi(r) r\, dr \tag{11.189a}$$

$$\psi(r) = -\frac{1}{8\pi} \int_\Sigma dt (1-t)^{-\frac{3}{2}} \int_0^{2\pi} \overline{A(\xi, r)} g(\xi) d\theta \tag{11.189b}$$

where

$$A(\xi, r) = \frac{2^{3/4}}{\pi^{1/4}} \frac{r^{-\frac{1}{2}}}{(1 - \xi)^{\frac{1}{2}}} \exp\left[-\frac{r^2}{4}\left(\frac{1 + \xi}{1 - \xi}\right)\right], \quad \xi = \sqrt{t}\, e^{i\theta}. \quad (11.189c)$$

The integral transform pair connecting $B_E(C)$ and $H_{1/4}$ also follows as before and is given by,

$$g(\xi) = \int_{\mathrm{Im}\, z > 0} e^{-\xi z^2/1} f(z) d\lambda(z) \quad (11.190a)$$

$$f(z) = -\frac{1}{8\pi} \int_\Sigma dt (1 - t)^{-3/2} \int_0^{2\pi} e^{-z^2 \bar{\xi}/2} g(\xi) d\theta \quad (11.190b)$$

11.3.3.3 *Arbitrary representations of* D_k^+: $\left(k = \frac{1}{2}, 1, \frac{3}{2}, \ldots\right)$

We now consider the Barut–Girardello and Perelomov coherent states and the $E(1)$ basis in $B_k(C)$:

$$f_\sigma^k = [I_{2k-1}(2|\sigma|)]^{-\frac{1}{2}} I_{2k-1}(\sqrt{2\sigma}\, z) \quad (11.191a)$$

$$f_a^k = \frac{(1 - |a|^2)^k z^{2k-1}}{2^{k-\frac{1}{2}}\sqrt{\Gamma(2k)}} \exp(az^2/2) \quad (11.191b)$$

$$f_r^k = \sqrt{2}\, e^{\frac{-r^2}{4}} e^{\frac{-z^2}{2}} I_{2k-1}(zr) \quad (11.191c)$$

where $r^2/4$ is the eigenvalue of $J_1 + J_3$, the positive Hermitian generator of the $E(1)$ subgroup.

The kernel of the integral transforms connecting $L_k^2(R^+)$ and H_k and $B_k(C)$ and H_k are given by the overlaps

$$(f_r^k, f_a^k) = \frac{2^{1-k}(1 - |a|^2)^k r^{2k-1}}{\sqrt{\Gamma(2k)}} e^{-\frac{r^2}{4}\left(\frac{1-a}{1+a}\right)} \quad (11.192a)$$

$$(f_\sigma^k, f_a^k) = \frac{[I_{2k-1}(2|\sigma|)]^{-\frac{1}{2}}}{\sqrt{\Gamma(2k)}}. \quad (11.192b)$$

The integral transform pair mapping $L_k^2(R^+)$ onto H_k is given by[8]

$$g(\xi) = \int_0^\infty A(\xi, r)\psi(r) r\, dr \quad (11.193a)$$

$$\psi(r) = \int_{|\xi| < 1} \overline{A(\xi, r)} g(\xi) d\sigma(\xi) \quad (11.193b)$$

where

$$A(\xi, r) = \frac{1}{\sqrt{\Gamma(2k)}} 2^{\frac{1}{2}-k} \frac{r^{2k-1}}{(1-\xi)^{2k}} e^{-\frac{r^2}{4}\left(\frac{1+\xi}{1-\xi}\right)} \tag{11.193c}$$

$$d\sigma(\xi) = \frac{2k-1}{\pi}(1 - |\xi|^2)^{2k-2} d^2\xi. \tag{11.193d}$$

Similarly the pair connecting $B_k(C)$ and H_k is given by,[8]

$$g(\xi) = \frac{2^{\frac{1}{2}-k}}{\sqrt{\Gamma(2k)}} \int_{\mathrm{Im}\, z>0} \bar{z}^{2k-1} e^{-\xi \bar{z}^2/2} f(x) d\lambda(z) \tag{11.194a}$$

$$f(z) = \frac{2^{\frac{1}{2}-k}}{\sqrt{\Gamma(2k)}} \int_{|\xi|<1} z^{2k-1} e^{-z^2 \bar{\xi}/2} g(\xi) d\sigma(\xi) \tag{11.194b}$$

where

$$d\lambda(z) = \frac{2}{\pi}|z|^2 K_{2k-1}(|z|^2) d^2z. \tag{11.194c}$$

References

[1] A. A. Kirillov, *Usp. Math. Nauk* **106** (1962) 57; *Russ. Math. Surv.* **17** (1962) 53.
[2] E. Schrödinger, *Naturwissenschaften* **14** (1926) 664.
[3] J. R. Klauder, *Ann. Phys. (NY)* **11** (1960) 123; J. R. Klauder and E. C. G. Sudarshan, *Fundamentals of Quantum Optics* (Benjamin, New York, 1968), (Dover, 2006); J. R. Klauder and B. S. Skagerstam, *Coherent States, Applications to Physics and Mathematical Physics* (World Scientific, Singapore, 1985).
[4] R. J. Glauber, *Phys. Rev. Lett.* **10** (1963) 84; *Phys. Rev.* **130** (1963) 2529; **131** (1963) 2766.
[5] E. C. G. Sudarshan, *Phys. Rev. Lett.* **10** (1963) 277.
[6] V. Bargmann, *Commun. Pure Appl. Math.* **140** (1961) 187; **20** (1967) 1.
[7] I. E. Segal, *Ill, J. Math.* **6** (1962) 500.
[8] D. Basu, *Proc. Roy. Soc. (Lond)* **A455** (1999) 975.
[9] R. M. Wilcox, *J. Math. Phys.* **8** (1967) 962.
[10] D. Basu, *J. Math. Phys.* **30** (1989) 1.
[11] M. Moshinsky and C. Quesne, *J. Math. Phys.* **12** (1971) 1772, 1780.
[12] A. O. Barut and L. Girardello, *Commun. Math. Phys.* **21** (1971) 41; D. Basu, *J. Math. Phys.* **33** (1992) 1.
[13] See Ref. 11.
[14] K. B. Wolf, *J. Math. Phys.* **15** (1974) 1295, 2102; C. P. Boyer and K. B. Wolf, *J. Math. Phys.* **16** (1975) 1493, 2215.